ELECTRONIC DESIGN AND CONSTRUCTION OF ALTERNATE ENERGY PROJECTS

No. 1672
$18.95

ELECTRONIC DESIGN AND CONSTRUCTION OF ALTERNATE ENERGY PROJECTS

R. ANDREW MOTES

FIRST EDITION

FIRST PRINTING

Printed in the United States of America

Library of Congress Cataloging in Publication Data

Motes, R. Andrew.
Electronics design and construction of alternate energy projects.

Includes index.
1. Electronics—Amateurs' manuals. 2. Renewable energy sources—Amateurs' manuals. I. Title.
TK9965.M68 1985 621.381 84-26891
ISBN 0-8306-0672-6
ISBN 0-8306-1672-1 (pbk.)

Contents

Introduction

It is obvious that in today's world the high cost of energy is unacceptable. The combination of oil shortages and the laws of supply and demand have caused the price of oil to more than quadruple in the last 10 years. Since the world oil reserves cannot last forever, the only recourse is to find new sources of energy. Coal is a good substitute for oil in electric power generation, but the resulting air pollution makes this undesirable for some. Nuclear power has no direct pollution properties, but the problem of nuclear waste disposal makes this alternative also unacceptable. Hydroelectric plants are nonpolluting, but many environmentalists are fighting these because they alter the natural waterways. The only types of alternate energies that are not being fought by special-interest groups are solar energy and wind energy.

All the energy sources just mentioned, except for nuclear, are merely sources of stored solar energy. Oil and coal are products of ancient vegetation that, in turn, received its energy from the sun. Hydroelectric plants generate power from rain and snow runoff, while the clouds, which produced the rain and snow, were actually formed by an evaporation process that received its energy from the sun and wind. Wind also receives its energy from the sun through differential heating of the earth's surface. Because all these sources of energy are indirect sources of solar energy, many people think it would be much simpler to bypass most of these natural processes and go directly to wind and solar energies.

Although there are now many people interested in solar energy and wind energy, few can afford the high cost of a factory-produced system, and even less have the technical knowledge to design and build their own. This book is basically a collection of research and design notes that I compiled while I designed my own alternate energy systems. It is written on a technical level and contains more than 70 electronic circuits that can be used for generating power or saving energy. All of the circuits were designed to use components available at any neighborhood electronics store and, hence, are not necessarily optimal circuit designs. Where possible, the circuits were designed to use automobile salvage parts. All circuits are complete with theory of operation, circuit diagrams, and enough information so that a person with some electronics training can build and operate them. For those lacking in electronics knowledge, several appendices have been added for instruction in basic electronics.

This book is primarily concerned with the electronics of alternate energy systems. There are many books available on the mechanics of home-built energy systems, but few, if any, on the electronics. Hopefully, this book fills that gap.

1

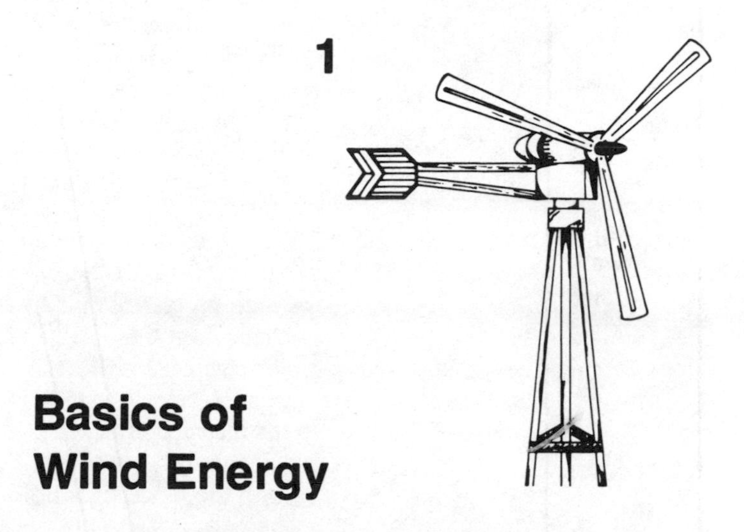

Basics of Wind Energy

A major advantage of solar energy and wind energy is the lack of pollution; however, at the present time direct conversion of solar energy into electricity is not very efficient. A recently developed gallium-arsenide photovoltaic cell has demonstrated efficiencies as high as 26 percent, but these cells are very expensive (Mims 1980, p. 99). The efficiency rating of most currently produced silicon photovoltaic cells is only about 10 percent. Hopefully, future research will make improvements in this area. In the meantime, wind energy looks like the most feasible way to convert solar energy to electricity, which is our primary form of usable energy.

POWER AVAILABLE IN THE WIND

The power density of wind, measured in watts per square foot of area impacted by the wind, is given by the following equation:

$$PD = d(WS)^3(2.14) \text{ watts/ft}^2 \qquad \textbf{Eq. 1-1}$$

where d is the air density measured in slugs per cubic foot or $(lb)(sec)^2/(ft)^4$ and is a function of altitude, and WS represents wind speed measured in miles per hour. Table 1-1 relates air density to altitude, and Fig. 1-1 shows a graph of power density versus wind speed for altitudes of 0 and 10,000 feet. To find the maximum wind power present at a particular wind turbine, multiply the power density by the effective surface area of the turbine. For a standard

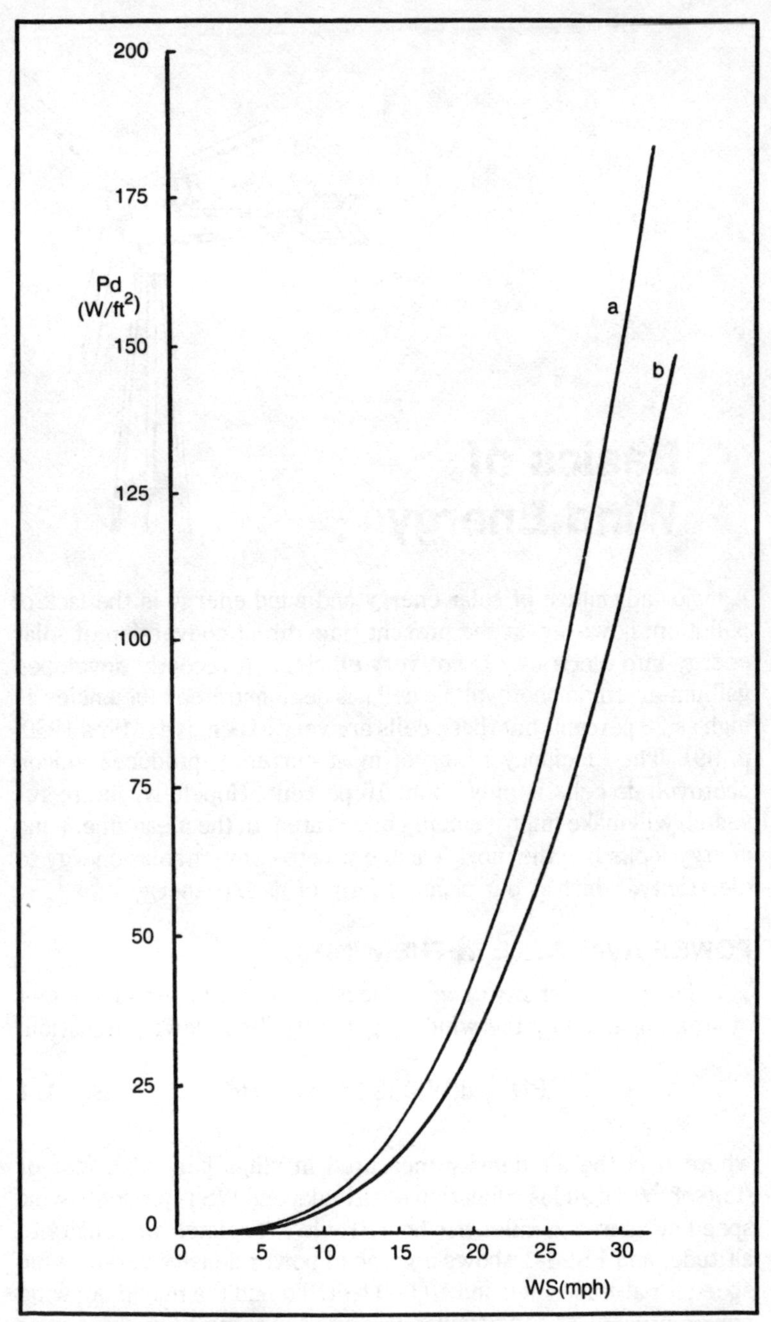

Fig. 1-1. Power density (PD) versus wind speed (WS) at altitudes of (a) 0 and (b) 10,000 feet.

Table 1-1. Air Density in Relation to Altitude.

Elevation in feet	Air Density in slugs/ft^3	Elevation in feet	Air Density in slugs/ft^3
0	0.0024	5500	0.0020
500	0.0024	6000	0.0020
1000	0.0023	6500	0.0020
1500	0.0023	7000	0.0019
2000	0.0023	7500	0.0019
2500	0.0022	8000	0.0019
3000	0.0022	8500	0.0018
3500	0.0022	9000	0.0018
4000	0.0021	9500	0.0018
4500	0.0021	10000	0.0018
5000	0.0021	10500	0.0017

propeller-type windmill, you can use Eq. 1-2 to calculate the maximum wind power (Pm):

$$\begin{aligned} Pm &= (PD)A \\ &= (PD)(\pi R^2) \end{aligned} \qquad \textbf{Eq. 1-2}$$

where R is the radius of the turbine or the length of one prop.

As an example, suppose the altitude is 1000 feet, the wind speed is 20 miles per hour, and the turbine radius is 6 feet. At this altitude (see Table 1-1) d is 0.0023 slugs per cubic foot. Using these values in Eq. 1-1 and 1-2, we calculate the maximum wind power present to be 4453 watts. However, maximum wind power present at the turbine and power available to the turbine are not the same.

The output power of a wind turbine is the product of maximum wind power at the turbine and the efficiency of the turbine. The maximum theoretical efficiency in converting wind energy into mechanical energy is 59.3 percent for any type of wind turbine (Daugherty and Franzini, 1977, p. 164). Therefore, power available to the turbine (Pa) is given by

$$\begin{aligned} Pa &= Pm(0.593) \\ &= (PD)(\pi R^2)(0.593) \end{aligned} \qquad \textbf{Eq. 1-3}$$

The efficiency in changing this mechanical energy into electrical energy depends on the generator or alternator used. The overall efficiency in changing wind energy to electrical energy could feasibly reach as high as 40 percent. Obviously, 40 percent is much more attractive than 10 percent, which is representative of photovoltaic

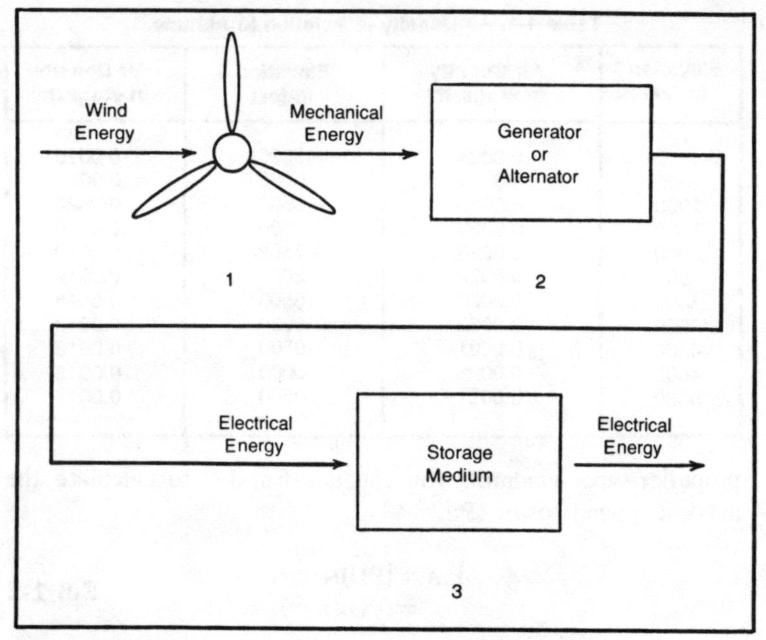

Fig. 1-2. Three basic components of a wind-power system.

cells. If the wind turbine of the previous example were capable of 40-percent efficiency, the output power would be 1781 watts when the maximum wind power present is 4453 watts.

COMPONENTS OF A WIND-ENERGY SYSTEM

A wind-energy system has three basic components. The first component, the *turbine*, converts wind energy into mechanical energy. There are many types of turbines, the most common of which is the familiar propeller type. While this turbine appears to be similar to an airplane propeller, it is actually quite different. The pitch and cross-sectional dimensions of the turbine blades are very different from those of the airplane propeller. Two other types of turbines are the *Savonius* and the *Darrieus*. The Savonius (S-rotor) turbine has excellent low-rpm-torque and poor high-rpm-torque, while the Darrieus (eggbeater type) has excellent high-rpm-torque and no starting torque at all. The propeller-type turbine is a good compromise because it has reasonable starting, low-rpm-torque, and good high-rpm-torque.

The second component, a *generator* or an *alternator*, converts the mechanical energy output of the turbine to electrical energy. A generator converts mechanical energy directly to direct current (dc) electricity, while an alternator first produces alternating current (ac) and then rectifies the ac to obtain dc. The advantage the alternator has over the dc generator is that it does not require a commutator. The brushes used in conjunction with the commutator of the dc generator tend to quickly wear out, causing the generator to require regular maintenance; however, if a permanent-magnet dc generator is available, it is an adequate substitute for an alternator if voltage regulation is not a consideration. If voltage regulation is desired, either an alternator or a dc generator with an accessible field winding is required.

The third basic component of a wind-energy system is the *energy storage medium*. Energy storage is necessary to provide power during periods of low wind. The chemical method, or batteries, is the most common way to store electrical energy. Batteries are not the most efficient means for storing energy, but they are easy to find, especially the 12-volt automobile type, and they are relatively inexpensive.

WINDWORKS of Mukwonago, Wisconsin has developed an inverter that changes dc to ac and then feeds all excess power back into the electric utility power lines, thus using the power lines as the energy storage medium (Lindsley 1975, p. 50). The interesting thing about this system is that it turns the wattmeter backwards when power is being fed into the power lines, and this, in effect, causes the utility company to buy customer-generated power at the same rate at which they are selling their own power. When the utility company discovers this, they usually insist on paying a discount rate. In 1975 the price of this sytem was $1200. With the high rate of inflation, the system is probably much more expensive now. As previously stated, batteries are relatively inexpensive and still recommended for the average homeowner who is trying to minimize costs.

Figure 1-2 shows a block diagram of the basic wind-power system just described. It is a very simple illustration, but it does effectively demonstrate how the three components fit together for producing electric power from the wind.

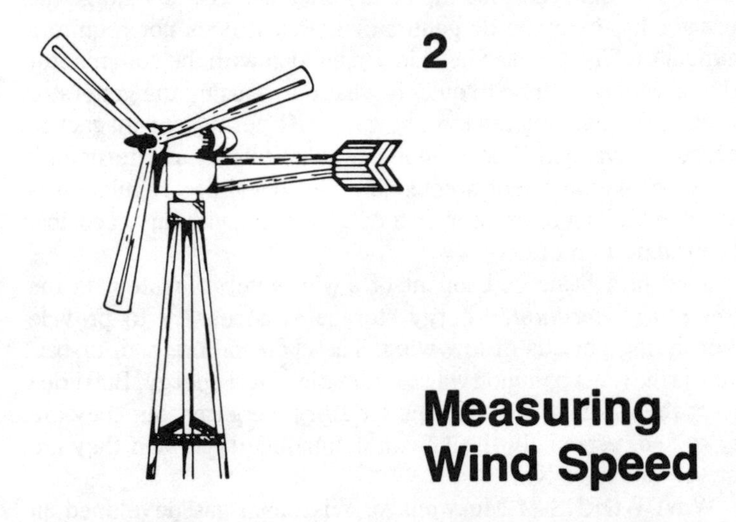

2

Measuring Wind Speed

The first project of a home-built wind-energy system should be an *average wind-speed detector*. Nothing would be more disheartening than to put a lot of time and money into building a wind-power system only to find that there is not enough wind to power the system. There is no set value for the minimum average daily wind speed required for efficient power production; however, a minimum average wind speed of 7 miles per hour is usually recommended.

Even though the average daily wind speed may be low, there is still the possibility of high wind gusts at certain periods of the day. Therefore, it is also recommended that you take actual wind-speed readings. An easy way to do this is to use the direct-readout wind-speed detector discussed in this chapter.

THE RECORDING ANEMOMETER

To measure average wind speed over some interval of time, you need a *recording anemometer*. This is simply a miniturbine device that records the number of revolutions of the turbine instead of giving a direct readout of wind speed. Figure 2-1 shows a simplified sketch of a miniturbine for recording average wind speed.

The number of revolutions of the miniturbine multiplied by some constant and divided by the number of hours of operation gives the average wind speed over that time period. Let C be this constant. The formula relating revolutions to average wind speed is then

$$AWS = (C)(Revs)/T \text{ miles per hour} \qquad \textbf{Eq. 2-1}$$

where AWS is the average wind speed, and Revs represents the number of revolutions of the miniturbine over the time interval of T hours. If we assume that the cups on the miniturbine are traveling at

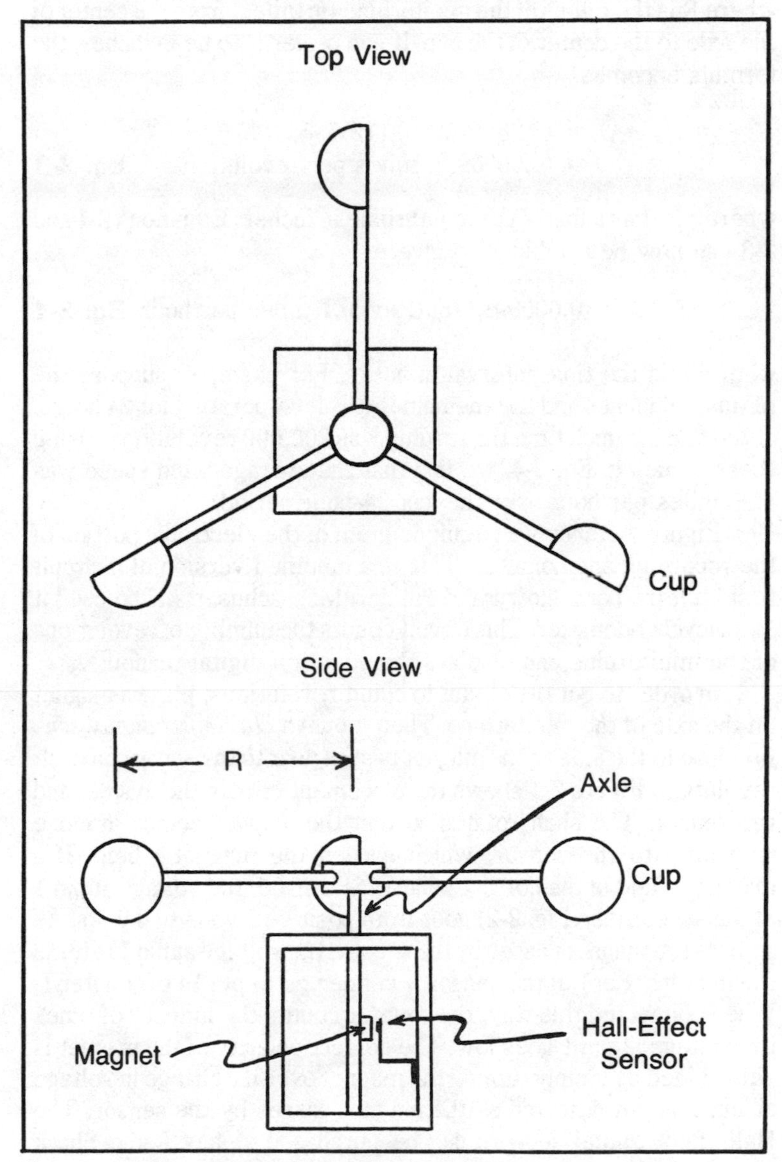

Fig. 2-1. Miniturbine with Hall-effect sensor for recording wind speed.

the same speed as the wind, then C is given by the following formula:

$$C = (6.28)R \text{ miles per revolution} \qquad \textbf{Eq. 2-2}$$

where R is the radius of the miniturbine (in miles) from the center of the axle to the center of the cup. If you prefer R to be in inches, the formula becomes

$$\begin{aligned} C &= (6.28)r/(5280)\ (12) \\ &= (0.000099)r \text{ miles per revolution} \qquad \textbf{Eq. 2-3} \end{aligned}$$

where r is the radius of the miniturbine in inches. Equations 2-1 and 2-3 can now be combined to give

$$AWS = (0.0000993\ (r)\ (Revs)/T \text{ miles per hour} \quad \textbf{Eq. 2-4}$$

where T is the time interval in hours. For example, suppose the radius is 6 inches and the anemometer is left operating for 24 hours, at the end of which time the reading was 200,000 revolutions. Using these values in Eq. 2-4, we find that the average wind speed was 4.95 miles per hour over the last 24-hour period.

Figure 2-2 shows a circuit diagram of the electronic portion of the recording anemometer. This is a modified version of a circuit first built by Tom Skowyra of Palmer, Massachusetts, who used it as a bicycle odometer. This circuit counts the number of revolutions of the miniturbine and displays the count on digital readouts.

In order to get the circuit to count revolutions, place a magnet on the axle of the miniturbine. Then mount a ***Hall-effect digital sensor*** close to the axle so the magnet passes near to the sensor on each revolution. Figure 2-1 shows the placement of both the magnet and the sensor. The shaft rotates so that the magnet comes in close proximity to the sensor, which senses the magnetic field. If a magnetic field of the correct polarity is sensed, the voltage at pin 1 of the sensor (see Fig. 2-2) goes from positive 5 volts to 0 volts. As soon as the magnet passes by the sensor, the voltage at pin 1 returns to +5 volts. Pin 1 of the sensor is connected to pin 14 of counter 1. When connected this way, the counter counts the number of times the voltage at pin 1 goes low. The correct polarity of the magnet is determined by simply turning the magnet over if a change in voltage at pin 1 is not detected as the magnet passes by the sensor. The Hall-effect digital sensors can be purchased at any Radio Shack store.

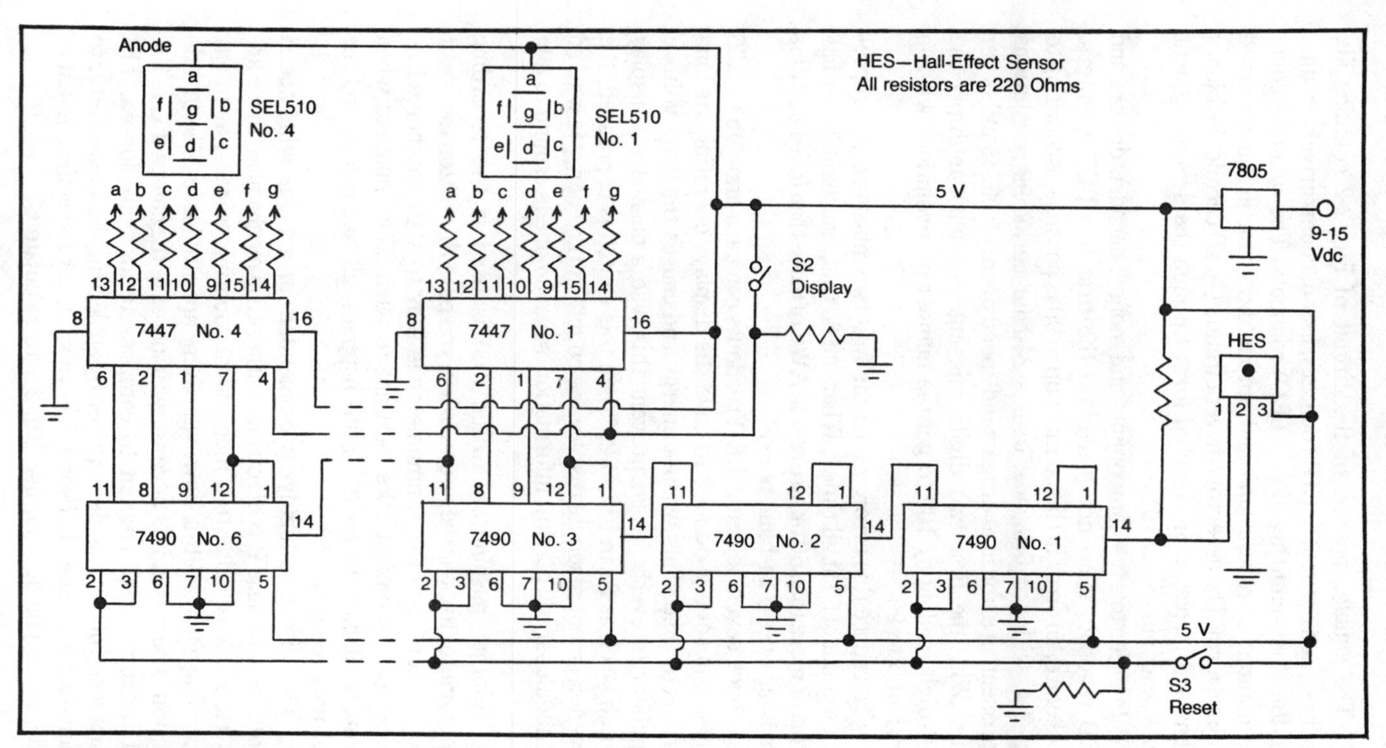

Fig. 2-2. Digital electronic recording anemometer.

The counter portion of the circuit of Fig. 2-2 contains six binary-coded decimal (BCD) counters, four decoder/driver circuits, and four light-emitting diode (LED) readouts. The counters count the number of revolutions and output the count in binary-coded decimal form. The detector/driver circuits (7447) decode the binary count and display it in decimal form through the seven-segment LED readouts (SEL510).

In order to save components and money, decoder/drivers and LED readouts were not placed on the first two BCD counters. Therefore, an error in the count of up to 99 is possible. According to Eq. 2-4, with r = 6 inches, over a 24-hour period 99 revolutions represent an error in average wind speed of only 0.00245 miles per hour. With the first two digits missing, you must multiply the four-digit reading by 100 to get the number of revolutions with an error of 0 to 99.

To reduce this error, we let the first two missing digits represent a count of 50 at all times. When we do this, the count never has an error more than 50 or an error in AWS of more than 0.00122 miles per hour over a 24-hour period.

Since seven-segment LED readouts can each draw up to 0.25 amps, it is not a good idea to have the displays operating for long periods of time. This wastes energy and causes the 7805 voltage regulator to overheat. To prevent this, use a pushbutton display switch, such as S2 in Fig. 2-2. When this switch is depressed, the seven-segment readouts display the count, and no energy is wasted by needlessly displaying information when no one is around to see it.

Another pushbutton switch (S3) is used in the recording anemometer to clear the count. When you push this button the count goes to zero, and the anemometer is reay to start another wind-speed test. You must take care not to push S3 by mistake when trying to display the count. If this happens, all data is lost, and the test must be repeated.

This circuit is highly recommended for average wind-speed detection. Because it is electronic rather than mechanical, there are no friction losses due to a mechanical counting process; as in any recording anemometer, however, there are still losses due to axle friction. Therefore, it is necessary to increase the constant C in Eq. 2-1 by some small amount to compensate for these losses. The exact value of C can be determined by subjecting the anemometer to a constant wind speed of some known value for a specific amount of time. After you do this, use Eq. 2-1 to calculate C.

A good way to obtain this known, constant wind speed is to hold the anemometer out of the window of a moving automobile on a calm day. If the friction losses are low, C will require only a small adjustment. For example, suppose the anemometer is held out of the window of a car going 30 miles per hour for 10 minutes and the reading on the display is 1700. Using Eq. 2-1 and solving for C, we see that

$$C = \frac{(AWS)T}{Revs}$$

where AWS is equal to 30 miles per hour, T is equal to 10/60 hours, and Revs is the display count. Putting these numbers in the equation gives a value for C of 0.00294. Substitute this in Eq. 2-1 to calculate the actual wind speed.

THE DIRECT-READOUT ANEMOMETER

Instead of displaying the number of revolutions, the *direct-readout anemometer* displays actual wind speed in miles per hour. It does this by counting revolutions over a small interval of time. If the correct time interval is chosen, the number of counts will equal the wind speed. Figure 2-3 shows a functional block diagram of the direct-readout anemometer circuit along with its digital timing pulses.

Using Eq. 2-4, with r equal to the radius of the miniturbine in inches and T equal to the time interval in hours, we can calculate the exact value of T that will give a direct readout in miles per hour. Because the digital readout of this circuit is the actual number of revolutions over the time interval T, to get this readout to also be equal to wind speed we must set AWS equal to Revs in Eq. 2-4 to find the value of T needed. Equation 2-4 then becomes

$$Revs = (0.000099)\,(r)\,(Revs)/T$$

or

$$T = 0.000099(r) \qquad \textbf{Eq. 2-5}$$

If r equals 6 inches, T should be equal to 0.00059 hours or 2.1 seconds. If friction is neglected, this is the time interval necessary for the 6-inch radius anemometer to be accurately calibrated in miles per hour.

Using the direct-readout anemometer circuit diagram (Fig.

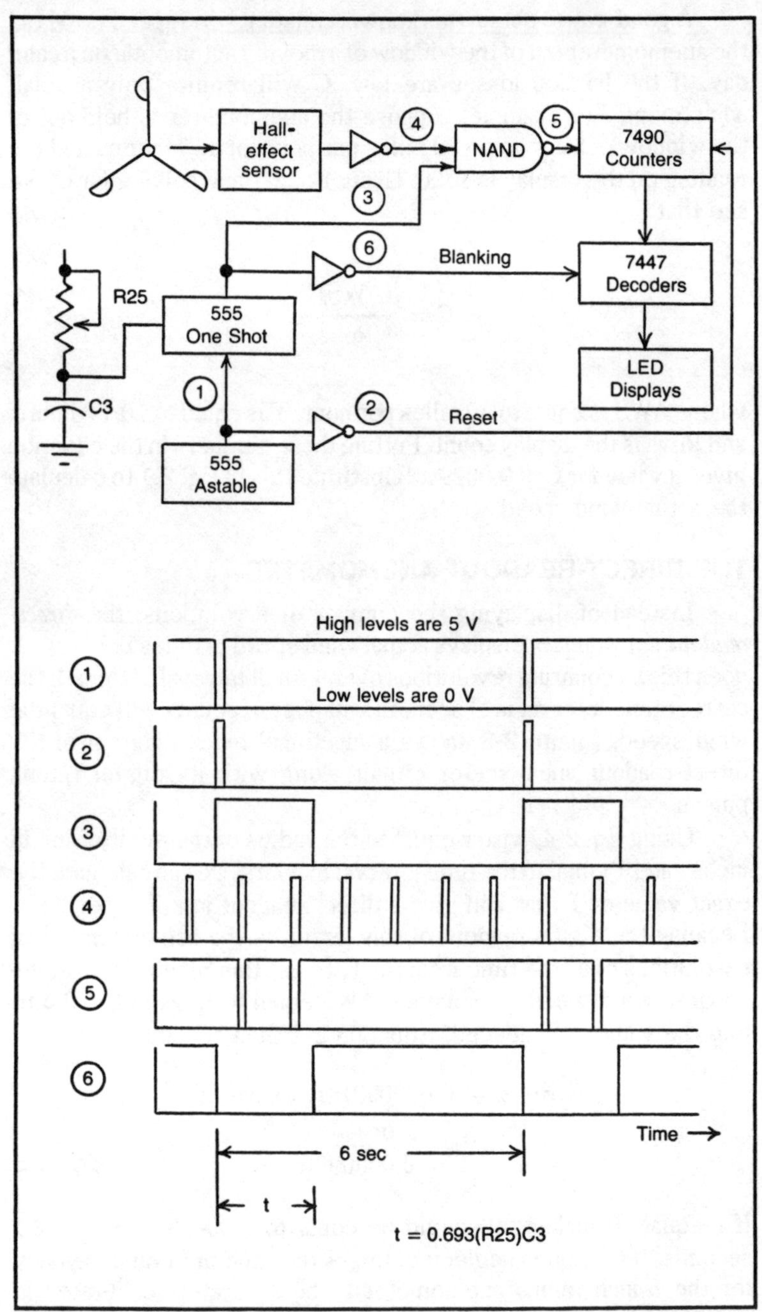

Fig. 2-3. Block diagram of the direct-readout anemometer along with timing pulses.

2-4), we can show that T is related to the resistance of potentiometer R25 by the formula

$$t = 0.693(R25)C3 \qquad \textbf{Eq. 2-6}$$

where t is the time interval T in seconds and C3 represents the capacitance in farads of capacitor C3. If a 5-μF capacitor is used for C3, then

$$t = 0.00000346(R25) \qquad \textbf{Eq. 2-7}$$

Substituting 2.1 seconds for t in Eq. 2-7 gives

$$2.1 = 0.0000346(R25)$$

or

$$R25 = 606{,}000 \text{ ohms}$$

If friction could be neglected, a resistor of this size could simply be used in place of R25, and the circuit would be calibrated. Because friction can never be entirely eliminated, the circuit must be calibrated by holding the anemometer out of the window of a moving automobile on a calm day. R25 can then be adjusted until the digital readout corresponds to the speedometer reading. Notice that in Eq. 2-4 we are calculating average wind speed over T. The smaller the time interval T, the closer AWS is to instantaneous wind speed.

The circuit diagram for this direct-readout anemometer is shown in Fig. 2-4. This circuit is a modified version of one presented in *Elementary Electronics* (Fox 1976, p. 46). Most of the components of this circuit are the same as those of the recording anemometer. The two circuits are basically the same. For the recording anemometer the counter reset and readout display functions are manually selected. Also, the time interval is selected manually. For the direct-readout anemometer all these functions are automatic once the length of the time interval is selected by adjusting potentiometer R25.

This circuit also utilizes the same Hall-effect sensor as the recording anemometer previously discussed. Before the negative pulses from the Hall-effect sensor reach the counter, they are first inverted and passed through a *NAND gate* (see Figs. 2-3 and 2-4), which allows the pulses to pass through to the counters only when a positive voltage (counting pulse) is applied to the other input of the gate. These counting pulses are received from the NE555 one-shot

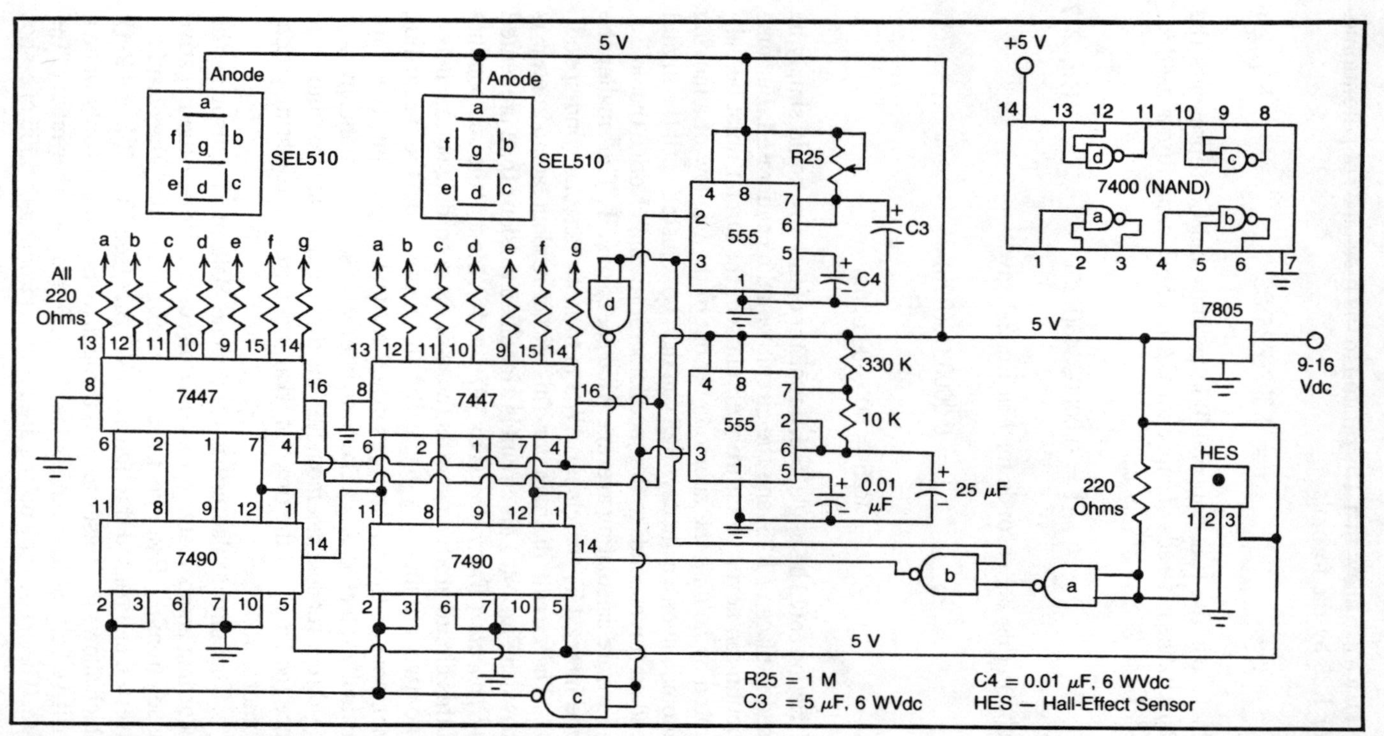

Fig. 2-4. Electronic circuit for the direct-readout anemometer.

and occur every 6 seconds. The duration of this positive counting pulse can be varied by adjusting R25. The inverse of the counting pulse is also applied to the decoders and causes the display to be blanked out while the counters are counting pulses from the sensor. The one-shot is triggered by short negative pulses from an NE555 astable multivibrator that occur every 6 seconds. These negative pulses are also inverted and used to reset the counters. Because a 7400 NAND gate integrated circuit chip has four NAND gates, in order to reduce the number of chips the two unused NAND gates were converted to inverters by connecting the two inputs.

Although these two anemometer circuits seem complicated, they are not difficult to build. Digital electronic circuits are easier to work with than analog circuits because there are only two voltage levels—0 volts and 5 volts. Hence, most digital circuits are compatible and can be directly connected without any interface circuitry.

POWER SUPPLIES FOR ELECTRONIC ANEMOMETERS

So far there has been no mention of where the +5-volts power for the two anemometers should come from. If 120-Vac power is available, the 120-Vac-to-5-Vdc power supply in Fig. 2-5 can be used. This circuit uses a 120-volts-to-12-volts (120:12) transformer with a 1.2-amp center-tapped secondary to drop the ac voltage level to 6 volts. The two general-purpose 1-amp diodes (1N4001) and the 1000-μF capacitor are then used to rectify and filter the 6 Vac into 8.48 Vdc. After the voltage is filtered, it is then regulated to 5 Vdc by the 7805 voltage regulator. This 5-Vdc output can be used to power the two previous anemometer circuits.

The 6-Vac becomes 8.48 Vdc when it is *rectified* and filtered

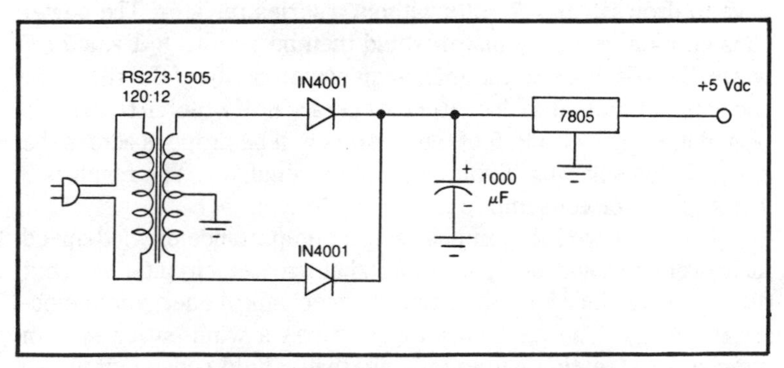

Fig. 2-5. A 120-Vac-to-5-Vdc power supply for the digital electronic anemometers.

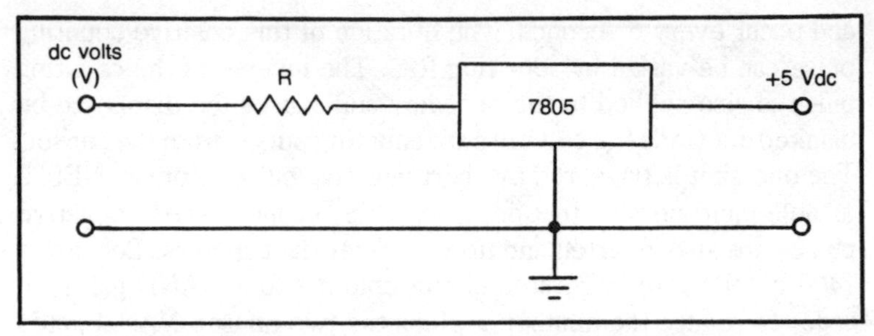

Fig. 2-6. A dc-to-5-Vdc power supply for the digital electronic anemometer.

because ac voltage is measured in *root-mean-square* (rms) voltage. The peak ac voltage is equal to 1.414 times the rms voltage. When the ac voltage is rectified and filtered, the capacitor charges to a dc level equal to the peak ac voltage level or 8.48 volts; however, this voltage level may decrease slightly when current is drained from the capacitor (for example, when a load is placed on the output).

If no 120-Vac power is available, a dc source can be used if it is regulated at 5 volts, as shown in Fig. 2-6. The resistance value R chosen for this circuit should be equal to

$$R = \frac{V - 7}{1} \qquad \textbf{Eq. 2-8}$$

The power rating Pr of the resistor should at least be equal to the resistance value of the resistor.

The purpose of the resistor is to absorb some of the voltage drop during heavy loads. For example, if the input voltage V was 12 volts and the output current was 1 amp, the 7805 regulator would have to drop 12–5 = 7 volts without a series resistor. The power consumption of the regulator would then be 7 volts × 1 amp or 7 watts. This is close to the maximum power rating of the regulator and could burn it up with continued operation. If a 5-ohm resistor is placed in series with it, 5 of the 12 volts will be dropped across the resistor, leaving only 2 volts across the regulator. The result is 2 watts of power consumption.

Now that we have emphasized the importance of wind-speed measurements and presented two anemometer circuits, we next discuss machinery for converting the mechanical energy into electrical energy. The next chapter describes a wind-power system simulator for testing automobile alternator field control circuits.

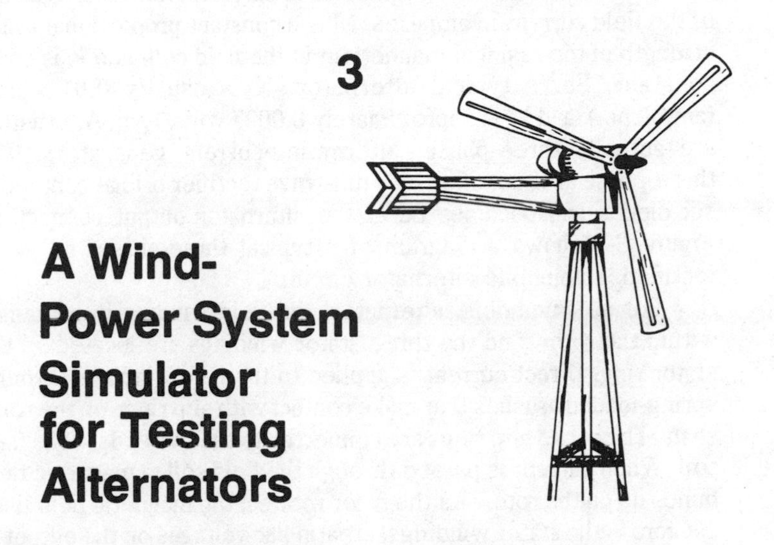

3

A Wind-Power System Simulator for Testing Alternators

Most homeowners who want to build their own wind-power systems cannot afford the high cost of generators and alternators specially designed for wind-power applications. For this reason, all the wind-power generating circuits described in this book were designed for use with automobile alternators that are easy to find and relatively inexpensive. They can usually be purchased at local salvage yards for about $15.

This chapter describes a wind-power system *simulator* that can be used to test the automobile alternator field control circuits before the actual system is constructed. Testing control circuits on a wind-energy system is especially difficult because the generator or alternator is usually located several feet in the air. Also, the wind is sometimes unpredictable and does not always spin the alternator/generator at the desired rpm for taking data. The system described in this chapter allows for a completely controlled and accessible environment for testing.

BASIC ALTERNATOR THEORY

The output voltage of automobile alternators is dc and varies in magnitude according to the field current and rotation rate of the rotor. This can be expressed in equation form as

$$Va = (rpm)\,[(If)(K)+M] \qquad \textbf{Eq. 3-1}$$

where rpm is revolutions per minute of the rotor. If is the magnitude of the field current in amperes, M is a constant proportional to the strength of the residual magnetism in the field coil, and K is some constant. For a typical alternator, K is usually 0.01 volts/(amp)(rpm), and M is approximately 0.0007 volts/rpm. Alternators are actually three-phase, alternating-current generators. The three-phase ac is rectified by a full-wave rectifier bridge containing six diodes and becomes dc at the alternator output connection. Figure 3-1 shows a diagram of a typical three-phase, full-wave rectified, automobile alternator circuit.

For an automobile alternator, the field winding is contained within the rotor, and the three stator windings are located on the stator ring. Direct current is applied to the field windings through spring-loaded brushes that make contact with slip rings on the rotor shaft. These two slip rings are connected to the two ends of the field coil. When current is passed through the field coil, a magnetic field builds up on the rotor. As the rotor rotates, the magnetic field lines cut across the stator windings, creating ac voltages on the output of the three stator coils. This ac is then rectified by a full-wave bridge containing six diodes to produce the dc output.

THE WIND-POWER SYSTEM SIMULATOR

Before building the full-sized wind-power system, it may be advantageous to test the alternator and alternator field control circuits on a wind-power system simulator. Figure 3-2 shows a

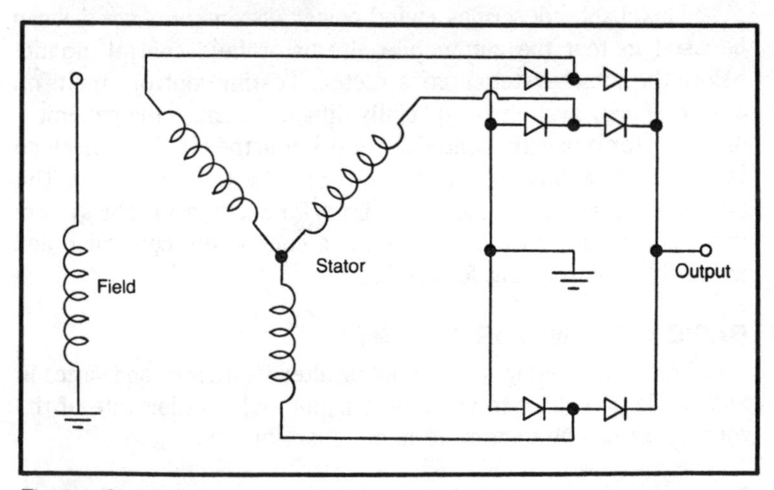

Fig. 3-1. Circuit diagram of a typical three-phase, full-wave rectified, automobile alternator.

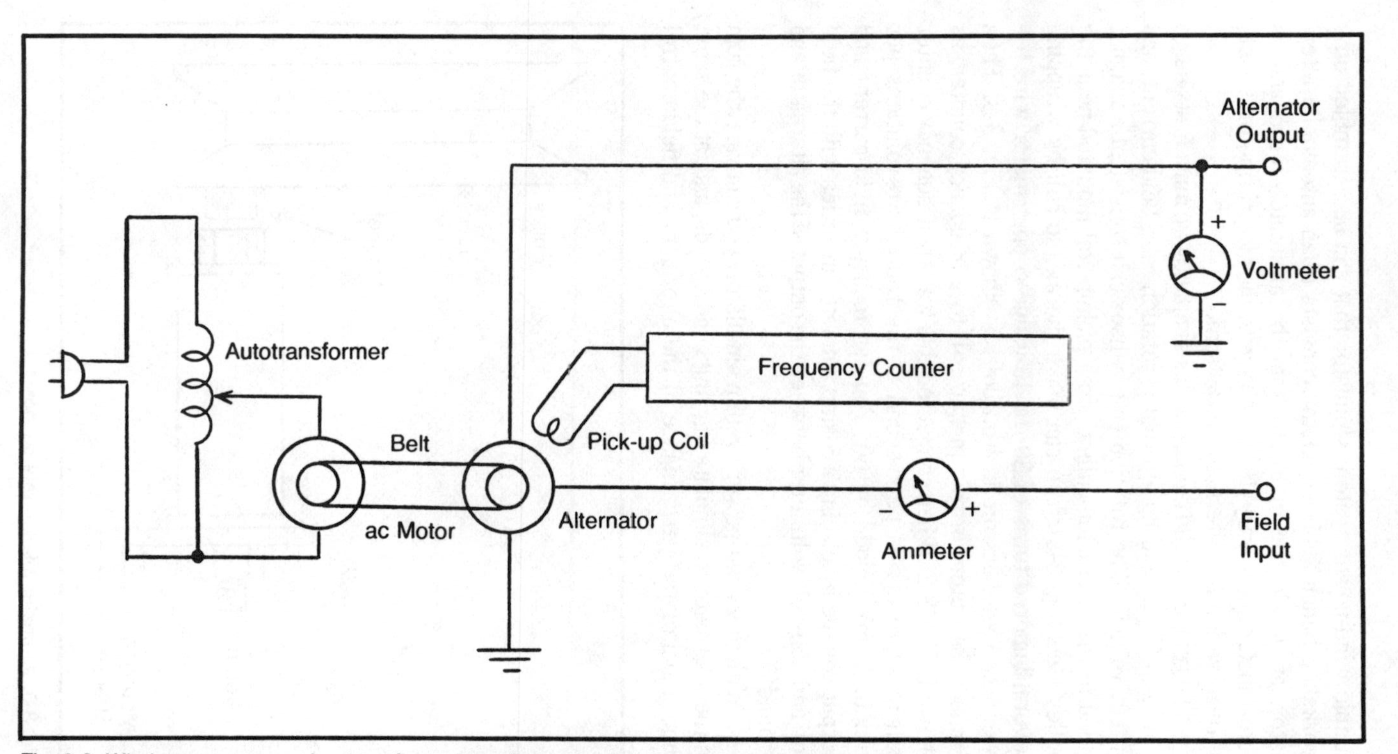

Fig. 3-2. Wind-power system simulator for testing alternators and alternator field control circuits.

simple wind-power system simulator that can be assembled on a workshop bench. This simulator consists of an automobile alternator, an ac electric motor to spin the alternator rotor, and an autotransformer to vary the ac voltage to the electric motor, which, in turn, varies the alternator rotor rpm.

To get meaningful data from this simulator we must first design a system for measuring alternator rotation rate, field current, and output voltage. The rotation rate sensor can consist of a magnet glued to the alternator pulley and a pickup coil mounted near the pulley (see Fig. 3-3). You can make the pickup coil by wrapping several hundred turns of 30- to 40-gauge copper motor wire between two washers on an iron bolt, as shown in Fig. 3-3. Then connect the output of this pickup coil to a frequency counter, as shown in Fig. 3-2. When connected this way the frequency counter will measure cycles per second of the rotor or revolutions per minute (rpm) divided by 60. You can measure field current and output voltage by placing a 5-amp ammeter in series with the field coil and a 50-volt voltmeter between the output of the alternator and ground.

With this wind-power system simulator, you can vary the rpm of the alternator to simulate varying wind speeds, and, at the same time, monitor the performance of the alternator and field control

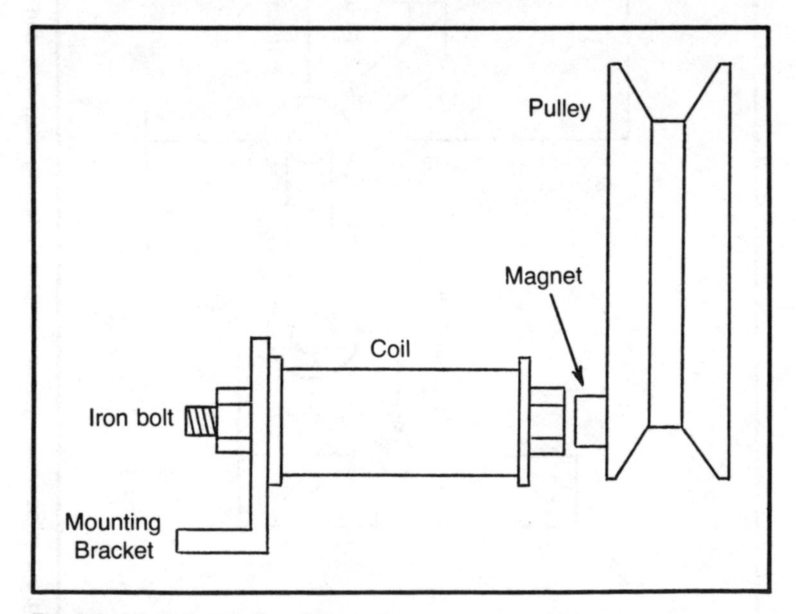

Fig. 3-3. Alternator rotation rate sensor.

circuits. Several possible field control circuits are discussed in the next three chapters.

AN ELECTRONIC FREQUENCY COUNTER

If no frequency counter is available for use on the wind-power system simulator, you can use the direct-readout anemometer circuit shown in Fig. 2-4. If you adjust the counting time interval t, discussed in Chapter 2, to exactly 1 second, the direct-readout anemometer will display input pulses per second. You may then place the Hall-effect sensor next to the magnet on the alternator pulley to produce one pulse per revolution. The output reading of the frequency counter (previous anemometer circuit) will be in revolutions per second and can be converted to rpm by multiplying by 60. Because there are only two seven-segment readouts, the maximum reading will be 99 revolutions per second or 5940 rpm. This should be more than adequate, because an alternator can begin producing power at less than 800 rpm.

Now the only problem is getting t to be equal to 1 second. You can do this with the circuit in Fig. 3-4. This circuit is a modified version of the anemometer power supply shown in Fig. 2-5 and allows the frequency-counter power supply to be a source of calibration. If you move the single-pole-double-throw switch in Fig. 3-4 to the upper position, a precise 60-cycle square wave is input to the 7490 counters. You can then adjust the calibration potentiometer R25 until the output on the display is 60. When this is done, t will be equal to 1 second and the circuit will be calibrated in revolutions per second.

Hence, the direct-readout anemometer can be used as both an anemometer and a frequency counter, and a three-pin connector can be used to plug in different Hall-effect sensors, as shown in Fig. 3-4. When the single-pole-double-throw switch is in the down position, the circuit can then be used as either an anemometer or frequency counter, depending on which Hall-effect sensor is connected. Once the circuit is calibrated as a wind-speed sensor, the calibration potentiometer's position can be marked. This will allow for alternate use of the circuit without having to recalibrate it as an anemometer each time it is used as such.

Note that if you use the circuit as a frequency counter, you must use the ac-to-dc (rather than dc-to-dc) power supply in order to obtain a 60-cycle square-wave signal for calibration; however, the 60-cycle square-wave generator shown in Fig. 3-5 does give a fairly accurate calibration if no utility power is available.

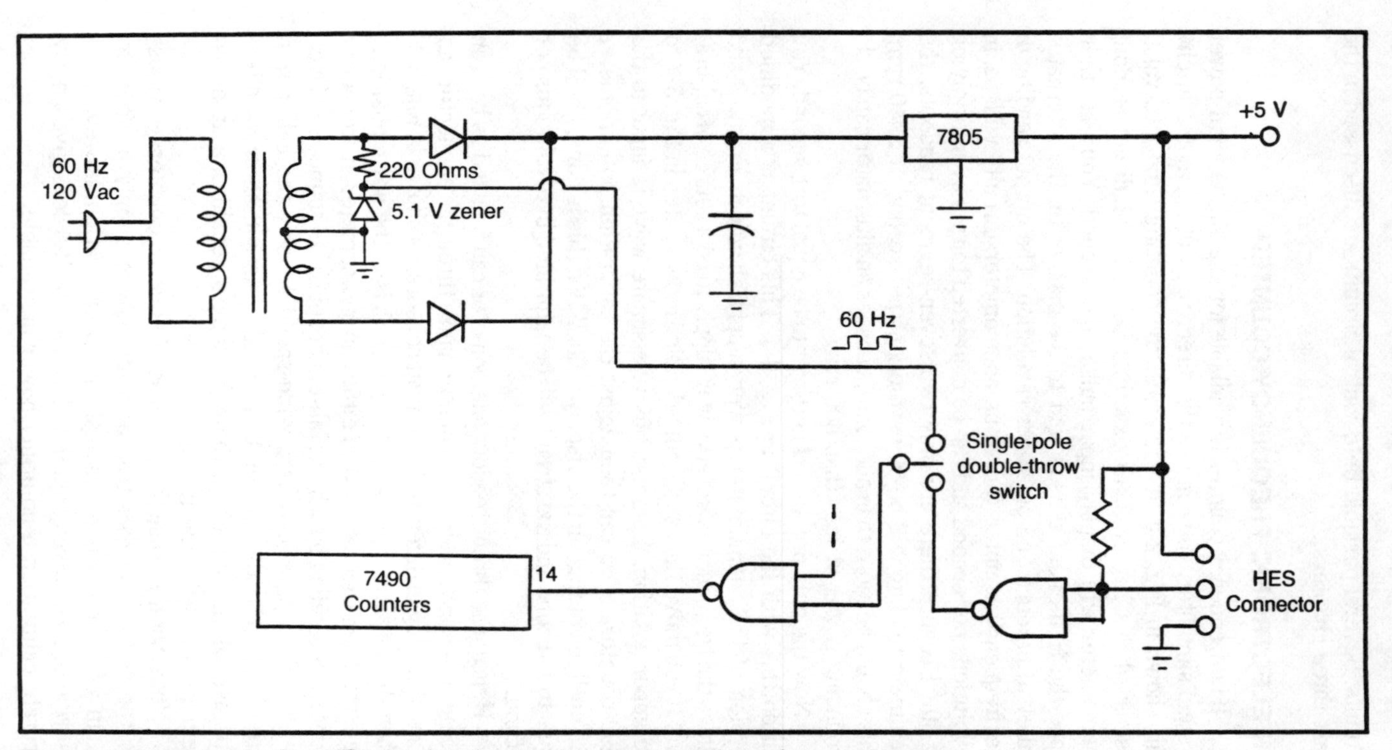

Fig. 3-4. Anemometer circuit modification that changes the anemometer into a frequency counter.

The circuit in Fig. 3-5 generates a square wave with a frequency of approximately 60 cycles per second. It is an astable multivibrator made from an NE555 timer and is similar to the timer circuit used in the direct-readout anemometer. This circuit can be built into the frequency counter with its output connected to the single-pole-double-throw switch in Fig. 3-4 instead of to the connection from the power supply. If you want a precise 60-cycle calibration source but no utility power is available, you can use the crystal-controlled 60-cycle source described in Chapter 8.

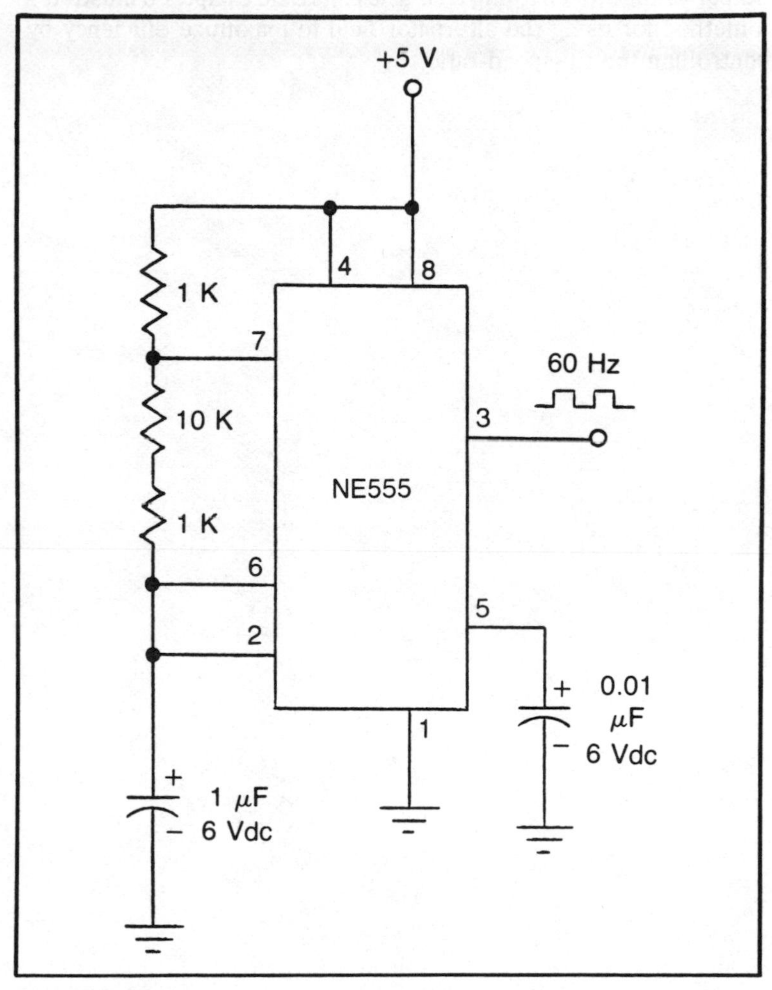

Fig. 3-5. A 60-cycle square-wave generator calibration source for the frequency counter shown in Fig. 3-4.

Now that the wind-power system simulator is complete with frequency counter, we can begin the alternator and control circuit testing. Using this simulator, we can design field control circuits to make the alternator's output voltage proportional to the rotor rpm, to make the output voltage remain constant regardless of rpm, or to make the output voltage vary for maximum system efficiency. Chapter 4 describes several field control circuits for charging nickel-cadmium batteries where the output voltage is proportional to rotor rpm. Chapter 5 contains two field control circuits that regulate output voltage for charging lead-acid batteries. Chapter 6 illustrates a method for using the alternator field to maximize efficiency by controlling the tip-speed-ratio.

4

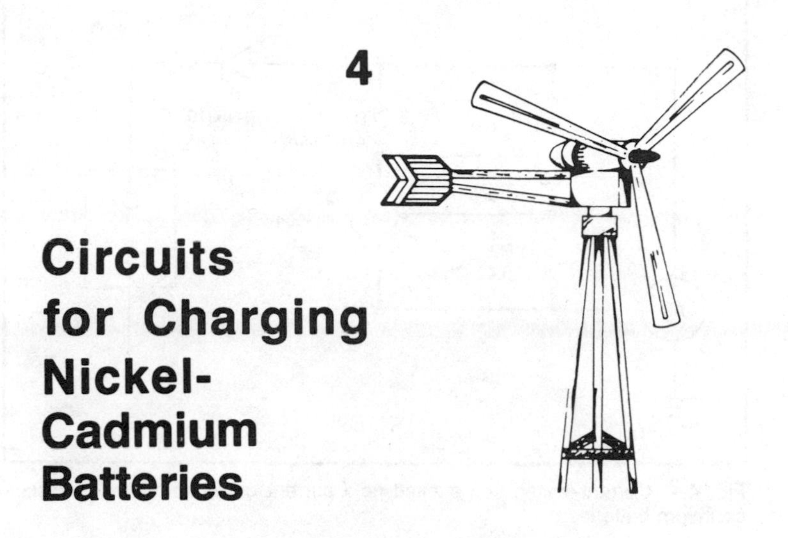

Circuits for Charging Nickel-Cadmium Batteries

Because nickel-cadmium batteries can withstand high charging currents, there is no need for voltage regulation when charging these batteries. This chapter describes three circuits that provide limited current to the alternator field coil and cause the alternator to simulate permanent-magnet dc generators without voltage regulation. With the field current held constant, the output voltage of the alternator is directly related to rpm.

A SIMPLE CURRENT-LIMITED FIELD CURRENT CIRCUIT

This section describes a simple field current circuit that utilitizes the residual magnetism in the alternator to produce its own field current. From Eq. 3-1 you can see that if M, the residual magnetism, is strong enough, the alternator will produce an output voltage even if the field current is zero (Moran 1976, p. 93). If the alternator output could be fed back into the field coil through a current-limiting circuit, this small output voltage could be used to energize the alternator. Figure 4-1 shows a simple circuit that does this.

In Fig. 4-1, the two transistors and two resistors form a current-limiting circuit between the alternator output and the field coil. This circuit was designed to limit field current to 1.27 amps. If no current-limiting circuit were employed here and the output of the alternator were routed directly into the field coil, a potentially dangerous situation would exist. At high rpm, the output voltage

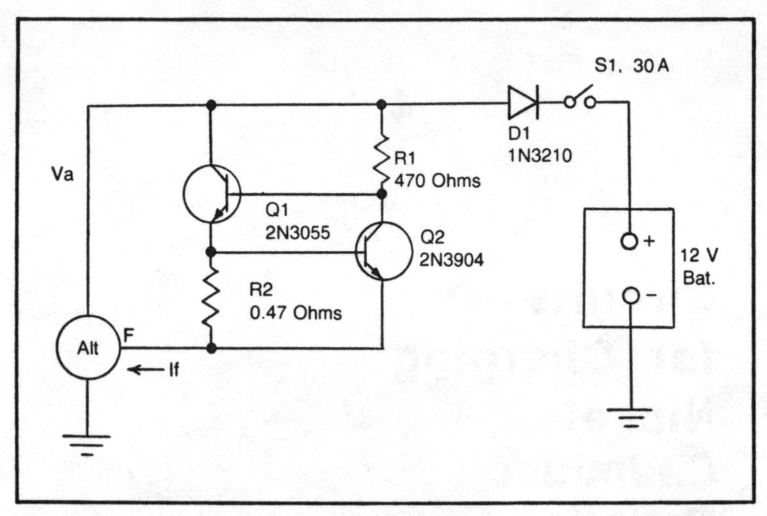

Fig. 4-1. Current-limited self-excited field current circuit for charging nickel-cadmium batteries.

would tend to increase rapidly in magnitude, and these high voltages could damage the field coil.

Diode D1 prevents battery current from flowing backward into the current-limiter circuit, while allowing alternator charging current to flow forward into the battery. Switch S1 was placed between the battery and the field current circuitry to provide a simple means for disconnecting the battery from the system. This switch could be used to prevent the battery from discharging through the field circuit if diode D1 were to become shorted during periods of low wind. Having a quick disconnect such as this could prevent damage to the battery in this type of situation.

The majority of the field current passes through transistor Q1 and resistor R2. The base current for Q1 is supplied through resistor R1. When the field current exceeds 1.27 amps, the voltage across R2 exceeds 0.6 volts, which is the base-to-emitter junction voltage of transistor Q2. At this point Q2 starts to drain the base current from Q1 and lowers the field current back to 1.27 amps.

By opening switch S1 and connecting a variable-voltage power supply to the alternator output, we generated the graph of Fig. 4-2, which illustrates the current-limiting properties of the circuit by graphing field current versus output voltage.

Some alternators have more residual magnetism than others. Usually, older ones, which have seen the most use, have higher values of residual magnetism. Experimentation has determined that

Va must reach at least 4 volts for the circuit to begin producing its own field current. From Eq. 3-1 you can see that the magnitude of M must be 0.005 volts/rpm for the output voltage to be 4 volts when If=0 and rpm=800. If the value of M were 0.0007 volts/rpm, the alternator would have to reach 5714 rpm before the output could reach 4 volts and the alternator could begin producing its own field current.

The circuit of Fig. 4-1 does not have a voltage-regulation capability and, therefore, is not recommended when voltage regulation is required. If you use automobile batteries, place enough in parallel to limit the individual charging current for each battery to a safe level. You can determine this by reading the voltage across the batteries while the alternator is operating at the maximum expected rpm. If the voltage exceeds 16 volts, place more batteries in parallel. This is an extremely simple circuit, and if it is used on an alternator with enough residual magnetism to produce the minimum 4 volts at a reasonable rpm and in a system where voltage regulation is not essential, it provides an inexpensive method for generating power.

A BATTERY-ASSISTED SELF-EXCITED FIELD CURRENT CIRCUIT

For alternators with low residual magnetism, the battery can be used to assist the field energizing process. The circuit of Fig. 4-3 does this by using the circuit of Fig. 4-1 with a 91-ohm resistor added between the battery and the field coil. This resistor allows

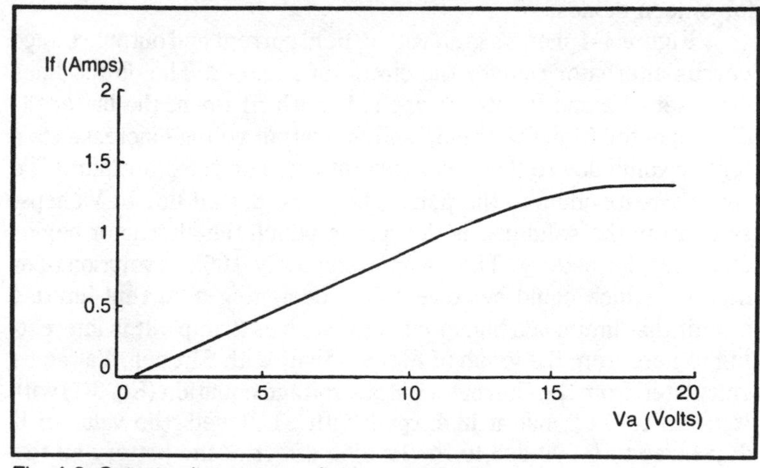

Fig. 4-2. Output voltage versus field current for the current-limiter circuit.

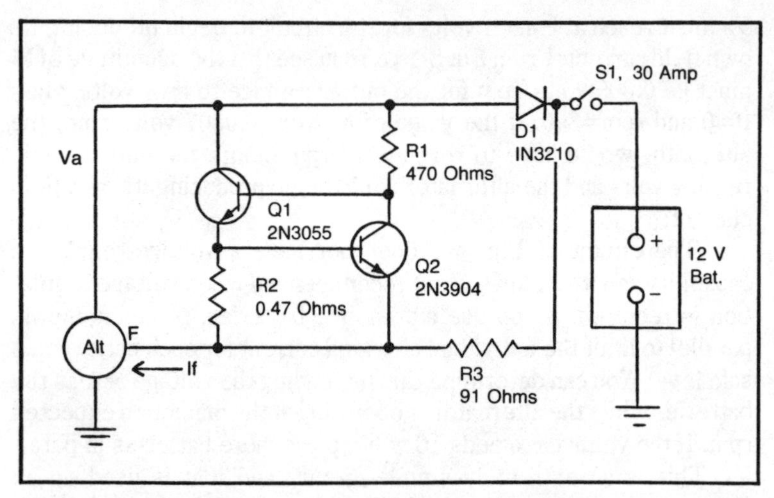

Fig. 4-3. Battery-assisted self-excited field current circuit for charging nickel-cadmium batteries.

0.13 amps of current to constantly flow through the coil. With the alternator spinning at sufficient rpm, this small amount of field current produces an output voltage, which, in turn, produces more field current through transistor Q1. More field current produces even more output voltage and, again, more field current. This process continues until the field current reaches 1.27 amps, or until the output voltage exceeds the value of the battery voltage. No current flows through diode D1 until the alternator voltage exceeds the battery voltage by 0.6 volts, which is the diode junction voltage for silicon diodes.

Figure 4-4 shows a graph of the field current and output voltage versus alternator rpm for the circuit of Fig. 4-3. The dotted lines represent Va and If with S1 opened. With S1 open, the battery is disconnected from the circuit, and the output voltage increases to a higher value due to the lower current drain on the alternator. The rpm corresponding to the point where the dotted line of Va separates from the solid line is the rpm at which the alternator begins charging the battery. This is approximately 1068 revolutions per minute, which could be lowered by designing a current-limiting circuit that limits at a higher current, such as 2 amps. It is interesting to note from the graph of Fig. 4-4 that, with S1 open, Va can be calculated from the alternator output voltage equation (Eq. 3-1) with K equal to 0.01, but at high rpm, with S1 closed, the value of K decreases to 0.009 due to the loading effect of the battery on the alternator.

The disadvantage of this circuit is the constant 0.13-amp current drain through resistor R3. This represents a constant 1.56-watt power consumption or 1.123 kilowatt-hours (kWh) per month. At 10 cents/kWh, this is insignificant. With a 150-amp-hour battery storage capacity (two typical car batteries in parallel), it would take more than a month and a half to discharge the batteries at this rate if no wind were blowing to keep the batteries recharged. The advantage of the circuit is its simplicity. For some people this may far outweigh the disadvantage.

A CURRENT-LIMITED FIELD CURRENT CIRCUIT WITH AUTOMATIC CUT-IN

One way to reduce the constant current drain characteristic of the previous circuit is to keep the field circuit shut down until an adequate charging rpm is reached. To do this, you need an rpm

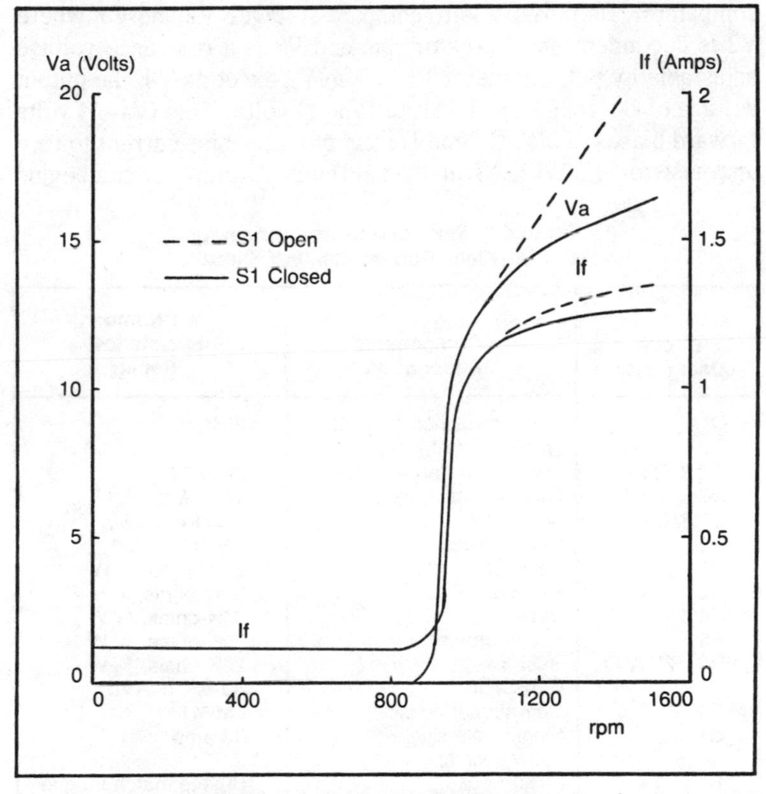

Fig. 4-4. Output voltage (Va) and field current (If) versus rpm for the battery-assisted self-excited field current-limiting circuit.

sensor. Figure 4-5 shows the field current-limiting circuit of Fig. 4-1 with an rpm sensor connected (parts list in Table 4-1). This rpm sensor keeps the field circuit shut off until a preset rpm is reached and then activates the field circuit through transistor Q3. The cut-in rpm is selected by adjusting potentiometer R5.

The rpm sensor utilizes the pickup coil configuration illustrated in Fig. 3-3. The ac output of the pickup coil, which is generated while the pulley is spinning, is rectified by diode D2 and filtered by resistor R4 and capacitor C1. The resulting dc output voltage of the filter is linearly related to rpm, as shown by the graph in Fig. 4-6. The slope of the pickup coil output voltage versus rpm graph depends on the number of turns of wire on the coil, the magnitude of the magnetic field, the permeability of the iron mounting bolt, and the spacing between the iron bolt and the magnet.

The operational amplifier (O1) in Fig. 4-5 acts as a voltage comparator. Its purpose is to compare voltages V2 and V3, where V2 is dependent on alternator rpm and V3 is a reference voltage adjustable by potentiometer R5. When V2 exceeds V3, the output voltage of O1 goes from 1 volt to (Va−1) volts. This (Va−1) volts forward biases diodes D3 and D4 and provides bias current to turn on transistor Q3. With Q3 on, the field current-limiter circuit begins

Table 4-1. Parts List for the Automatic Cut-in, Field Current-Limiting Circuit.

Circuit Designator	Component Description	Part Number, Resistance, Rating
Q1	High-power npn (mounted on a heat sink)	2N3055
Q2 & Q3	Low-power npn	2N3904
D2	Germanium diode	1N34A
D3 & D4	Silicon diode	1N914
D5	Silicon diode	1N5400
R1	Resistor	270 ohms, ½ W
R2	Resistor	0.47 ohms, 1 W
R4 & R9	Resistor	33K-ohms, ¼ W
R5	Potentiometer	20K-ohms, ¼ W
R6, R7, & R8	Resistor	20K-ohms, ¼ W
C1	Capacitor	30 μF, 5 WVdc
O1	Operational amplifier	LM741
S1	Single-pole-single-throw switch	30 amp
H1	Pickup coil	Built to match magnet

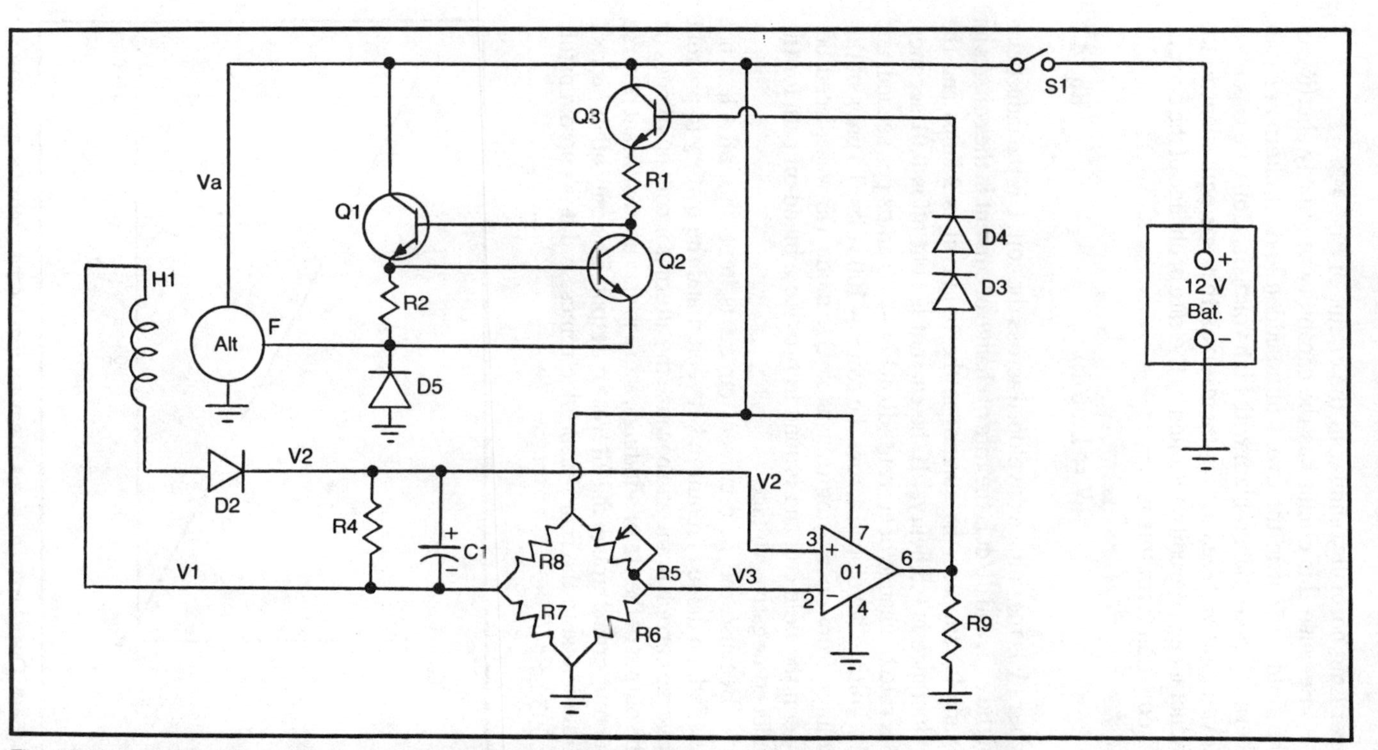

Fig. 4-5. Automatic cut-in, field current-limiting circuit for charging nickel-cadmium batteries.

operating normally, similar to the circuit of Fig. 4-3.

Because this circuit has the capability of quickly shutting off the field current in the event of a sudden loss of alternator rpm, there is a potential problem with high voltages. Coils are capable of developing voltages across themselves that are equivalent to the inductance of the coil multiplied by the rate of change of the current. In equation form this is

$$V = L(dI/dt) \qquad \textbf{Eq. 4-1}$$

where V is the voltage developed across the coil, L is the inductance of the coil, and dI/dt is the rate of change of current in the coil. As an example, suppose the current in the field coil is 2 amps and the inductance is 0.5 henrys. If the current is shut off in 0.01 seconds, the rate of change of current (dI/dt) is −200 amps per second, and the voltage developed across the coil is −100 volts. To prevent this voltage from destroying transistor Q1, diode D5 was placed between the field coil and ground to provide a bleed-off path for the induced negative voltage.

Because the field current circuit is kept shut off when not needed, no diode is required between it and the battery to prevent reverse current drain; and because the alternator contains diodes on the output of the stator windings, there is also no significant reverse current drain through the alternator output. The only other current drain is due to the rpm sensor electronics, and it is approximately 0.003 amps.

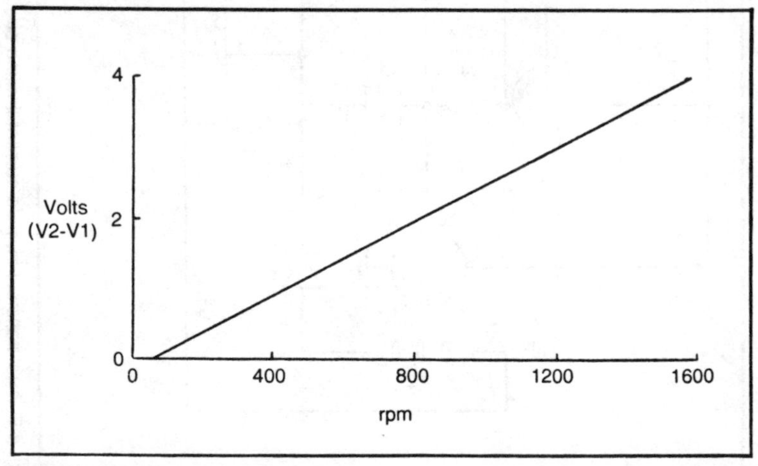

Fig. 4-6. Output voltage versus alternator rpm for the alternator rpm sensor.

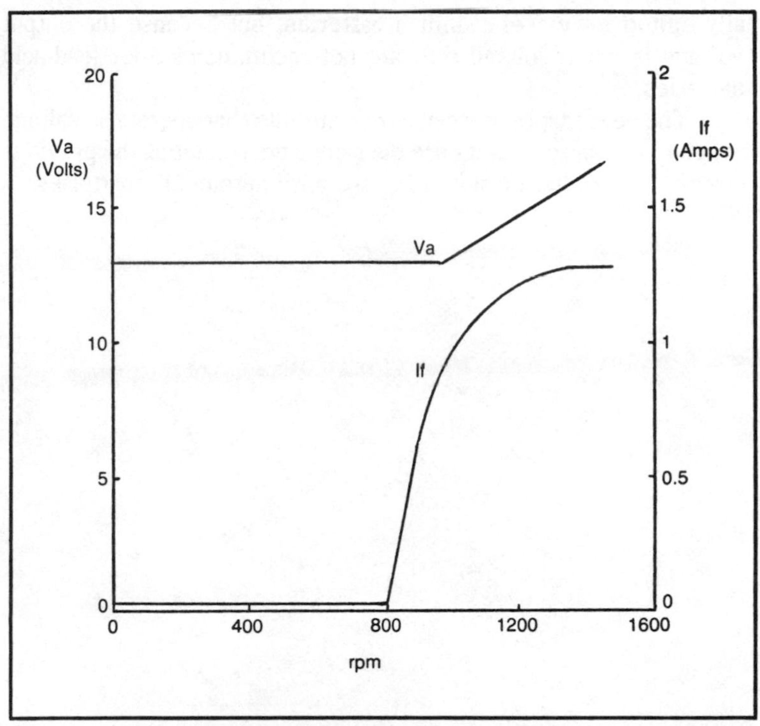

Fig. 4-7. Alternator output voltage (Va) and field current (If) versus rpm for the automatic cut-in, field current-limiting circuit.

A graph of the output voltage and field current versus rpm for the circuit of Fig. 4-5 is shown in Fig. 4-7. From the field current curve in Fig. 4-7 you can see that potentiometer R5 is adjusted for the field current to start cutting in at 800 rpm. At approximately 1000 rpm the field current is high enough for the alternator to begin charging the battery. At 0 rpm the field current is only 0.01 amps, as compared to 0.13 amps for the current-limiting circuit of Fig. 4-3. This 0.01 amps, plus the 0.003-amp current drain of the rpm sensor, is equivalent to the total of 0.013 amps of low-rpm current drain for the entire circuit of Fig. 4-5.

Notice from Fig. 4-7 that Va at low alternator rpm is 13 volts. Because there is no diode between the battery and alternator, the battery voltage is always present at the alternator output connector. The actual voltage of the battery connected to the alternator when this graph was generated was 13 volts.

The circuits discussed in this chapter cause the automobile alternator to simulate permanent-magnet generators. They are ideally suited for nickel-cadmium batteries, but because the output voltage is not regulated they are not recommended for lead-acid batteries.

The next chapter describes two circuits that operate as voltage regulators. These circuits use the same rpm sensor as the previous circuit and are better suited for use with automobile batteries.

5

Circuits for Charging Lead-Acid Batteries

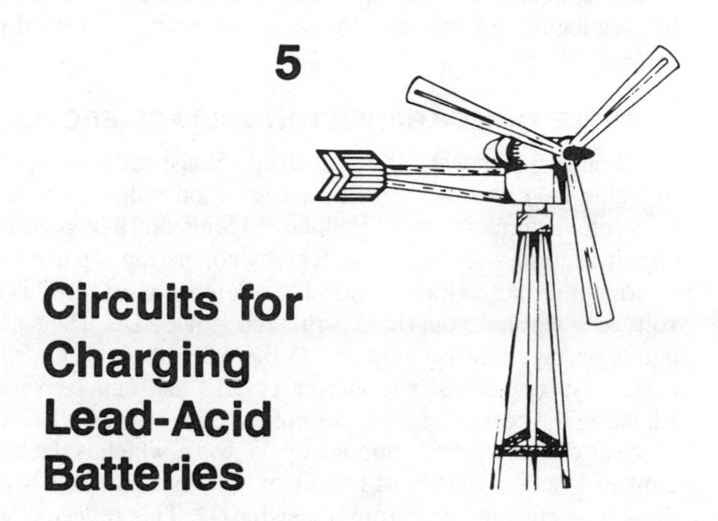

To keep costs to a minimum, some people may prefer to use lead-acid automobile batteries as the energy storage medium for their wind-power systems. In this case, to prevent damage to the batteries, the charging voltage must not be allowed to exceed 16 volts. Automobiles use *voltage regulators* connected to the alternator field coil to perform this function. These devices regulate alternator output voltage by limiting field current according to the magnitude of the output voltage. Remember from Eq. 3-1 that output voltage is directly proportional to field current and alternator rpm. Because the rpm is constantly changing, the only other parameter available for control is field current.

Although automobile voltage regulators can be used on wind-power systems for controlling voltage, they are not recommended for several reasons. First, to prevent the batteries from being discharged through the field coil, there must be some device for disconnecting these regulators during periods of low wind. Because relays and mechanical switches are unreliable, the only dependable way of doing this is to use an electronic sensing and switching device. After all this effort, it would be just as simple to build the electronic sensor and regulator in a single unit. Second, most older automobile voltage regulators have moving parts that tend to wear out quickly, especially with continuous use. If protected from weather, an electronic regulator could last 20 years or more. Third, automobile voltage regulators have a preset regulation voltage. A home-built electronic voltage regulator can be built with an adjustable regulation voltage, and thus add versatility to a wind-power system.

A SIMPLE THREE-TRANSISTOR VOLTAGE-REGULATOR

Figure 5-1 shows a transistorized voltage regulator (parts list in Table 5-1) with an adjustable regulation voltage and an rpm sensor for automatic cut-in (Halkias and Millman 1972, p. 700). This circuit regulates alternator voltage by comparing a portion of the alternator output voltage to a stable reference voltage. This stable voltage is derived from the 6.2-volt zener diode D6. The portion of output voltage that is compared to the 6.2 volts is taken from the center tap (wiper) of potentiometer R13 and can be varied by adjusting the potentiometer. When the voltage at the wiper of R13 exceeds the zener-diode voltage by 0.6 volts, which is the base-to-emitter junction voltage of transistor Q2, transistor Q2 begins to draw base current away from transistor Q1. This reduces the field current and forces the output voltage down until the voltage at the wiper of potentiometer R13 is no longer larger than (6.2+0.6) volts by any significant amount. In this way the output voltage of the

Table 5-1. Parts List for the Transistorized Voltage Regulator.

Circuit Designator	Component Description	Part Number, Resistance, Rating
Q1	High-power npn (mounted on a heat sink)	2N3055
Q2 & Q3	Low-power npn	2N3904
D2	Germanium diode	1N34A
D3 & D4	Silicon diode	1N914
D5	Silicon diode	1N5400
D6	Zener diode, 6.2 V	1N4735
R4 & R9	Resistor	33K-ohms, ¼ W
R5	Potentiometer	20K-ohms, ¼ W
R6, R7 & R8	Resistor	20K-ohms, ¼ W
R10	Resistor	100 ohms, ½ W
R11	Resistor	470 ohms, ¼ W
R12	Resistor	4700 ohms, ¼ W
R13	Potentiometer	10K-ohms, ¼ W
C1	Capacitor	30 μF, 5 WVdc
O1	Operational amplifier	LM741
S1	Single-pole-single-throw switch	30 amp
H1	Pickup coil	Built to match magnet

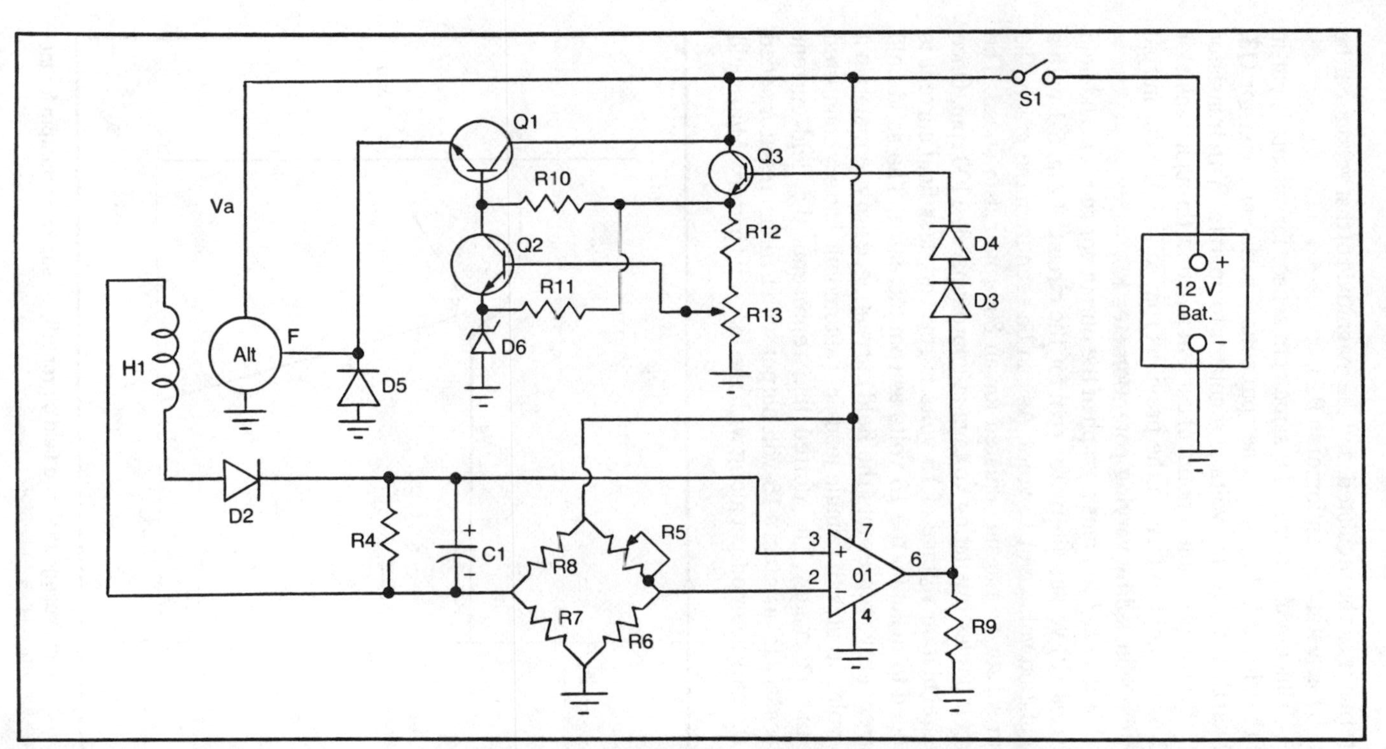

Fig. 5-1. Transistorized voltage regulator with an adjustable regulation voltage and an rpm sensor for automatic cut-in.

alternator can be regulated, and the magnitude of that voltage can be set by adjusting potentiometer R13.

This circuit utilizes the same rpm sensor for automatic cut-in as the circuit in Fig. 4-5. The rpm sensor turns on transistor Q3, which then allows the voltage regulator to operate. With transistor Q3 off, no appreciable amount of current will flow through transistor Q1 or the field coil. As in the previous rpm sensor, the cut-in rpm can be adjusted by varying potentiometer R5.

Figure 5-2 contains a graph of the output voltage (Va) and field current (If) versus alternator rpm for the circuit of Fig. 5-1 with a regulation voltage of 15 volts. As can be seen from the graph, the rpm sensor was again adjusted for an 800-rpm cut-in speed. The circuit begins regulating voltage at approximately 1100 rpm. Once the regulation voltage of 15 volts is reached, less field current is needed to maintain the 15 volts as rpm increases. The solid field current curve represents the field current necessary to maintain a 15-volt output with a single lead-acid automobile battery connected to the alternator. The dotted line represents the field current necessary to maintain a 15-volt output with a 5-ohm resistor added to the alternator output along with the lead-acid battery. More field

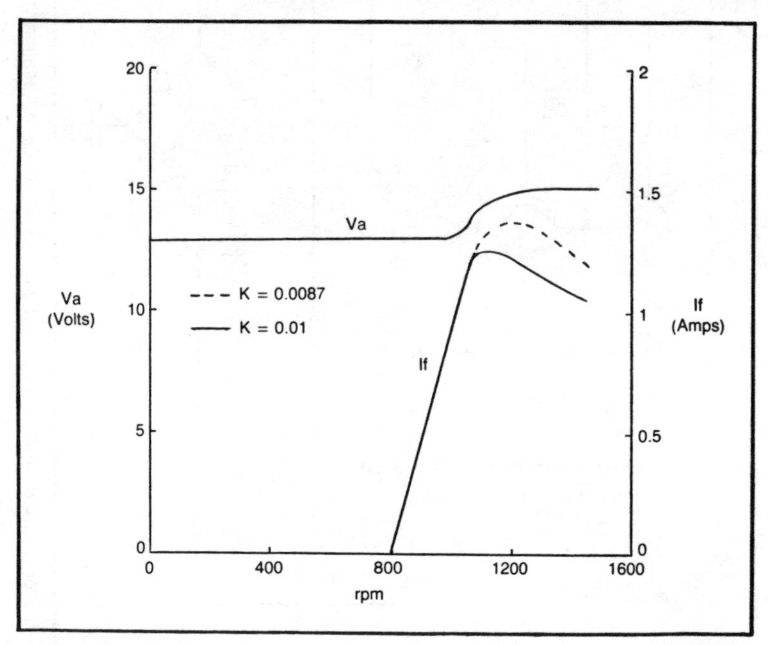

Fig. 5-2. Output voltage (Va) and field current (If) versus alternator rpm for the transistorized voltage regulator.

current is needed to maintain the 15 volts when the alternator is loaded because the constant K decreases with increased load. Using the values of Va and If (solid line) at 1400 rpm, we calculate K to be 0.01 volts/(amp)(rpm). Using If from the dotted line at 1400 rpm, we get K = 0.0087 volts/(amp)(rpm). These results show that the regulator automatically supplies more field current to maintain the 15 volts when the alternator has an increased load on the output.

There is one disadvantage to this particular regulator circuit. Because of its design, when transistor Q3 is on, the voltage across the field coil will never go lower than the zener-diode (D6) voltage minus the base-to-emitter junction voltage of transistor Q1. If a 6.2-volt zener is used, the voltage across the field will not go below (6.2–0.6) volts. Since the field resistance is approximately 6.3 ohms, Ohm's law dictates that the field current never goes below (6.2–0.6) volts/6.3 ohms or 0.888 amps. From Eq. 3-1, with K = 0.01 volts/(amp)(rpm), we can see that, if field current can never go below 0.888 amps, the circuit can never regulate at 15 volts once the rpm exceeds 1689. When the rpm exceeds 1689, the voltage must begin to increase. Using a lower-voltage zener diode would help, but it would not eliminate the problem. Suppose a 3.9-volt zener was used. In this case, the field current cannot go below 0.524 amps. With this field current, a 15-volt output regulation can be maintained up to an rpm of 2862. Remember, however, that when the alternator output is loaded, the magnitude of K is reduced. This also will extend the maximum rpm for voltage regulation. Even with this disadvantage, the transistorized regulator's worth should not be overlooked. It is simple to build and regulates voltage very well at moderately low rpm. For those systems that will be subjected to higher rpm, the circuit described in the next section is required.

A PRECISION OPERATIONAL AMPLIFIER VOLTAGE REGULATOR

The voltage regulator described in this section is an extremely precise regulator that has no maximum rpm for voltage regulation. At any rpm this circuit has the capability of completely shutting off the field current if necessary. The schematic for this regulator circuit is shown in Fig. 5-3 (parts list in Table 5-2). Notice again that the same rpm sensor was used in this circuit as in all the other automatic cut-in circuits. This time, however, instead of providing bias current for a transistor, the rpm sensor provides the reference voltage developed across the 6.2-volt zener diode D6.

This regulator, like the previous one, regulates alternator

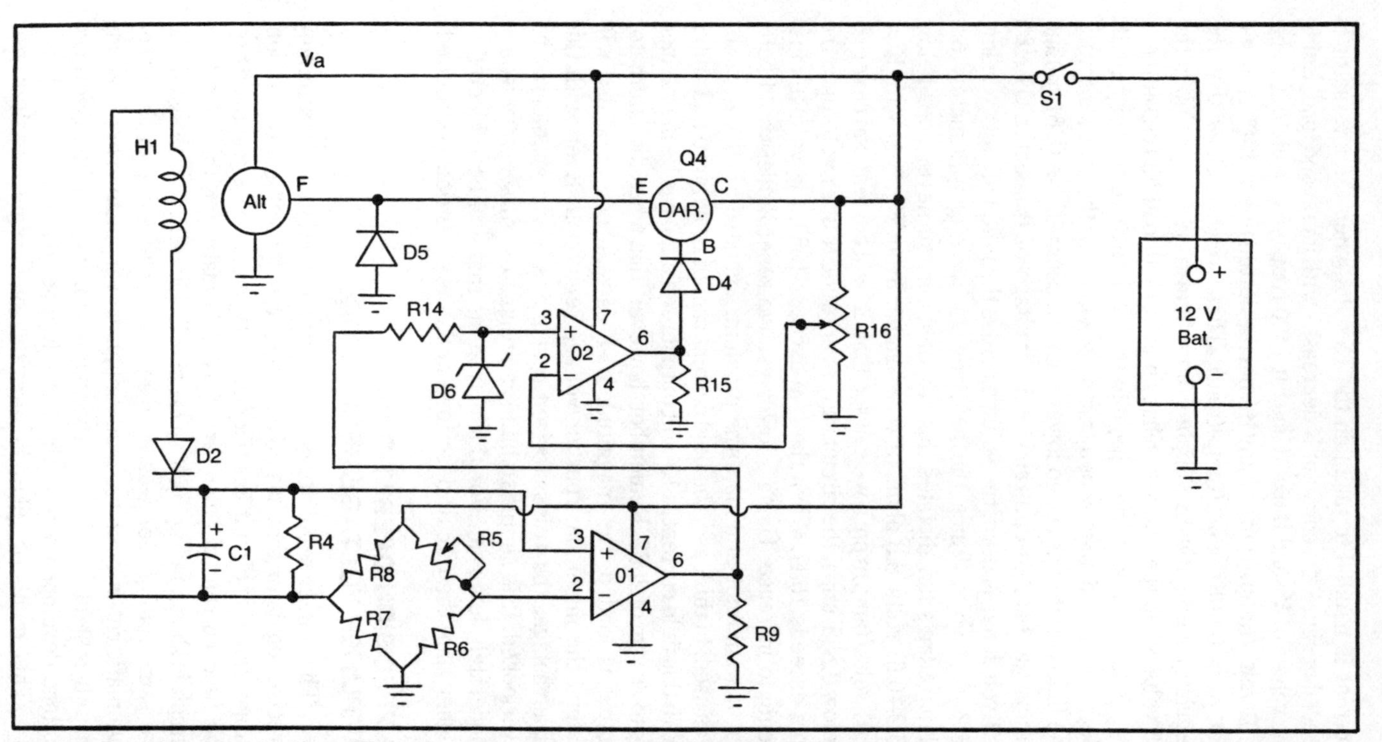

Fig. 5-3. Precision, operational amplifier voltage regulator with an adjustable regulation voltage and an rpm sensor for automatic cut-in.

Table 5-2. Parts List for the Precision, Operational Amplifier Voltage Regulator.

Circuit Designator	Component Description	Part Number, Resistance, Rating
Q4	High-power npn Darlington (mounted on a heat sink)	2N6576
D2	Germanium diode	1N34A
D4	Silicon diode	1N914
D5	Silicon diode	1N5400
D6	Zener diode, 6.2 V	1N4735
R4 & R9	Resistor	33K-ohms, ¼ W
R5	Potentiometer	20K-ohms, ¼ W
R6, R7 & R8	Resistor	20K-ohms, ¼ W
R14	Resistor	330 ohms, ¼ W
R15	Resistor	33K-ohms, ¼ W
R16	Potentiometer	50K-ohms, ¼ W
C1	Capacitor	30 μF, 5 WVdc
O1 & O2	Operational amplifier	LM741
S1	Single-pole-single-throw switch	30 amp
H1	Pickup coil	Built to match magnet

output voltage by comparing the output voltage to the reference voltage developed across the zener diode D6. The desired regulation voltage can be selected by adjusting potentiometer R16. When the voltage at the wiper of R16 exceeds the voltage across D6 (6.2 volts), the output voltage of amplifier O2 decreases. Because transistor Q4 is placed in the emitter-follower configuration and diode D4 is in the base circuit, the voltage across the field must be approximately equal to the output voltage of amplifier O2 minus 1.8 volts. Because of this, when the output of amplifier O2 decreases, the voltage across the field also decreases, which, in turn, decreases the field current and alternator output voltage. Similarly, when the voltage at the wiper of potentiometer R16 goes below 6.2 volts, the output of amplifier O2 increases, causing the field current and output voltage to increase. Because the gain of the LM741 operational amplifier O2 is extremely high, this regulator circuit will regulate voltages to about the same accuracy as the zener diode D6 regulates its 6.2 volts.

Zener diode D6 receives its voltage from the rpm sensor. Once alternator rotation rate is below the cut-in rpm, the voltage across D6 decreases from 6.2 volts to the minimum output voltage of amplifier O1, which is about 1 volt for the LM741. When this happens, the output voltage of amplifier O2 also drops to its minimum voltage. Because the field voltage is always 1.8 volts less than the output voltage of amplifier O2, the field voltage must be at most 0 volts when the output of O2 drops to 1 volt.

Figure 5-4 shows a graph of output voltage and field current versus rotation rate for the precision voltage regulator. As in Fig. 5-2, the solid lines represent field current and output voltage with a single lead-acid automobile battery connected to the alternator output, and the dotted lines represent field current and output voltage with a 5-ohm resistor added to the output with the lead-acid battery. As in the previous all-transistor regulator, this regulator circuit automatically supplies more field current to maintain the preset output voltage when a load is added to the alternator output. Also, by comparing this graph to that in Fig. 5-2, you can see, by the straight lines and sharp curves, that this circuit is much more precise.

This precision regulator is highly recommended for use with lead-acid batteries. It regulates voltage at any rpm greater than 1100 and produces minimum low-rpm current drain. The low-rpm current drain for the entire circuit of Fig. 5-3 is only 0.04 amps.

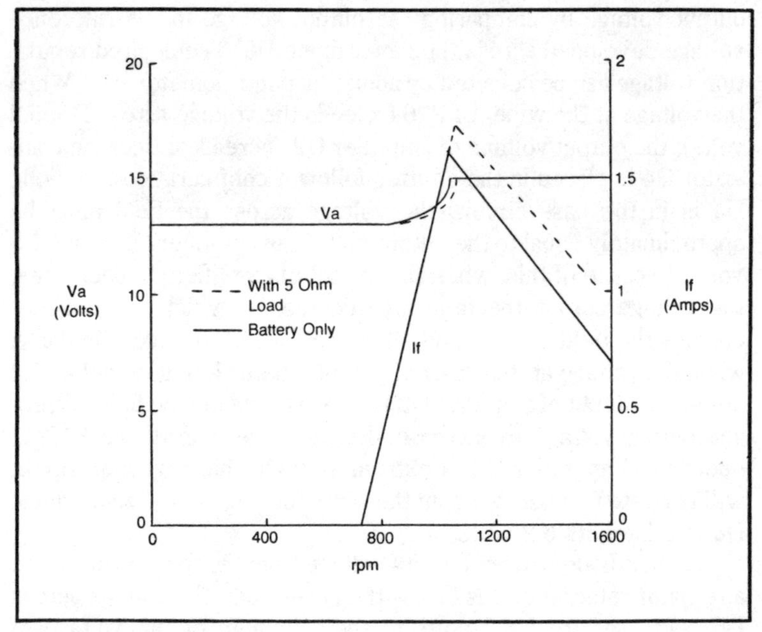

Fig. 5-4. Output voltage (Va) and field current (If) versus alternator rpm for the precision, operational amplifier voltage regulator.

The two preceding voltage-regulator circuits, though good for protecting lead-acid batteries from overcharging, are not the most efficient means for generating power. During high winds these circuits will not extract all available power, due to the higher priority of protecting the batteries. The next chapter discusses a circuit with the single priority of extracting the most power possible from the wind.

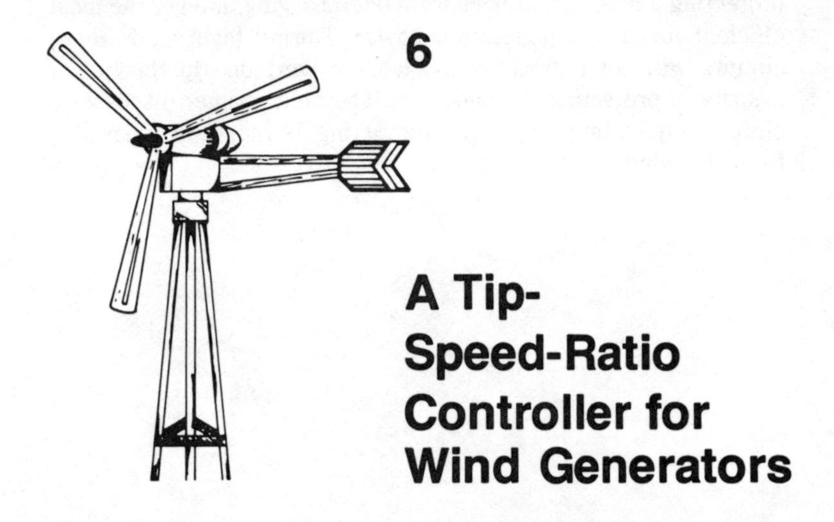

6

A Tip-Speed-Ratio Controller for Wind Generators

Tip-speed-ratio is the speed of the turbine tip divided by the speed of the wind. Any wind turbine will reach a maximum efficiency at some particular tip-speed-ratio (TSR), and this optimum TSR varies with different blade designs. This chapter describes a TSR controller with a variable tip-speed-ratio to accommodate these various designs.

THE TIP-SPEED-RATIO CONTROLLER

The TSR controller circuit is shown in Fig. 6-1 (parts list in Table 6-1). This circuit, a modified version of one first designed by Rob Smith of Jamestown, Rhode Island (Smith 1977, 38), controls the TSR by simultaneously monitoring wind speed and tip speed and using the difference to control field current. Wind speed is measured directly with an anemometer (or miniturbine), but tip speed is measured indirectly by measuring alternator rpm. Tip speed (TS) is directly related to alternator rpm (Arpm) according to the following equation:

$$TS = (Arpm)\,(R)\,(0.0714)/(GR) \qquad \textbf{Eq. 6-1}$$

In Eq. 6-1, TS is the tip speed measured in miles per hour, R is the turbine radius measured in feet, and GR is the gear ratio between the alternator and the turbine. The relationship between GR, Arpm, and turbine rpm (Trmp) is shown in Eq. 6-2.

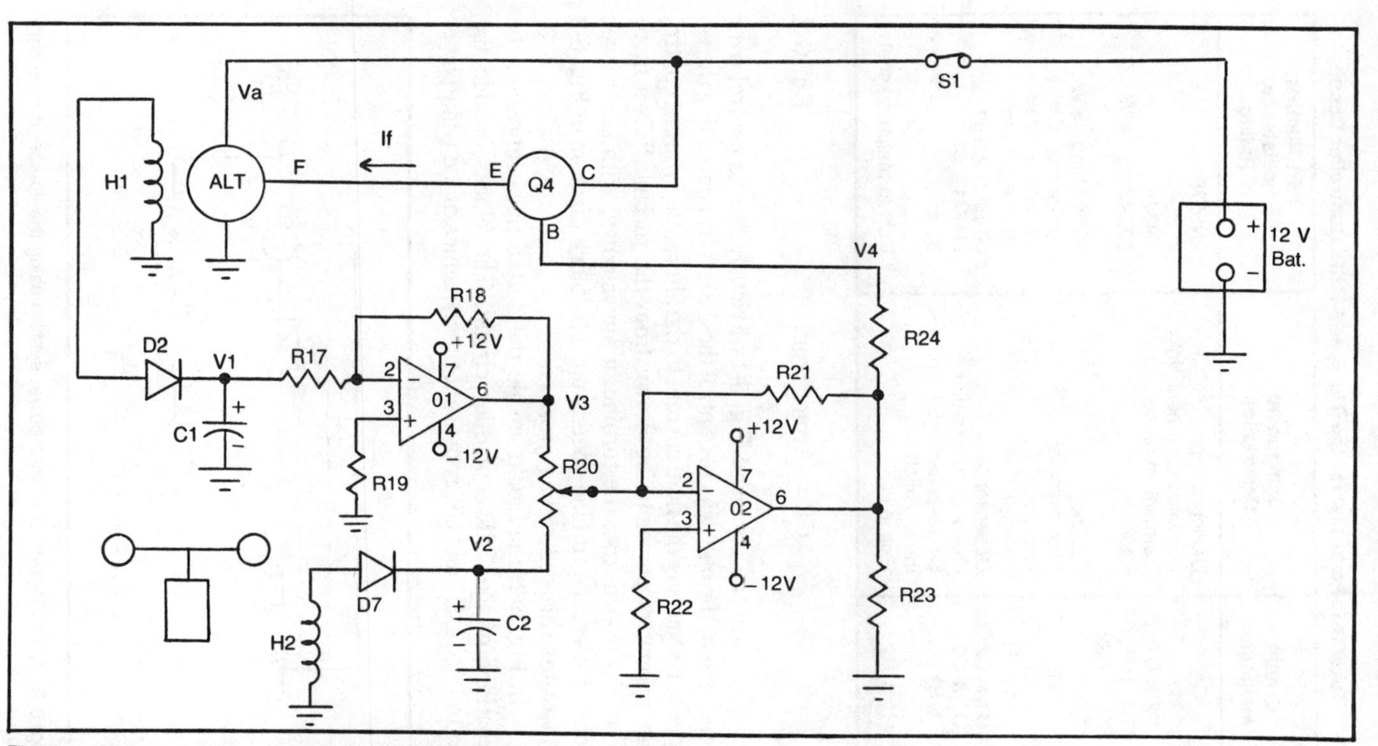

Fig. 6-1. Tip-speed-ratio controller circuit for maximizing output power.

Table 6-1. Parts List for the Tip-Speed-Ratio Controller Circuit.

Circuit Designator	Component Description	Part Number, Resistance, Rating
Q4	Darlington npn (mounted on a heat sink)	2N6576
D2 & D7	Germanium diode	1N34A
R17, R18, R19 & R22	Resistor	10K-ohms, ¼ W
R21	Resistor	100K-ohms, ¼ W
R20	Potentiometer	10K-ohms, ¼ W
R23	Resistor	33K-ohms, ¼ W
R24	Resistor	1K-ohms, ¼ W
C1 & C2	Capacitor	30 μF,5 WVdc
O1 & O2	Operational amplifier	LM741
S1	Single-pole-single-throw switch	30 amp
H1 & H2	Pickup coil	Built to match magnet

$$GR = Arpm/Trpm \qquad \textbf{Eq. 6-2}$$

Because the circuit of Fig. 6-1 obviously has no control over wind speed, the only way to control the TSR is by controlling the tip speed. By varying the field current to the alternator, you can control the amount of power being drawn from the turbine. When more power is drawn from the turbine, it slows down; with less power being drawn, the turbine speeds up. The block diagram of Fig. 6-2 illustrates this controlling process.

In Fig. 6-2, the blocks inside the dotted lines represent the functions of the TSR controller circuit. The blocks outside the dotted lines, G7 and G8, represent the dynamics of the wind-power

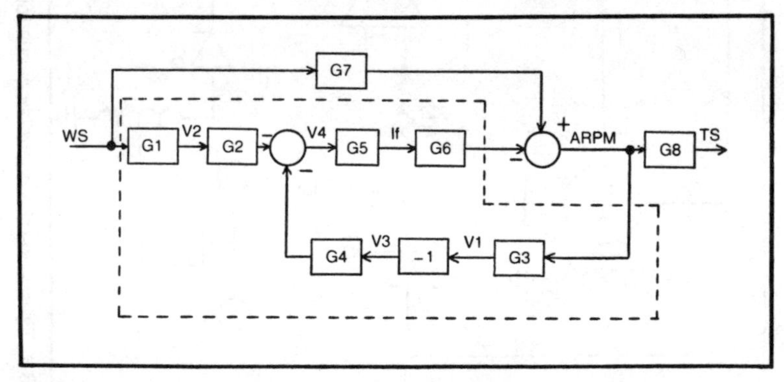

Fig. 6-2. Block diagram of a wind-power system using the tip-speed-ratio controller.

system and the wind itself. G7 represents the no-load dynamic gain of the turbine and the alternator. In other words, this represents the gain in Arpm for a given increase in wind speed when there is no field current supplied to the alternator. The magnitude of G7 will largely depend on the size and design of the turbine. G8 is simply a constant relating tip speed to Arpm and can be derived from Eq. 6-1 as follows:

$$TS = (Arpm)\,(R)\,(0.0714)/(GR) \qquad \textbf{Eq. 6-1}$$

$$G8 = TS/Arpm = (R)\,(0.0714)/(GR) \qquad \textbf{Eq. 6-3}$$

In Fig. 6-2, blocks G1 through G6 represent the functions of the TSR controller circuit. G1 and G3 represent the gains of the wind miniturbine and Arpm sensors, respectively. The rpm sensor operates exactly as the one discussed in Chapter 3 and includes diode D2, capacitor C1, and coil H1 of Fig. 6-1. The placement of the pickup coil H1 is illustrated in Fig. 3-3. The magnitude of G3 varies, depending on the construction of the pickup coil, but should be about 0.0025 volts/rpm. The wind-speed sensor operates much the same as the Arpm sensor and will be discussed later. The magnitude of G1 should be about 0.1 volts/mph.

The block after G3 simply shows a negative "1." This represents the unity gain inverting amplifier O1. G4 and G2 represent the gain of amplifier O2 for their particular paths of the block diagram. The two gains are adjustable by varying potentiometer R2O and are interdependent. When R2O is varied, either G4 increases and G2 decreases, or G2 increases and G4 decreases. The actual relationship between the two gains is given by Eq. 6-4:

$$G4 = G2(100)/(G2(10) - 100) \qquad \textbf{Eq. 6-4}$$

The purpose of amplifier O2 is to sum the two signals (wind speed and negative Arpm) and then invert the resulting sum. The total effect at this point is that we have a signal proportional to the difference of Arpm and wind speed. This voltage, V4, which is proportional to (Arpm-WS), is then applied to the Darlington transistor Q4, where it is converted to field current.

G5 represents the gain of Q4 with the units of amps/volt. Because the field resistance is 6.3 ohms and Q4 is in the emitter-follower configuration, G5 is equal to 1/6.3, or 0.1587, amps/volt. The units of G6 are rpm/amp. G6 simply represents the change in Arpm when a given amount of field current is supplied to the alternator. The actual value of G6 depends on the size of the load connected to the alternator output when field current is applied. An estimate of G6 with a light load on the alternator output is approximately 200 rpm/amp.

Now that all the elements of the block diagram have been explained, we can draw another block diagram (see Fig. 6-3) containing the actual gain values. Using this block diagram, we obtain a complete system equation relating TS to WS. First, however, we need to assign values to G7 and G8. For this example let the turbine radius be 5 feet and the gear ratio 9. With these values and Eq. 6-3, G8 = 0.03967. Also, suppose that G7 is approximately equal to 176.4 rpm/mph. Using these values in Fig. 6-3, we can derive the system equation:

$$TS = WS \left[\frac{(0.0396)\,[(0.01587)(G2)(G6) + 176.4]}{1 + \dfrac{(G2)(G6)(0.039675)}{(G2)(10) - 100}} \right] \qquad \textbf{Eq. 6-5}$$

If a constant load is assumed, G6 will be constant. If a value for G6, such as 200, is inserted, the equation becomes a function of G2 only. If the function in brackets is now called f(G2), Eq. 6-5 can be written as

$$TSR = TS/WS = f(G2) \qquad \textbf{Eq. 6-6}$$

As long as G6 is constant, the tip-speed-ratio is a function of G2 only, which can be set by adjusting potentiometer R20.

The only disadvantage of this circuit is that TSR is a function of load, because G6 changes with load. If G6 is 200 and G2 is 20 then

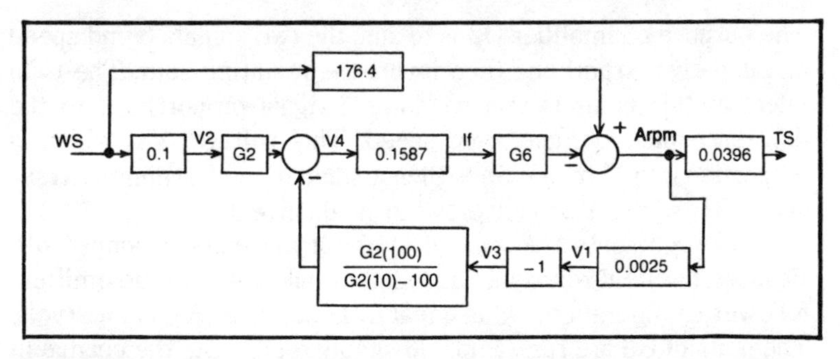

Fig. 6-3. Block diagram of a wind-power system using the tip-speed-ratio controller and having a turbine radius of 5 feet and a gear ratio of 9.

TSR is 3.67. If G6 changes to 300 (larger load), the tip-speed-ratio changes to 4.518. This could be a problem if the output load is constantly changing.

One advantage of this TSR controller circuit is that it does not require a separate cut-in circuit to prevent current drain during periods of low wind. When voltage V1 (the Arpm sensor output) drops, the output voltage of amplifier 02 also drops and stops field current from flowing. If no wind is blowing, neither sensor will produce a voltage, and the output of amplifier 02 will again be a minimum. When the wind beings to blow, the miniturbine starts first and the wind-charger turbine follows. The alternator field current does not flow until the tip speed of the wind-charger turbine catches up with the wind speed according to the TSR previously set. This provides minimum starting torque and allows the main turbine to reach optimum rpm before the alternator turns on.

MAKING THE TSR CONTROLLER INSENSITIVE TO LOAD CHANGES

If the circuit of Fig. 6-1 is modified by placing an integrator between amplifier 02 and transistor Q4, the tip-speed-ratio will no longer be a function of output load. This circuit modification is shown in Fig. 6-4.

Because each operational amplifier is also an inverter, you must use two stages in this circuit modification to prevent having a sign change. Figure 6-4 shows the integrator connected immediately following amplifier 02. The operational amplifier 03 is converted to an integrator by placing two capacitors in the feedback path. Two capacitors need to be used back-to-back because the polarity of the voltage across the capacitor can be either positive or negative. Placing the capacitors back-to-back prevents leakage current during reverse voltage. The second stage of the circuit modification is simply an inverter to change the sign of the output voltage back to its original polarity before it is input to transistor Q4.

With the integrator modification, the block diagram of Fig. 6-3 can be redrawn as shown in Fig. 6-5. The output of the integrator is a function of time and can be represented as kt for simplicity, where k is some constant and t is time in seconds. The system equation now becomes

$$TS = WS \left[\frac{(0.0396)\,[(kt)\,(0.01587)(G2)(G6) + 176.4]}{1 + \dfrac{(kt)(G2)(G6)(0.039675)}{(G2)(10) - 100}} \right] \qquad \textbf{Eq. 6-7}$$

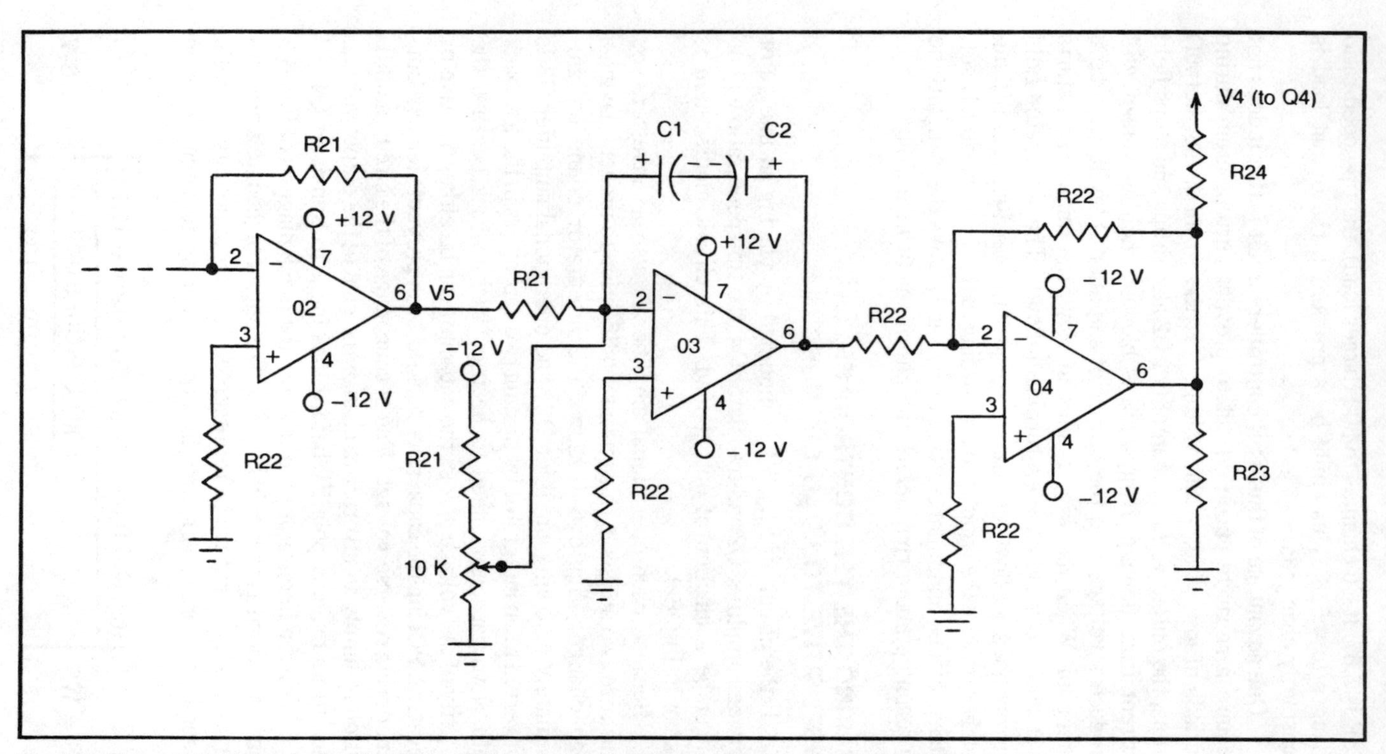

Fig. 6-4. Tip-speed-ratio controller circuit modification for making tip speed insensitive to load changes.

The limit of TS as t goes to infinity is

$$\lim_{t \to \infty} TS = WS\,[0.1584\,(G2) - 1.584] \qquad \textbf{Eq. 6-8}$$

This can also be written as

$$TS/WS = 0.1584(G2) - 1.584 \qquad \textbf{Eq. 6-9}$$

where TS/WS is the tip-speed-ratio. If G2 = 10 (the minimum value possible), TS/WS = 0. If G2 = 50, TS/WS = 6.336. Equation 6-9 shows no upper limit on TS/WS; however, the maximum value is determined by mechanical limitations of the system.

One disadvantage with this circuit is the possibility of integrator drift during periods of low wind. This problem can be eliminated by placing the 10K-ohm potentiometer between −12 volts and the input of amplifier O3 in Fig. 6-4 and adjusting it to force the integrator output to drift high when no input is applied by amplifier O2. When the integrator output drifts high, the output of O4 goes low, and no field current will flow to discharge the batteries.

There is also a possibility that the system can become unstable once the circuit modification is installed in the tip-speed-ratio controller. This is a situation where the output of the TSR controller circuit continually oscillates between minimum and maximum voltages. Fortunately, this problem can easily be corrected by using larger feedback capacitors on amplifier O3. The instability of this system with the integrator modification is a function of the turbine's moment of inertia. In other words, if the turbine is heavy and responds slowly to changes in wind speed or alternator field current, the greater is the chance for instability. Placing larger capacitors in the feedback of amplifier O3 simply slows the integrator down until it responds more slowly than the turbine. Once this is done, the output of the integrator will not overshoot, and the system is stable.

Also notice from Figs. 6-1 and 6-4 that amplifiers O1, O2, O3, and O4 all require positive and negative 12 Vdc to operate. In this case one additional battery is needed to supply the −12 volts. The battery configuration required for positive and negative 12 volts is shown in Fig. 6-6. Because the −12-volt battery can not be charged by the alternator during normal operation, the batteries must be switched periodically to prevent discharging of the negative 12-volt battery.

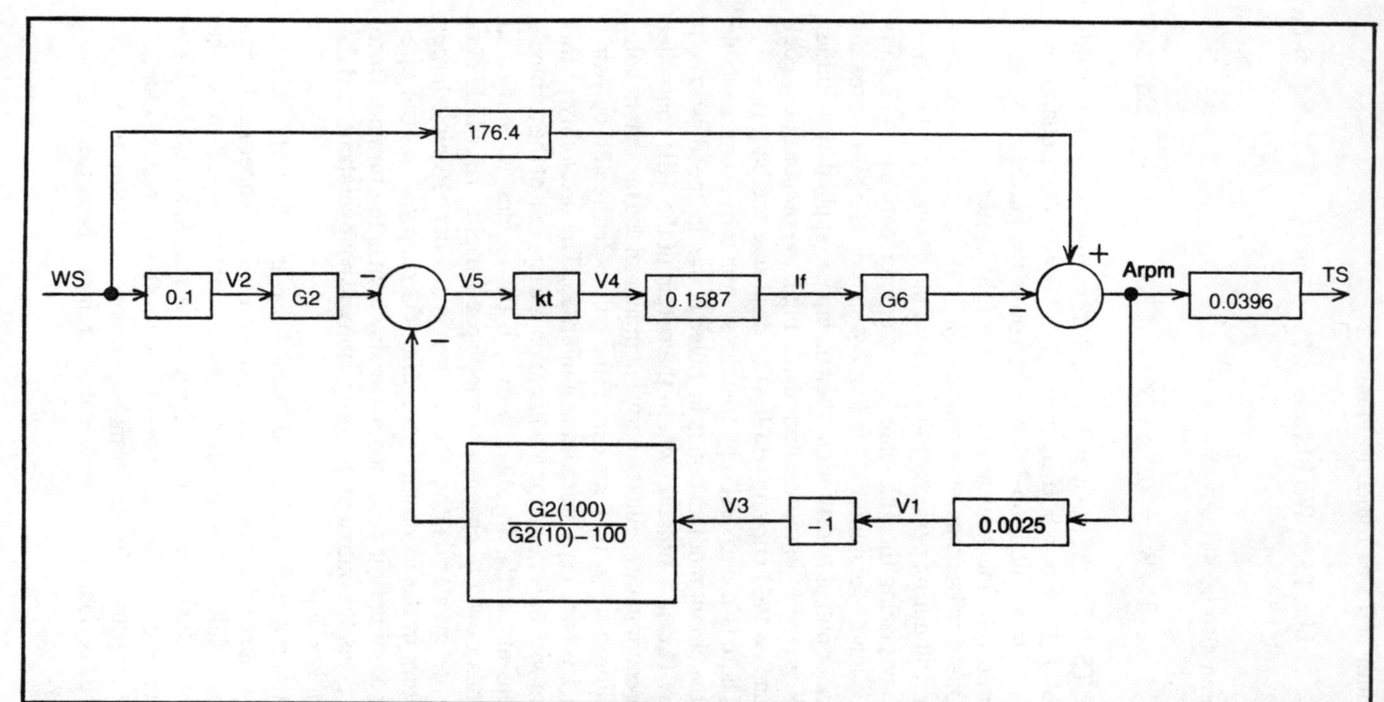

Fig. 6-5. Block diagram of a wind-power system using the modified tip-speed-ratio controller circuit and having a turbine radius of 5 feet and a gear ratio of 9.

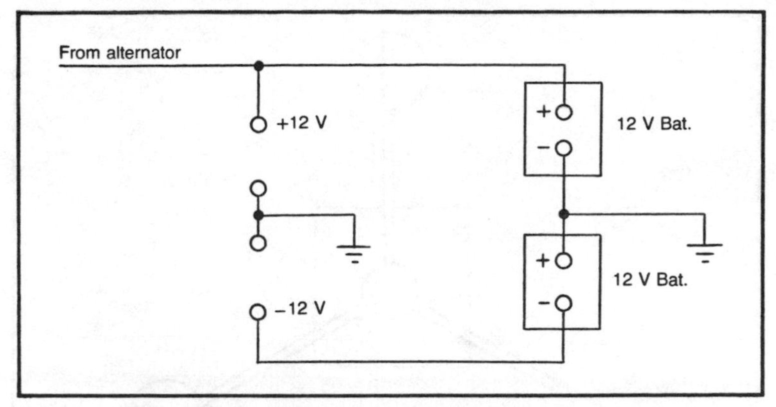

Fig. 6-6. Battery configuration for supplying positive and negative 12 Vdc.

THE WIND-SPEED SENSOR

Any wind-speed sensor can be used with this circuit as long as it produces an output voltage proportional to the wind speed and limited to about 5 volts. Figure 6-7 shows a sketch of a simple wind-speed sensor that can be built for use with this TSR controller circuit. The pickup coil and magnet placement for this sensor are similar to that of the alternator rpm sensor discussed in Chapter 3, except there are four magnets on the wheel instead of one. Also, the magnetic field polarity on each successive magnet is reversed to get the effect of one continuous cyclic magnetic field. The gain, G1, of the wind-speed sensor can be of any reasonable value and does not have to be known. As long as G1 is not too large, R20 can be adjusted to give the desired TSR.

Utilizing this tip-speed-ratio controller is an excellent way to maximize the efficiency of a wind-power system that uses the automobile alternator. In addition to maximizing efficiency at high Arpm, it also creates efficiency by automatically turning off the field current when the alternator stops spinning. This maximizes efficiency of the system by preventing wasteful current drain at low Arpm. The only disadvantage of this field current controller is that there is no provision for voltage regulation. Therefore, it may not be appropriate to use lead-acid batteries as the energy storage medium unless you place enough in parallel to reduce charging current to a safe level.

All the circuits discussed so far have been for use with 12-volt systems. The next chapter describes ways of converting the automobile alternator for use with higher-voltage systems.

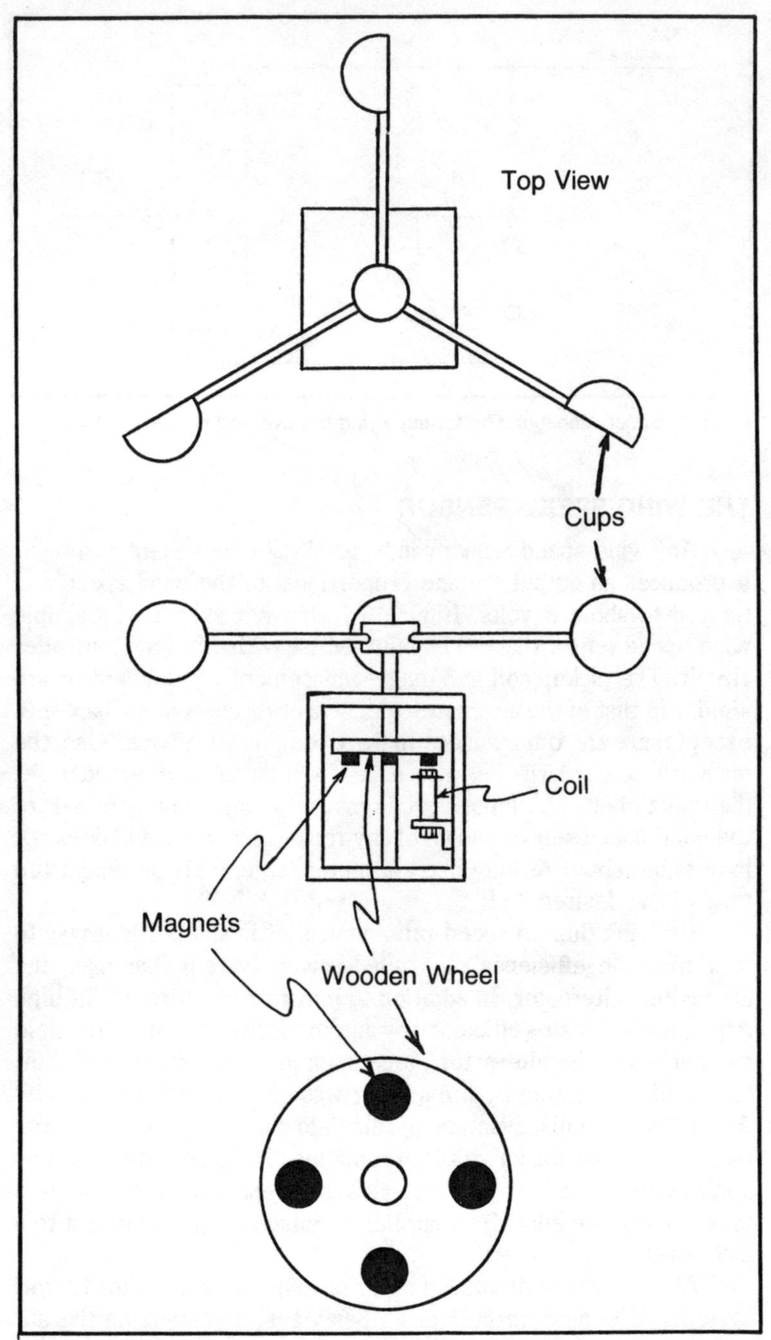

Fig. 6-7. Wind-speed sensor designed for use with the tip-speed-ratio controller.

7

Automobile Alternator Modifications for Higher Output Voltage

Due to the recent popularity of recreational vehicles, there are presently many 12-Vdc appliances on the market. This makes the 12-Vdc systems of the previous three chapters very practical. In some cases, however, it may still be necessary to provide higher dc or ac voltages. This chapter describes two ways of modifying an automobile alternator so that higher ac or dc regulated voltages can be supplied.

CASCADE TRANSFORMER METHOD

One method for increasing the output voltage of an automobile alternator is to connect a three-phase "Y"-connected stepup transformer to the stator windings, as shown in Fig. 7-1. This three-phase transformer can be made from three single-phase transformers by connecting the primary and secondary windings as shown in the figure. The primary (side facing the alternator) of each of these transformers must be able to handle at least 40 amps of current, and the secondary must be rated for at least 5 amps.

If the turns ratio of secondary windings over primary windings for each of the three transformers is 10, the output ac voltage will be 115 volts rms (root-mean-square) when the original dc output is 15 Vdc. The reason for this can be seen in Fig. 7-2, which shows the ac waveforms of the three stator windings. These particular sine waves have a peak value of 16.26 volts and an rms voltage of 11.5 volts. When these sine waves are applied to a transformer with a

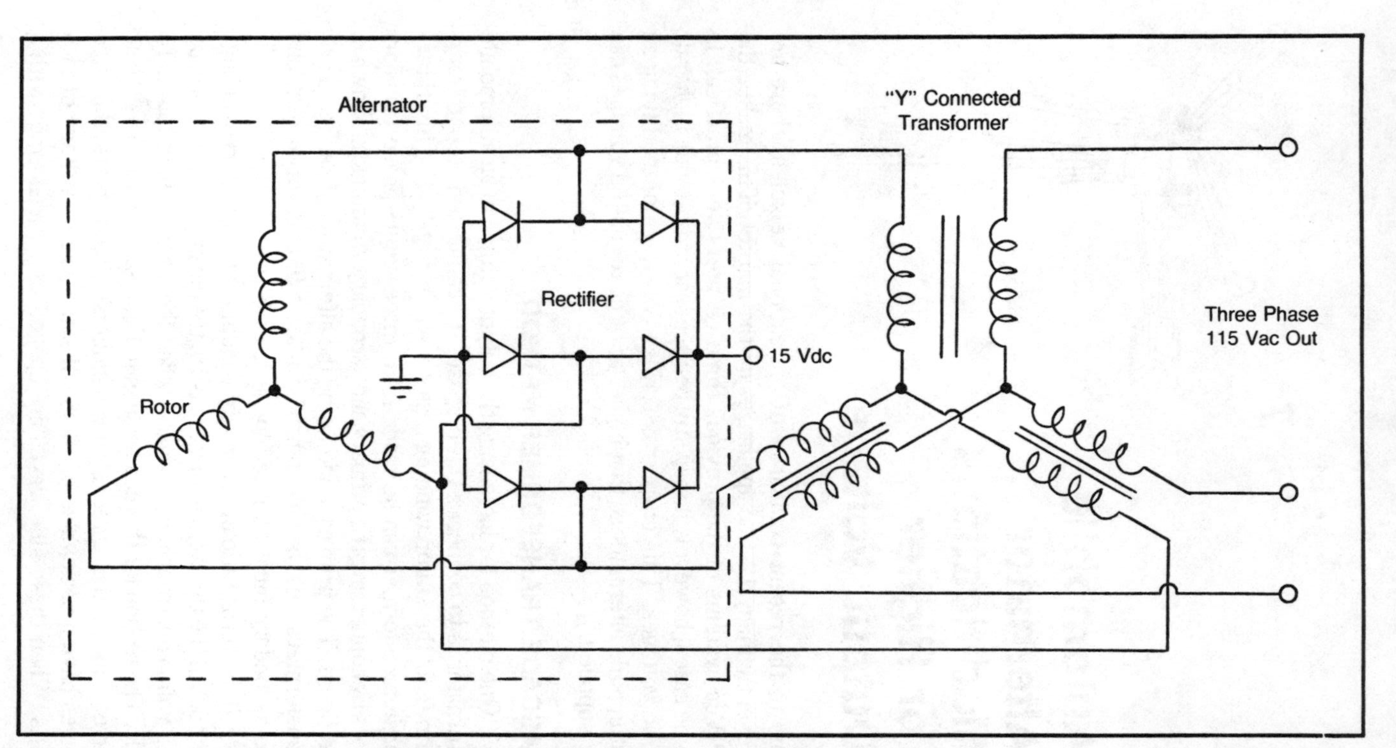

Fig. 7-1. Cascade transformer method of producing high voltage with an automobile alternator.

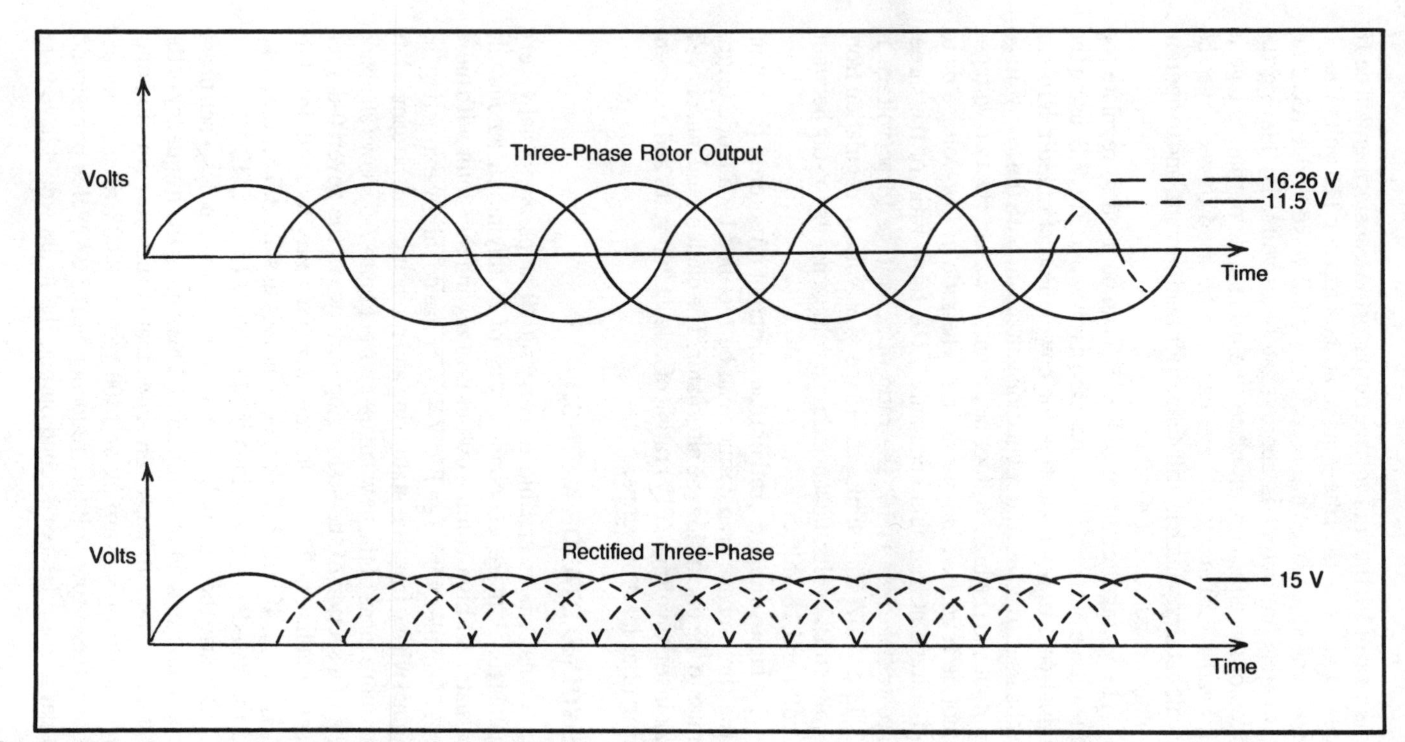

Fig. 7-2. Output voltages of alternator rotor and full-wave rectifier of Fig. 7-1.

turns ratio of 10, the rms voltage output of the transformer will be 10 times as great as the rms voltage at the input, or 115 volts rms. If these same sine waves are rectified by a full-wave bridge rectifier, the average dc output is approximately 15 volts. By leaving the diodes of the alternator in place, the original dc output voltage is still available. This dc voltage can then be used to recharge the 12-volt battery, which is still needed for field current, and to operate a voltage regulator, if desired.

The voltage regulator of Fig. 5-3 can be used to regulate the high-voltage output if it is connected as shown in Fig. 7-3, using the original dc output connection for power. Potentiometer R16 can now be used to adjust the regulation voltage for both the low-voltage dc output and the high-voltage ac output; however, the two voltages cannot be regulated independently. The ratio of low-voltage dc to high-voltage ac remains constant. If the turns ratio of the three-phase transformer is 10, the ratio of dc voltage to ac voltage is 115/15 or 7.67. This limits the range of ac voltage regulation from 92 volts to 123 volts, because the dc voltage needs to stay between 12 volts and 16 volts.

If a high-voltage dc, rather than ac, output is required, the output of the three-phase transformer can be rectified by a diode bridge similar to the one inside the alternator. The diodes used must have a peak inverse voltage (PIV) rating of at least 200 volts and a current rating of at least 10 amps.

REWOUND STATOR METHOD

Transformers capable of carrying 40 amps are usually very expensive. A more economical way of modifying an automobile alternator for high-voltage output is to rewind the stator with more turns of smaller wire (see Fig. 7-4). You can do this by extracting the original wiring from the stator ring and replacing it in the same slots with more turns of the new smaller wire. Figure 7-5 shows an inside view of a section of the stator ring with two of the seven 0-degree phase windings. The 0-degree windings are located in slots 1,4,7,10, . . ., 40; the 120-degree windings in slots 2,5,8,11, . . ., 41; and the 240-degree windings in slots 3,6,9,12, . . ., 42.

Because the automobile alternator has seven poles, each phase of the stator must have seven coils. This causes the frequency of the ac stator voltage to be seven times that of the rotor rotation rate. When the rotor is turning at 1400 rpm or 23.33 revolutions per second, the output voltage frequency is 163.33 cycles per second. To get a 60-cycles-per-second output, the Arpm would have to be

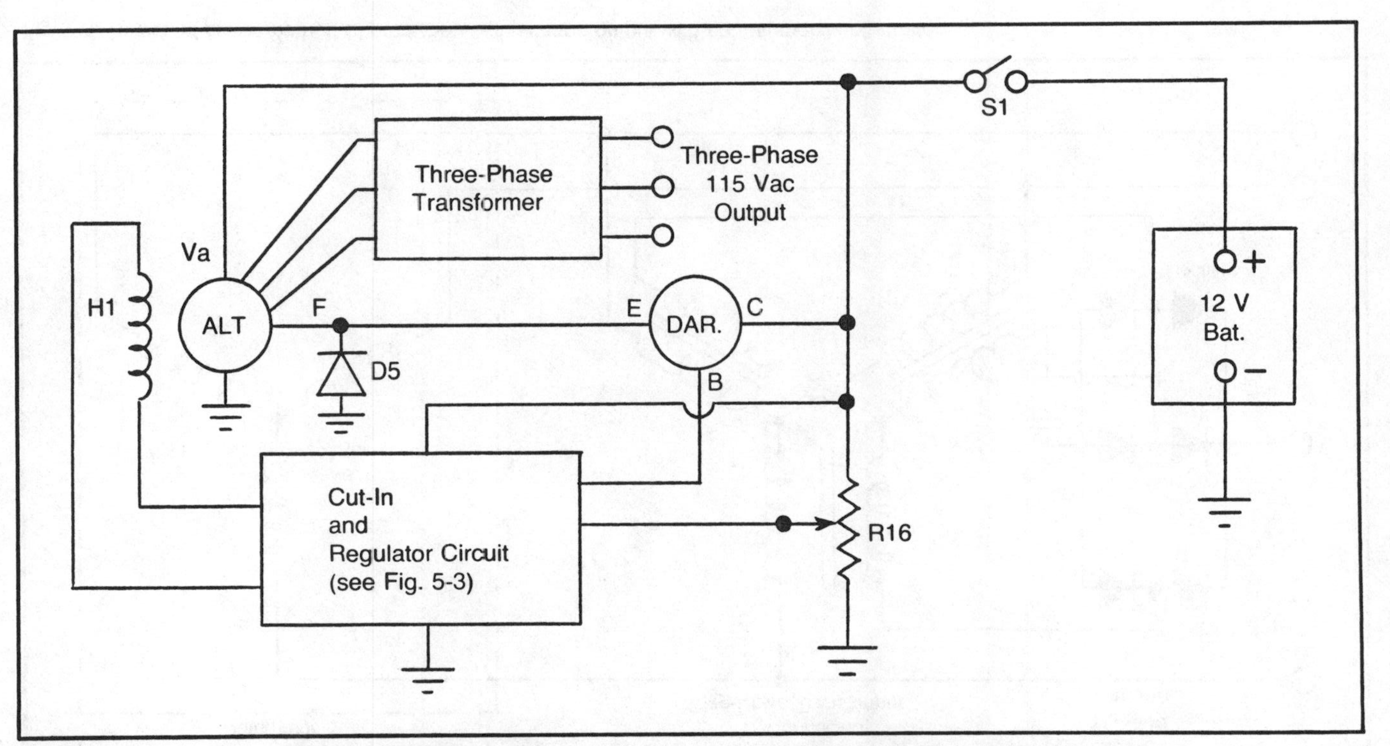

Fig. 7-3. Control circuitry for obtaining regulated three-phase 115-Vac output voltage from a standard automobile alternator.

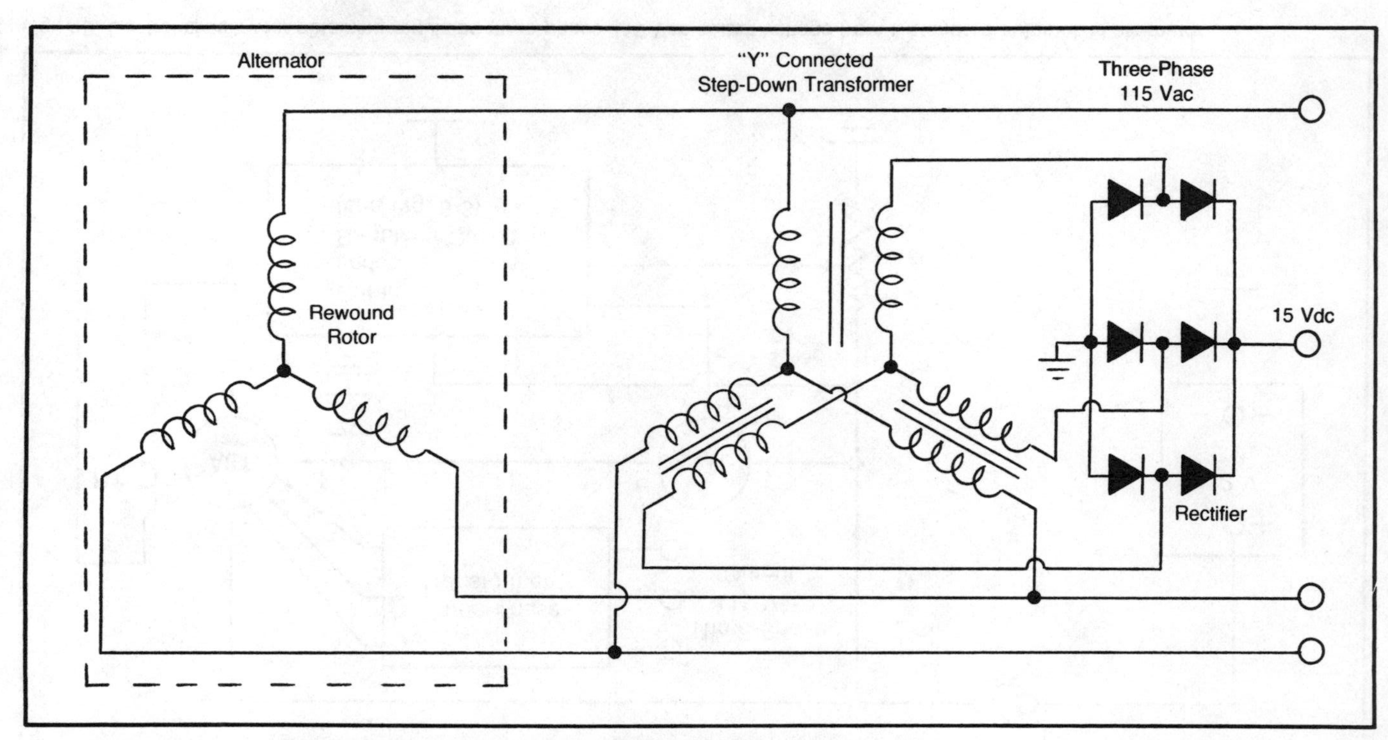

Fig. 7-4. Rewound stator method of producing high-voltage output with an automobile alternator.

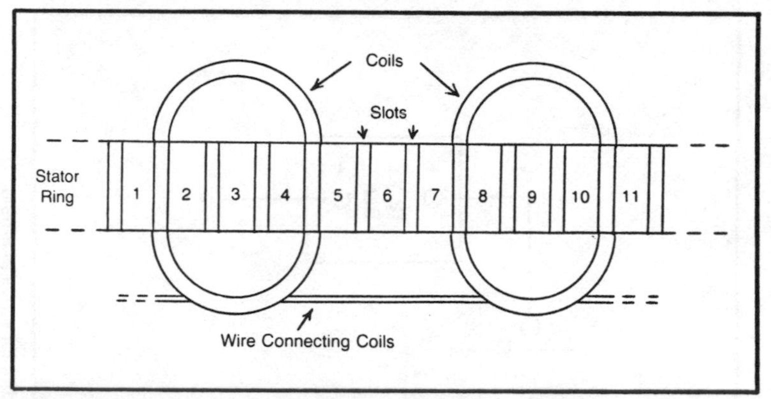

Fig. 7-5. Inside view of a section of the alternator stator ring.

514.28. Because an automobile alternator will not produce any appreciable power at this low rpm, there is no need to waste time trying to produce 115 Vac at 60 cycles per second.

To be able to achieve an ac voltage of 115 volts rms at a minimum Arpm of 1400, each pole of the stator must be wound with 110 turns of 30-gauge copper wire. This would require about 200 feet of wire per phase. Because 30-gauge copper wire has a resistance of about 0.103 ohms per foot, the total resistance per phase is 20.6 ohms. Because the power dissipated in the stator wire is equal to the wire resistance multiplied by the square of the current, the 30-gauge wire tends to waste much of the hard-earned power. With an output current of 5.58 amps per phase and an output voltage of 115 Vac, the power dissipated in the stator is equal to the alternator output power.

If 25-gauge wire were used, each pole could be wound with 55 turns. This is five times as many turns as the original pole windings and one-half as many windings as possible with 30-gauge wire. Because 25-gauge copper wire has a resistance of 0.0324 ohms per foot, 100 feet of wire per phase results in a resistance of 3.24 ohms per phase. At 5.58 amps, each phase of the stator will now dissipate only 100 watts, while the output power per phase is 641 watts. With this larger wire the minimum rpm for reaching an ac output voltage of 115 volts rms is about 2800.

It is interesting to note that an unmodified standard automobile alternator is capable of producing 115 volts if it is rotated fast enough. From Eq. 3-1, with K = 0.01 and If = 2 amps, we see that the output voltage of the alternator is 115 volts when the rpm is 5750. Because a typical automobile alternator can produce 40 amps,

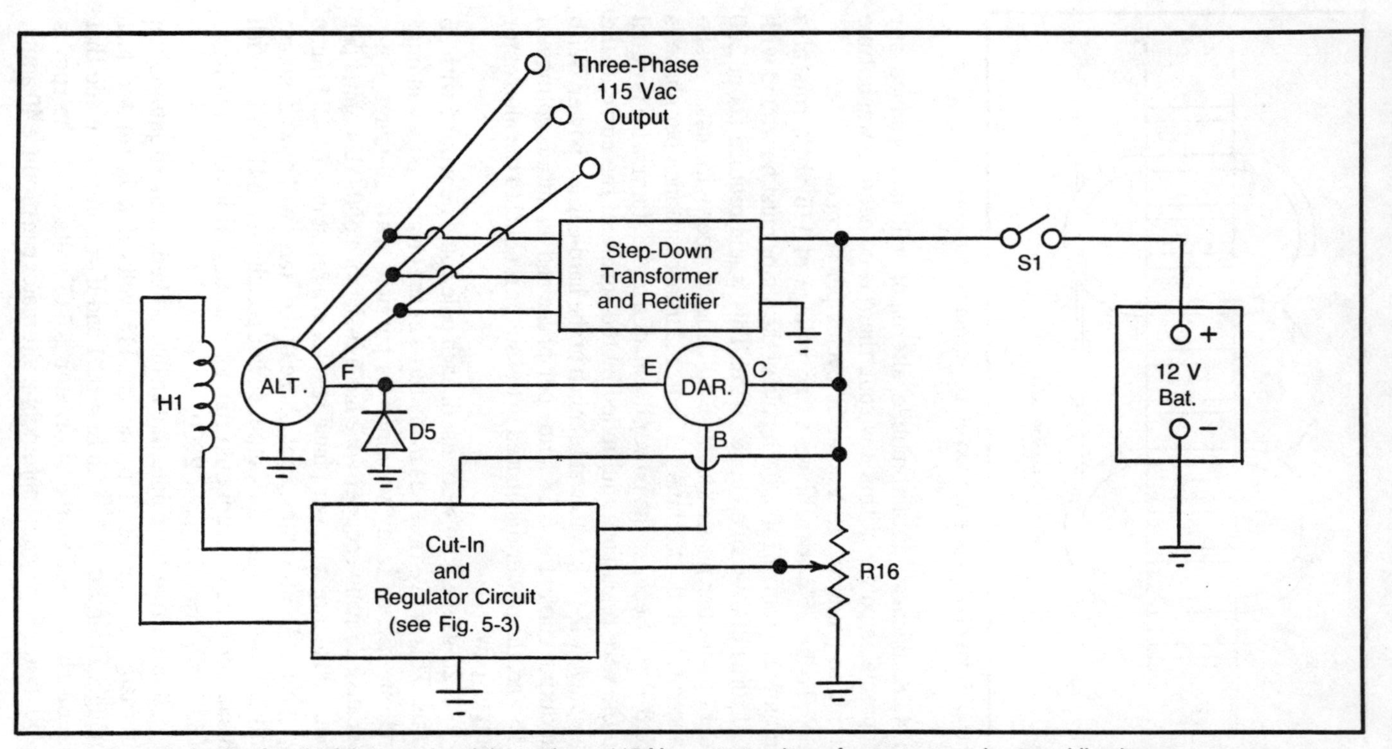

Fig. 7-6. Control circuitry for obtaining regulated three-phase 115-Vac output voltage from a rewound automobile alternator.

this implies that it is also capable of producing 4600 watts of power. Actually, this is possible, but not recommended for extended periods. Another interesting note is that the output voltage frequency would be 671 cycles per second when the input rpm is 5750. This could cause problems with line losses if the power is transmitted over long distances.

If the voltage regulator of Fig. 5-3 is used with the rewound alternator, a stepdown transformer must be connected, as shown in Fig. 7-4. The turns ratio of this transformer must be 1:10. The primary will have 10 times as many windings as the secondary, where the primary is the side facing the 115-volt output of the alternator. This Y connected transformer can also be built from three separate single-phase transformers; however, these transformers will not be as expensive as those of the last section. The secondary of these transformers need only handle a maximum of 5 amps. The full-wave bridge rectifier connected to the secondary of the transformer can be built from the diodes taken out of the alternator. The dc output voltage of the rectifier can now be used to recharge the field current battery and to power the voltage-regulator circuit.

Figure 7-6 shows how the voltage-regulator circuit discussed in Chapter 5 can be used with rewound alternator to regulate the three-phase 115-Vac output. In this circuit the output of the three-phase stepdown transformer is used to recharge the 12-volt battery and to power the voltage-regulator circuit. Actually, the voltage being regulated is the dc output of the three-phase transformer; however, this indirectly regulates the 115 Vac.

We have described two excellent ways of producing high-voltage ac power directly from the alternator. This power, however, is not a constant frequency source. Its frequency is directly related to the rotation rate of the alternator. If a high dc, rather than ac, voltage is required the outputs of the two previous circuits can be rectified with a bridge similar to the one originally built into the alternator. The diodes of the rectifier must be able to handle at least 10 amps and must have a PIV rating of more than 200 volts. The next chapter describes several electronic circuits that produce high-voltage ac at a constant frequency and may be suitable for applications that require precise 60-cycle power.

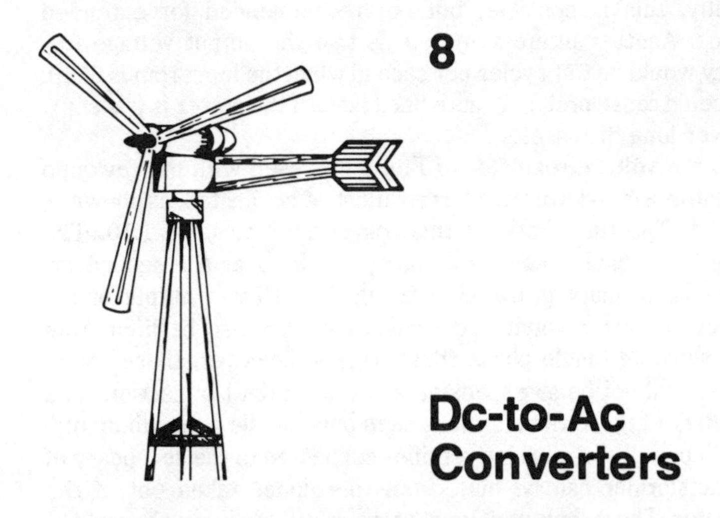

8

Dc-to-Ac Converters

The last chapter described ways of generating high-voltage ac power at no specific frequency. This chapter describes several ways of producing high-voltage ac power at a frequency of 60 cycles per second. There are basically two types of dc-to-ac converters that can be used to generate this constant-frequency power: *electronic converters* and *motor-generator sets*. Most electronic converters change dc to ac by varying the direct current at the input (primary) of a transformer and extracting the resulting alternating current at the output (secondary) of the transformer. A motor-generator set changes dc to ac by using a dc motor to drive an ac generator or alternator.

The term *inverter* is often used when referring to machines that convert dc to ac. This term is sometimes misleading because inverter is also used to refer to an electronic circuit that changes a zero dc voltage level to a higher dc voltage level and vice versa, depending on which is on the input. For the remainder of this chapter the term *converter* is used to represent devices that change dc to ac.

The advantages of electronic converters are their light weight, smaller size, and lack of rotating parts to wear out. The disadvantages of electronic converters are their inability to produce an efficient sine wave and their sensitivity to sudden load changes. Motor-generator sets are bulky, but they do supply sine waves, and, due to the large amounts of stored inertial energy in the rotating parts, they are not as sensitive to sudden load changes.

Most electronic converters produce a square-wave output, because it is more efficient to operate transistors in a switching mode than a linear mode, which is necessary to produce a sine wave. It is possible to produce a sine wave with an electronic converter, but these units are very expensive and very wasteful.

Because a square wave contains many high-frequency harmonic components, it is not as suitable for operating induction-type equipment as the sine wave. These distortion components tend to cause excessive heating in equipment such as transformers and induction motors. Therefore, square waves are not recommended for most household appliances, such as refrigerators, televisions, and washing machines.

SELF-COMMUTATED CONVERTERS

Self-commutated converters, as opposed to *line-commutated* converters, do not require a separate triggering source. These converters are completely self-contained and do not rely on the 60-cycle line frequency for timing. They are well suited for applications, such as remote cabins, where utility power lines are not accessable.

Basic Square-Wave Converters

A block diagram of a basic self-commutated 12-Vdc-to-120-Vac square-wave converter is shown in Fig. 8-1. It is simply a power circuit driven by two 60-cycle square waves that are 180 degrees out of phase.

A 60-watt power circuit that can be used with this basic converter is shown in Fig. 8-2. It is basically a center-tapped transformer

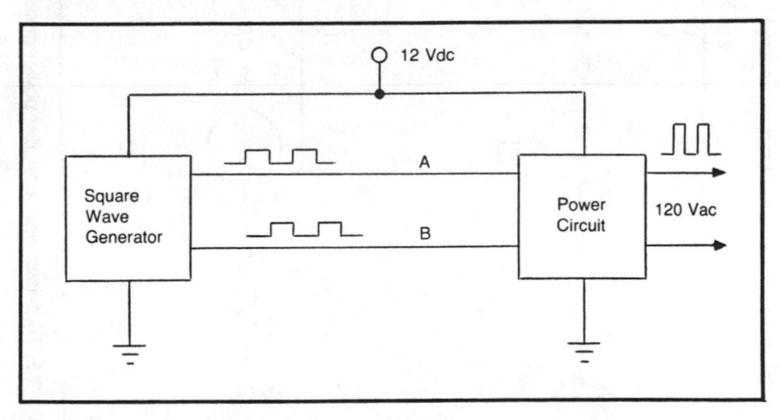

Fig. 8-1. Block diagram of a basic square-wave converter.

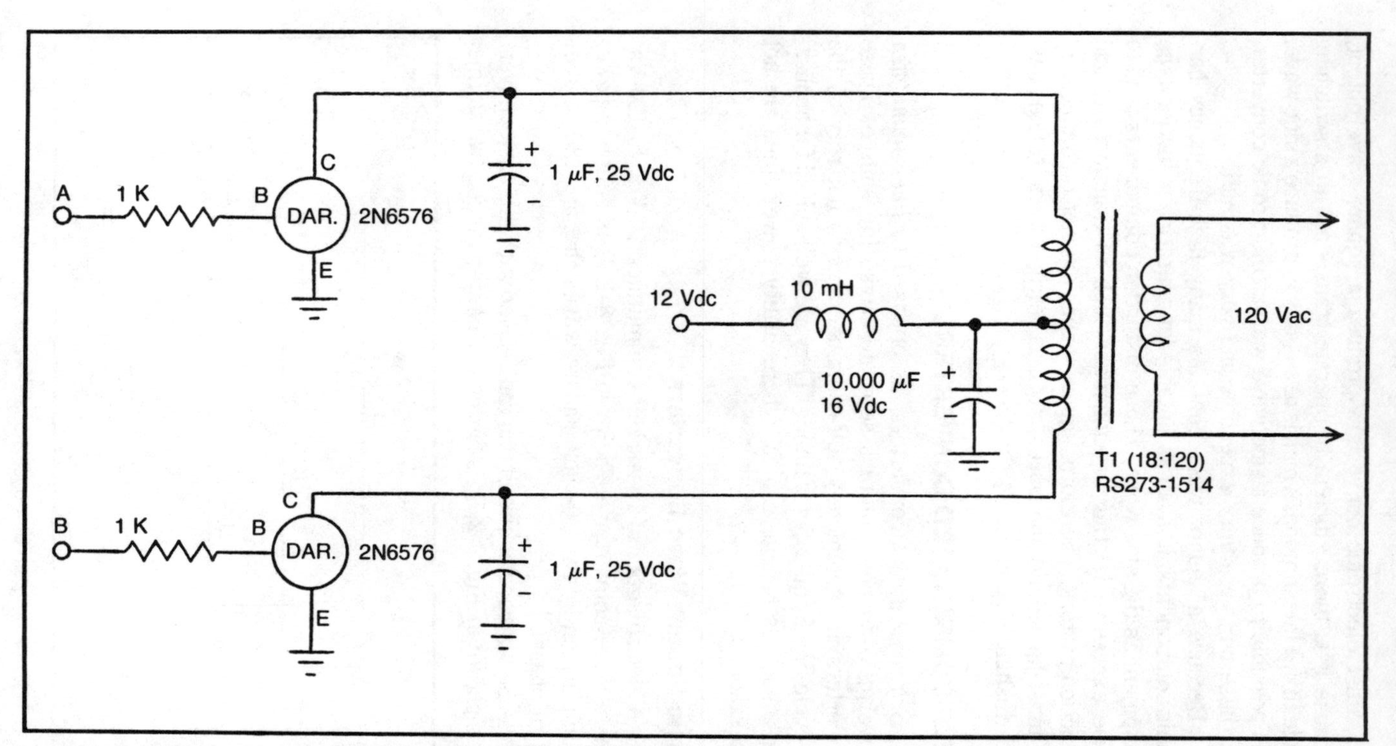

Fig. 8-2. Transistorized converter power circuit rated at 60 watts output.

with the center tap connected to positive 12 Vdc and the outside ends of the primary connected to ground through transistors. The two transistors are driven by the two out-of-phase square waves. When operated this way, the transistors alternately draw current through their respective halves of the primary coil.

The voltage at the secondary is a square wave with a peak no-load voltage value of twice the input dc voltage multiplied by the turns ratio (TR) of the transformer. For example, if the input voltage is 12 Vdc and the turns ratio is 6.66 (turns ratio for an 18-volt center-tapped transformer with a 120-volt secondary, such as transformer T1 in Fig. 8-2), then the peak no-load output voltage is 160 volts. If a load is placed on the output of the transformer, the output voltage varies according to the amount of power being drawn. If we assume that the secondary current is low enough (less than 1 amp) for the secondary coil resistance to be ignored, then Eq. 8-1 gives the relationship between output voltage Vs and output power Ps.

$$Vs = (TR)Vp + \sqrt{(TR)^2\,(Vp^2 - 4(Ps)Rp)} \qquad \textbf{Eq. 8-1}$$

Where Vp is the input voltage to the primary center tap, and Rp is the resistance of the primary coil. If TR = 6.66, Vp = 12 Vdc, and Rp = 0.45 ohms, which is the case for the converter power circuit of Fig. 8-2, then Eq. 8-1 reduces to

$$Vs = 80 + \sqrt{6400 - Ps(80)} \qquad \textbf{Eq. 8-2}$$

If no load is placed on the output, then Ps = 0 and Vs = 160 Vac, as previously determined. If a 60-watt load is placed on the output, then Vs = 120 Vac. Therefore, this converter is rated at 60 watts. Note that for a square-wave ac voltage the peak and rms voltages are the same. This is not true for a sine-wave ac voltage, where the rms voltage is 70.7 percent of the peak voltage.

Even though the converter of the previous example is rated at 60 watts, it is capable of producing 80 watts at a reduced output voltage. Equation 8-3 gives a formula for the maximum output power available.

$$Ps(max) = Vp^2/4Rp \qquad \textbf{Eq. 8-3}$$

Thus, if Vp = 12 Vdc and Rp = 0.45 ohms, the maximum output power is 80 watts. From Eq. 8-2, with Ps = 80 watts, the output voltage is 80 Vac.

The 1-μF capacitors (see Fig. 8-2) connected between the outer ends of the primary and ground absorb the voltage spikes caused by the sudden change in current through the primary coils. The 10-mH coil and 10,000-μF capacitor were placed between the input voltage source and center tap to smooth the input current supplied by the batteries. This is not only good for the batteries, but it also prevents noise from entering other electronic equipment that might be connected to the batteries, such as the square-wave generator.

If you need more output power from the converter, you can connect several power circuits in parallel or use the power circuit shown in Fig. 8-3. This circuit uses 60 Vdc instead of 12 Vdc. It also uses two 2N5578 transistors in each primary leg to achieve higher current capabilities. Each 2N5578 transistor is capable of drawing 30 amps, giving this circuit a power capability greater than 3000 watts.

The wires connecting each of the power transistors to the transformer primary must be of equal length. At these high currents, wire resistance becomes a factor, and if the resistance in one leg is lower than the resistance in the other leg, one transistor will tend to hog the current. Actually, it is good to have some resistance in the lines to keep one transistor from drawing all the current; however, if too much resistance exists, a large amount of power is wasted. The resistance should be between 0.05 and 0.1 ohms. Because the power dissipated in the wiring can be as much as 90 watts, the wire should be coiled and placed in a location that allows it to cool.

The transformer used on this high-power circuit must have a current rating of at least 60 amps on the primary and 30 amps on the secondary. The primary must be center tapped and the turns ratio should be 1. With this turns ratio, the peak output voltage is 120 volts. This does not leave much room for voltage drop with increased load, but, with 3000 watts behind it, not much is needed.

The 10,000-μF coil and the three capacitors should also have 60-amp ratings. Capacitors of this size might be hard to find; therefore, you might have to place several in parallel. Just remember that capacitance adds when capacitors are in parallel.

The square-wave generator for the basic square-wave converter is shown in Fig. 8-4. The NE 555 timer circuit shown generates a square wave whose frequency can be adjusted by varying the 50 K-ohm potentiometer. This single square-wave output is then tapped and inverted by the 2N3904 transistor to create

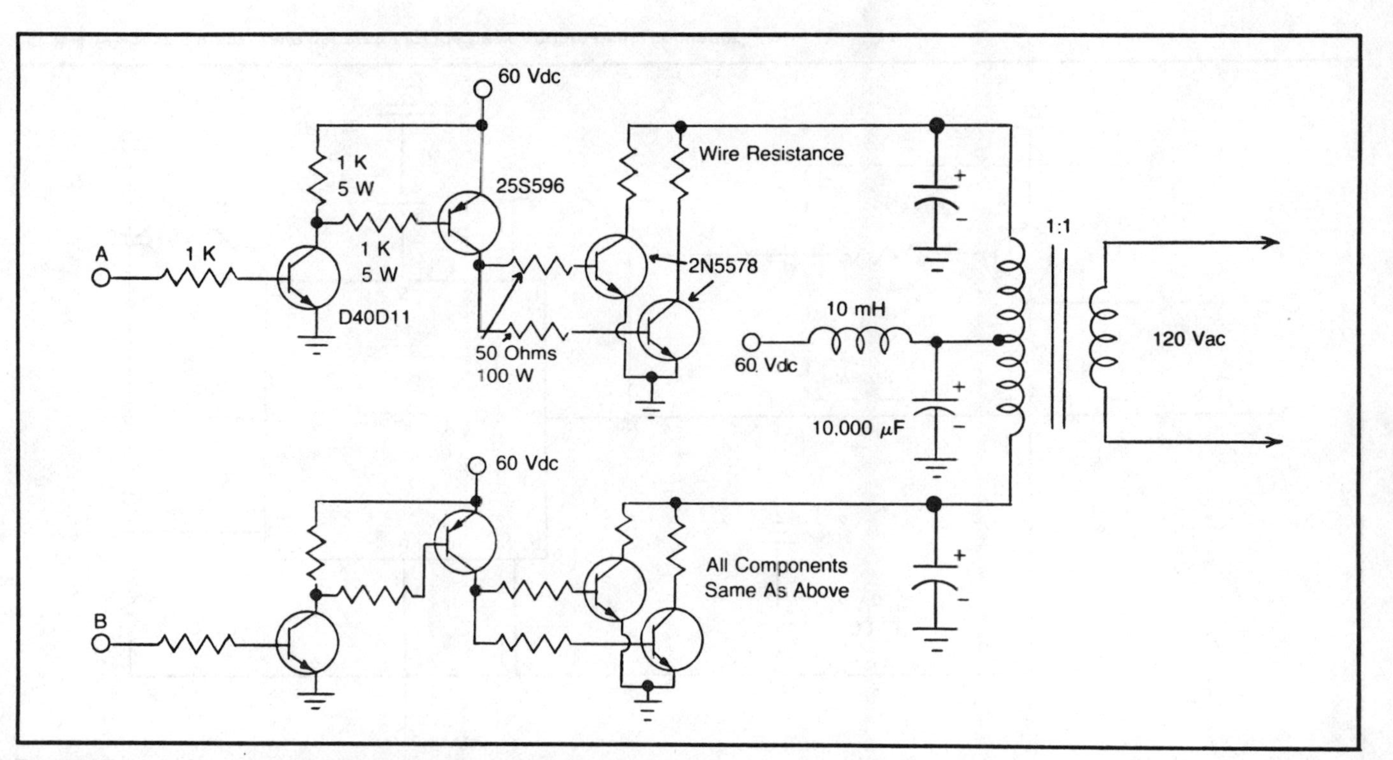

Fig. 8-3. Transistorized converter power circuit rated at 3000 watts output.

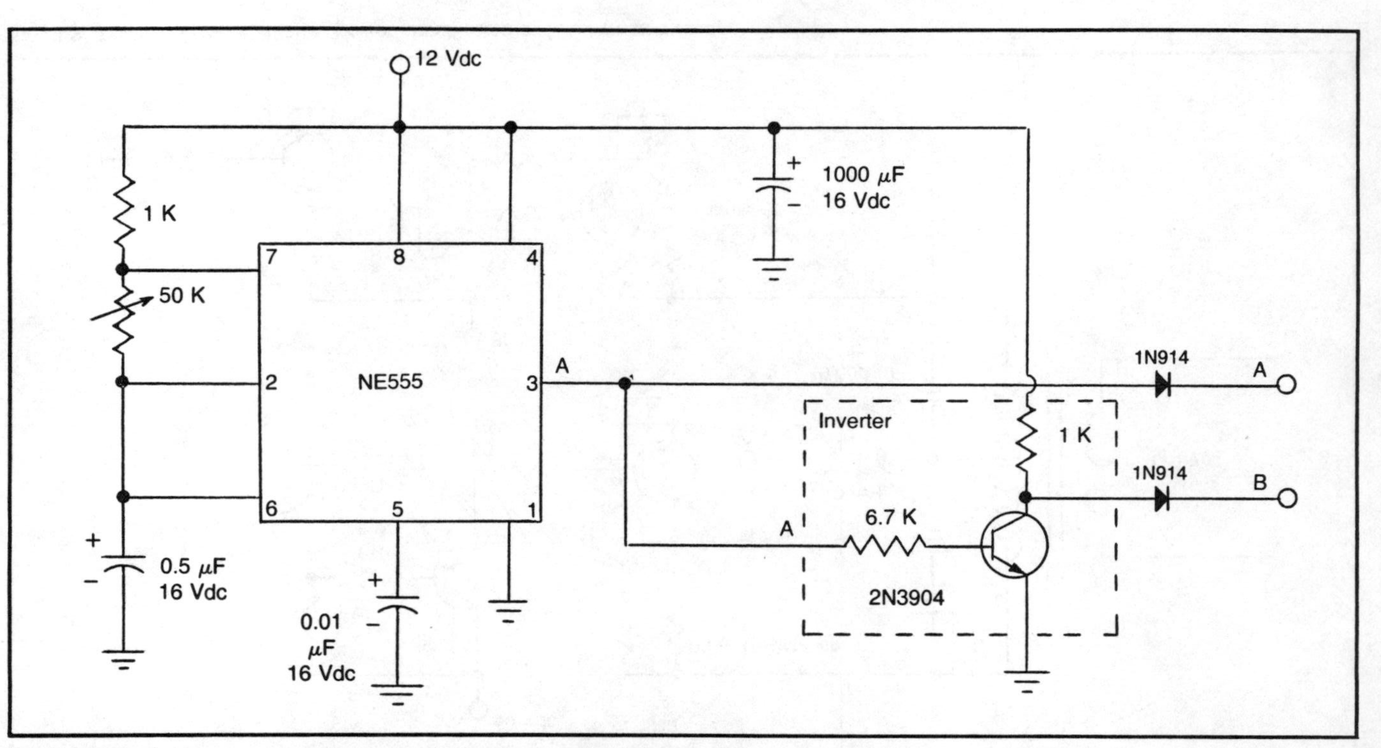

Fig. 8-4. 60-cycle square-wave generator for the basic square-wave converter.

another square wave that is 180 degrees out of phase with the original one. Both signals are then applied to the converter power circuit.

This square-wave generator can be calibrated by connecting one of the outputs to a frequency counter and adjusting the 50 K-ohm potentiometer until the frequency counter output reads 60 cycles per second. If no frequency counter is available, the digital anemometer circuit in Chapter 2 can be converted for that use, as discussed in Chapter 3. The circuit in Fig. 8-4 is not an overly precise piece of equipment, but a simple square-wave generator designed for approximate 60-cycle operation. For applications requiring a precise 60-cycle converter, you can use the one described in the next section.

Precision-Frequency Square-Wave Converters

Some appliances in the home, such as clocks, televisions, timers, and record player turntables, require an accurate 60-cycle power input for correct operation. If any of these appliances are operated with an electronic converter, some method of accurately controlling the frequency of the converter is necessary.

Figure 8-5 shows a block diagram of a precise 60-cycle square-wave generator that can be used to control a converter power circuit. This circuit uses a 6-megacycle crystal oscillator as shown in Fig. 8-6 with its output square-wave frequency divided by 100,000 to get an extremely precise 60-cycle output. This output is also inverted to provide a second square wave 180 degrees out of phase with the original output.

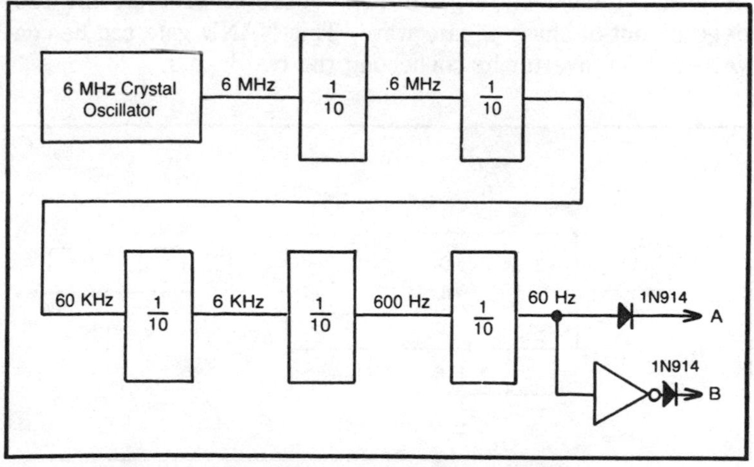

Fig. 8-5. Crystal-controlled, 60-cycle square-wave generator.

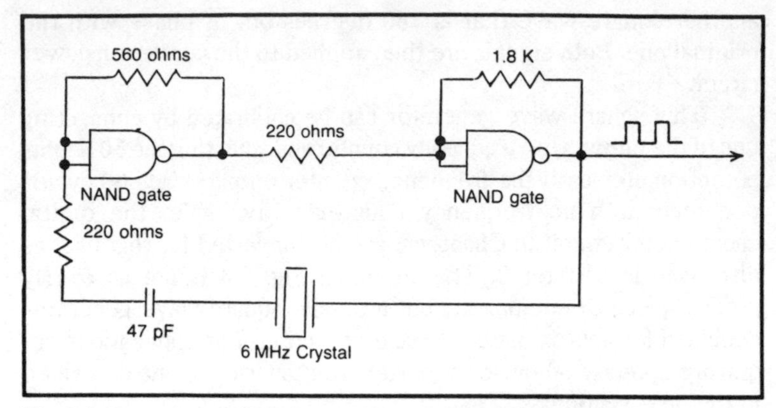

Fig. 8-6. Crystal-controlled, 6-megacycle square-wave oscillator.

To get the frequency divided by 100,000, you must use five divide-by-10 frequency-divider circuits. One of these circuits, shown in Fig. 8-7, uses a 74LS90 decade counter as the divider circuit and requires a 5-volt square-wave input. If five of these circuits are wired in series at the output of the 6-megacycle crystal oscillator, a 60-cycle square wave is obtained. This square-wave output, along with its inverted form, can be used in place of the square-wave generator output of the previous converter to drive the power circuit.

The crystal oscillator shown in Fig. 8-6 uses two NAND gates in its circuit. The 7400 quad NAND gate chip shown in Fig. 8-8 can be used to supply these two gates. A third gate in the 7400 chip can be used for the inverter, which is needed to generate the 180-degrees out-of-phase square wave. This NAND gate can be converted to an inverter by connecting the two inputs.

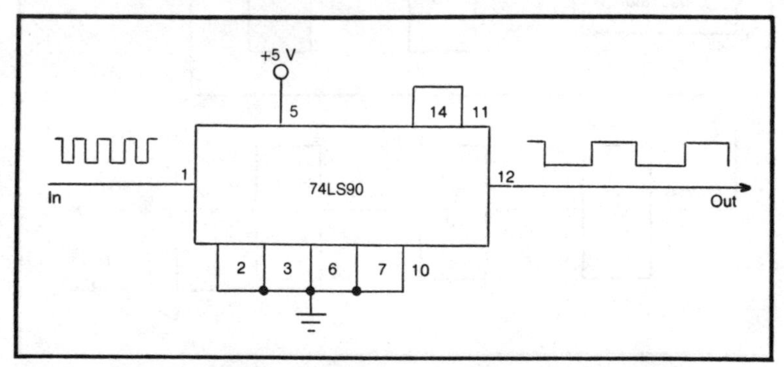

Fig. 8-7. Divide-by-10 frequency divider made from a decade counter.

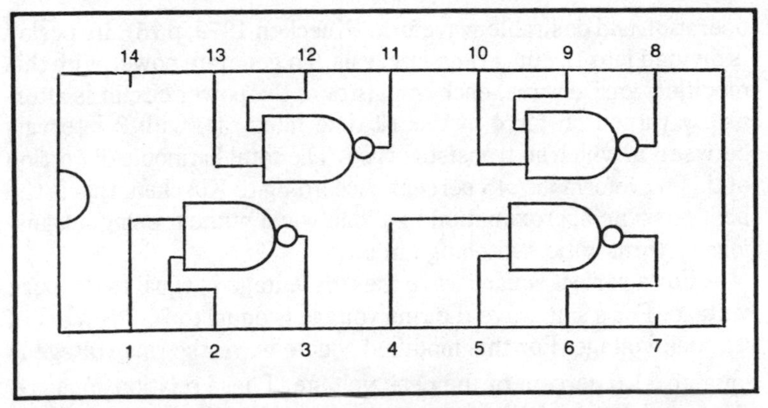

Fig. 8-8. Pinout diagram for a 7400 NAND gate integrated circuit chip.

Remember that all the digital circuits in this 60-cycle square-wave generator require 5 volts instead of 12 volts. A 7805 voltage-regulator chip can be used between a 12-volt source and the digital circuits to provide the needed 5 volts. Since the power supply voltage for these digital circuits cannot be greater than 5 volts, the output square-wave voltage cannot be greater than 5 volts; however, 5 volts is still capable of driving the power circuits previously described.

Sine-Wave-Approximated Square-Wave Converters

As stated earlier, sine waves are more desirable for most applications, but electronic converters are more efficient when operated in the switching mode. The waveform in Fig. 8-9 was suggested by John Kuecken as a compromise between efficient

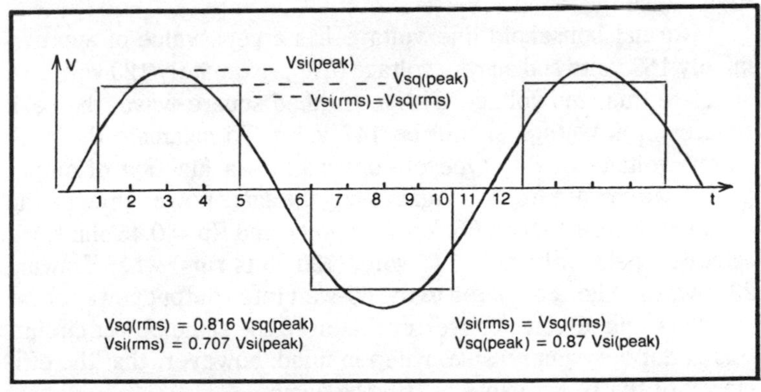

Fig. 8-9. Square wave used to approximate a sine wave.

operation and desirable waveform (Kuecken 1979, p 75). Its period is divided into 12 equal time intervals. To generate power with this modified square wave, each transistor of the power circuit is alternately turned on for 4 of the 12 time intervals, with 2 intervals between in which no transistor is on. The total harmonic distortion of this waveform is 7.45 percent. According to Kuecken, this is the best possible approximation to a sine wave without using a transformer turns-ratio switching circuit.

For a perfect square wave the rms voltage is equal to the peak voltage. For a sine wave the rms voltage is equal to 70.7 percent of its peak voltage. For this modified square wave the rms voltage is equal to 81.6 percent of the peak voltage. These relationships are illustrated by the following equations:

$$Vsi(rms) = 0.707Vsi(peak) \quad \textbf{Eq. 8-4}$$

$$Vsq(rms) = 0.816Vsq(peak) \quad \textbf{Eq. 8-5}$$

Where Vsi(rms) and Vsq(rms) represent the rms voltages of the sine wave and the modified square wave, respectively, and Vsi(peak) and Vsq(peak) represent their peak voltages. If your goal is to have the modified square-wave rms voltage equal to the sine-wave rms voltage it is approximating, then Eq. 8-4 and 8-5 can be equated to give:

$$Vsq(peak) = 0.87Vsi(peak) \quad \textbf{Eq. 8-6}$$

Therefore, the peak voltage of the modified square wave should be 87 percent of the peak voltage of the sine wave for the rms voltages to be equal.

Normal household line voltage has a peak value of approximately 169 volts and an rms voltage of approximately 120 volts. To simulate this rms voltage with a modified square wave, the peak square-wave voltage should be 147 volts. To calculate the peak output voltage of this type of converter as a function of output power, we can use Eq. 8-1. Again using the same power circuit as an example, where TR = 6.66, Vp = 12 volts, and Rp = 0.45 ohms, we achieve a peak voltage of 147 volts (120 volts rms) when drawing 23.9 watts. When compared to the 60-watt rated output power of the previous square-wave converter that used the same power circuit, this is not very impressive. Keep in mind, however, that the efficiency of the two circuits is still the same.

The maximum output power available from this converter is determined from Eq. 8-3, using the rms equivalent of Vp rather than the peak value. Like the output voltage, the rms value of the input voltage is 81.6 percent of the peak value. For the converter of this example the maximum output power is 53.3 watts at a peak output voltage of (using Eq. 8-2) 126.2 volts or 103 volts rms.

A block diagram of the sine-wave-approximated square-wave converter is shown in Fig. 8-10. The power circuit used can be the 60-watt version in Fig. 8-2 or the high-power version shown in Fig. 8-3. The square-wave modifier circuit is shown in Fig. 8-11.

The square-wave modifier circuit can be built using three integrated circuits, excluding the 7805 5-volt regulator. One of these circuits is a 7492 divide-by-12 counter used to divide the period into 12 equal parts. The other two are 7402 NOR gates used to create the two alternating pulses that drive the power transistors. A pinout diagram of the 7402 NOR gate chip is shown in Fig. 8-12.

To get a 60-cycle converter output frequency, the frequency of the square-wave generator must be 12 times 60 cycles per second or 720 cycles per second. This 720-cycle square wave can be generated by an NE 555 timer, such as in Fig. 8-13, or by a crystal-controlled oscillator similar to the one in Fig. 8-5 with a 7.2-megacycle crystal. If a 7.2-megacycle crystal is used, only four divide-by-10 frequency dividers are needed in series on the output. Also, no inverter circuit is required to produce a second 180-degree square wave.

Both the square-wave modifier and the square-wave generator

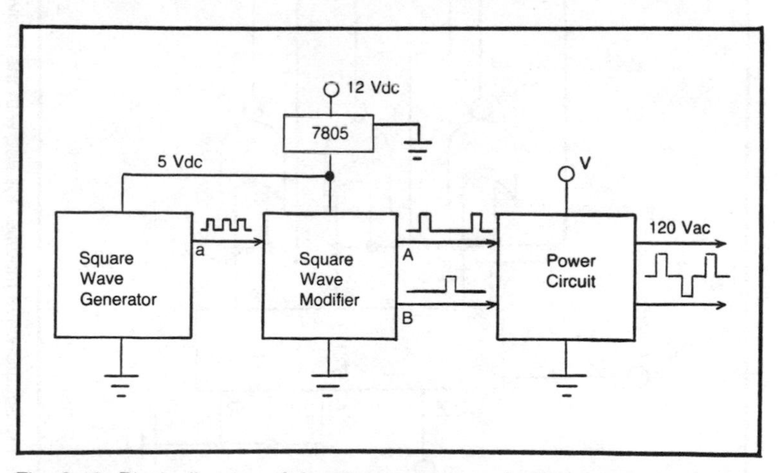

Fig. 8-10. Block diagram of the sine-wave-approximated square-wave converter.

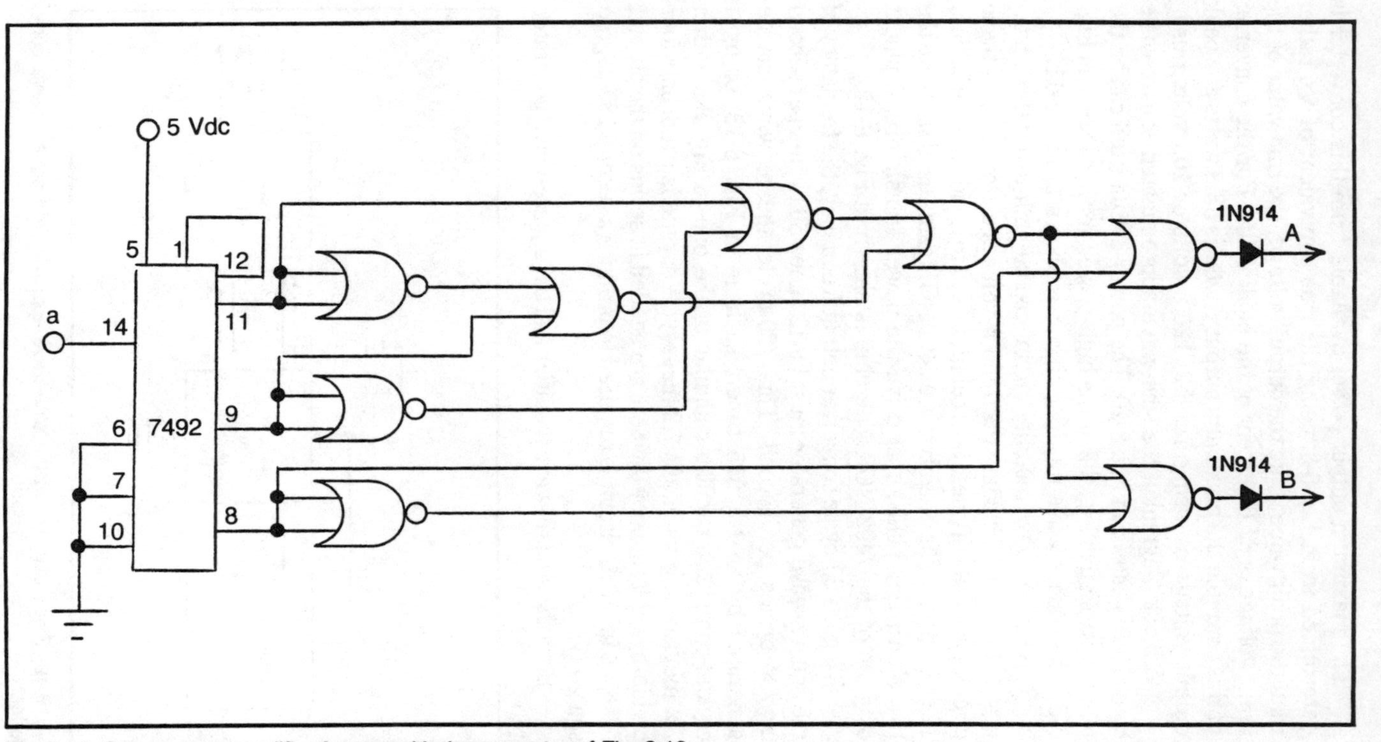

Fig. 8-11. Square-wave modifier for use with the converter of Fig. 8-10.

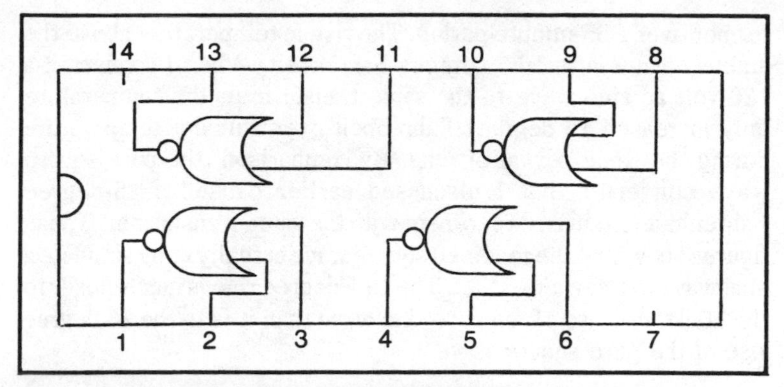

Fig. 8-12. Pinout diagram of the 7402 NOR gate integrated circuit chip.

require +5 volts for operation. The 7805 regulator chip can supply power for both these circuits.

You can determine how closely the modified square wave of this converter simulates a sine wave by measuring the amount of reactive losses the waveform experiences in a transformer as compared to a sine wave of similar magnitude. I measured the reactive losses of the modified square by connecting a transformer to the converter output. I measured the rise in temperature of the trans-

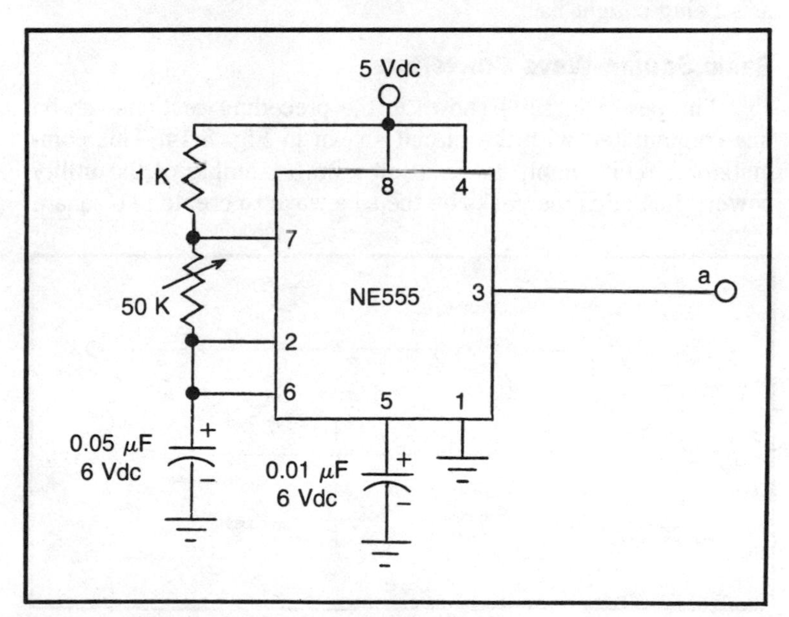

Fig. 8-13. 720-cycle square-wave generator.

former over a 35-minute period. The rise in temperature above the ambient value was 19.5 degrees Fahrenheit. When I connected a 120-volt ac sine wave to the same transformer, the temperature only increased 16 degrees Fahrenheit over ambient temperature during the same period of time. By comparison, the pure square wave converter that I discussed earlier caused a 26-degree Fahrenheit rise in an experiment with the same transformer. These figures show that the modified square wave actually does simulate a sine wave to a certain extent. The 19.5 degree rise is much closer to the 16 degree rise of the pure sine wave than it is to the 26 degree rise of the pure square wave.

LINE-COMMUTATED CONVERTERS

Line-commutated converters do not have their own triggering source. They use, instead, the 60-cycle frequency of the utility power lines. This is not only convenient but also lends itself to the possibility of synchronizing the converter output with the utility power lines so power from the converter can be fed back into the utility grid. The advantage of this type system is that batteries are no longer necessary. The utility power lines are effectively used as an infinite source of energy storage. Actually, this excess energy is being sold to the power company, and when the wind isn't blowing, it is being bought back.

Basic Square-Wave Converters

The power circuits shown in the preceding sections can be line-commutated with the circuit shown in Fig. 8-14. This commutator circuit simply takes a low-voltage sample of the utility power, then clips the peaks off the sine wave to create two square

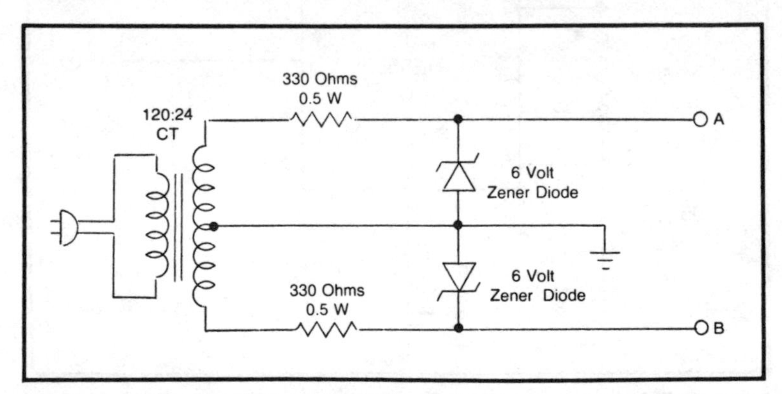

Fig. 8-14. Line-commutator circuit for the basic square-wave converter.

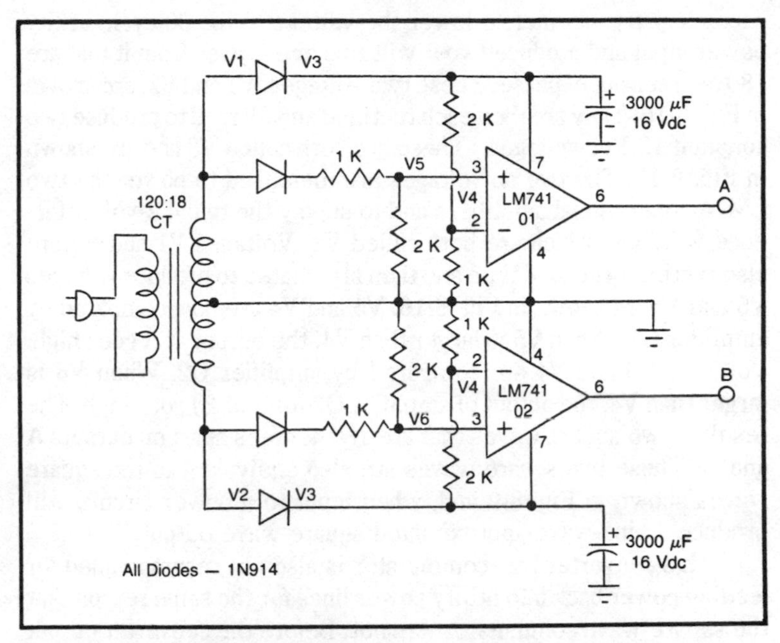

Fig. 8-15. Line-commutator circuit for the sine-wave-approximated square-wave converter.

waves that are 180 degrees out of phase. These two square waves are then used to trigger the converter power circuit.

This is a very simple method of line commutating the converter, but it is not recommended for feeding power back into the utility power lines. When each transistor is turned on by an increase in line voltage, it must then rely on a decrease in line voltage to turn it off. At the same time, this transistor is trying to keep the line voltage high. This creates a problem because the converter is fighting against the line voltage. Later, a converter will be described that does not have this problem.

A converter using this commutator could be used to power noninduction type equipment or appliances that require precise 60-cycle power but not necessarily a perfect sine wave.

Sine-Wave-Approximated Square-Wave Converters

A more acceptable line-commutator circuit is shown in Fig. 8-15. This circuit uses the 60-cycle line voltage to create a sine-wave-approximated square wave similar to the one shown in Fig. 8-9.

The circuit in Fig. 8-15 first uses an 18-volt center-tapped-secondary transformer to lower the voltage of the 60-cycle utility power input and produce two 9-volt rms sine waves from it that are 180 degrees out of phase. These two voltages, V1 and V2, are shown in Fig. 8-16. They are then each rectified and filtered to produce two constant 12-Vdc voltages. These are both called V3 and are shown in Fig. 8-16. The two dc voltages are then used to power the two LM741 operational amplifiers and to supply the two 4.2-volt reference voltages, which are both called V4. Voltages V1 and V2 are also rectified a second time and then attenuated to produce voltages V5 and V6, as shown in Fig. 8-16. V5 and V4 are now compared by amplifier O1. When V5 is larger than V4, the output at A goes high. Voltages V6 and V4 are compared by amplifier O2. When V6 is larger than V4, the output of amplifier O2 (output B) goes high. The result is two square waves that are 180 degrees apart on outputs A and B. These two square waves are also equivalent to the square waves shown in Fig. 8-9 and, when input to a power circuit, will produce a sine-wave-approximated square-wave output.

This converter line-commutator is also not recommended for feeding power back into utility power lines for the same reason that the square-wave commutator was not. Before the converter output is connected to the utility lines, some circuit must be placed between the line-commutator of Fig. 8-15 and the power circuit to cause the power transistor shutoff to be independent of line voltage. In other words, the width of the commutator square waves should be controlled by timers that turn the power transistors off at a present time. The circuit shown in Fig. 8-17 will accomplish this task.

In Fig. 8-17 the outputs of the commutator circuit are first passed through a capacitor/transistor combination. These are actually inverted-pulse generators that produce short negative pulses every time the commutator outputs go from low to high. These short negative pulses cause the outputs of the two NE 555 timers to go high. The NE555 timers are preset so the outputs will automatically go low in 0.0055 seconds after they go high. The new commutator outputs A′ and B′ now look exactly like the previous outputs A and B. The only difference is that the positive to negative transitions of the square waves are independent of the line voltage.

The commutator circuit shown in Fig. 8-14 should not be used with this square-wave timer circuit to feed power back into utility power lines. The pure square-wave output of this commutator would cause unnecessary degradation of the utility power line sine wave. Because the harmonic distortion of the modified square wave is only 7.45 percent, the commutator circuit shown in Fig. 8-15

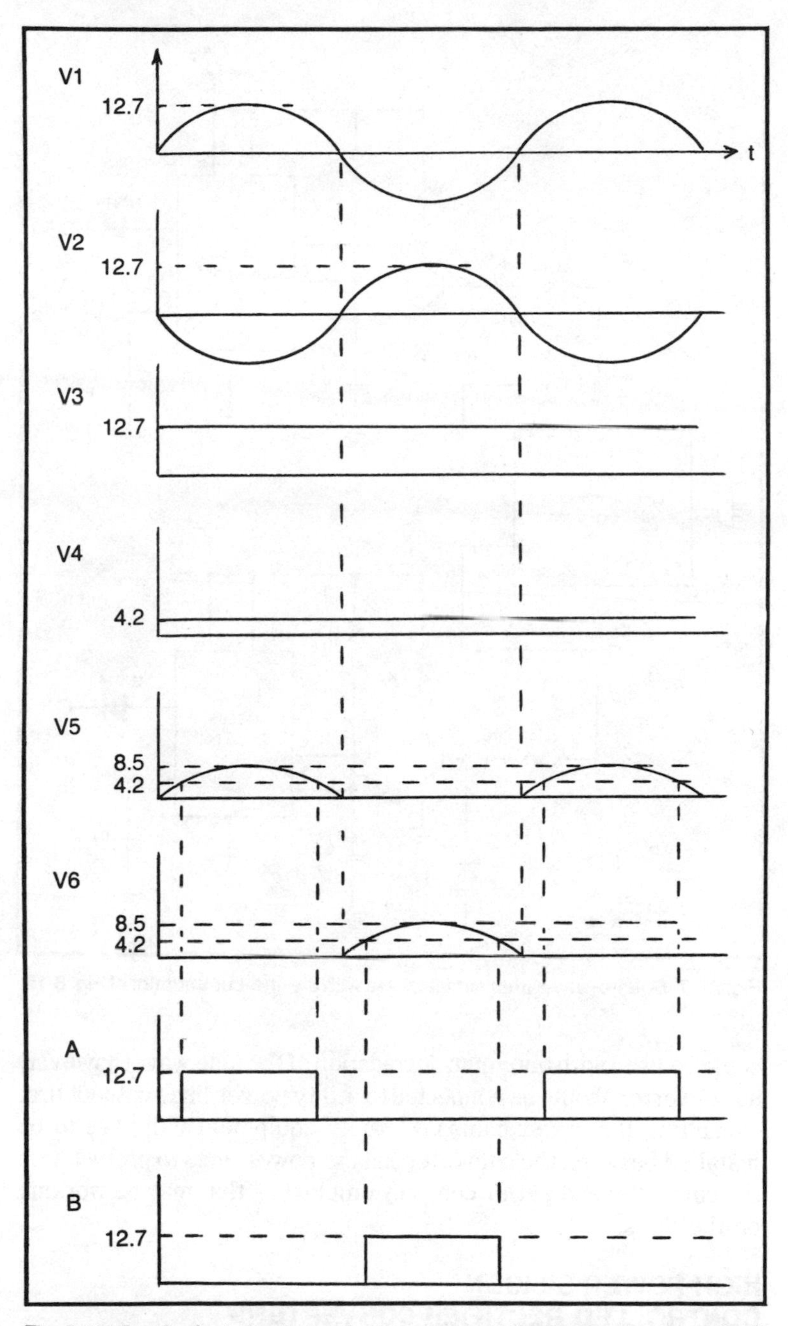

Fig. 8-16. Graph of relative timing for the voltages in the line-commutator circuit shown in Fig. 8-15.

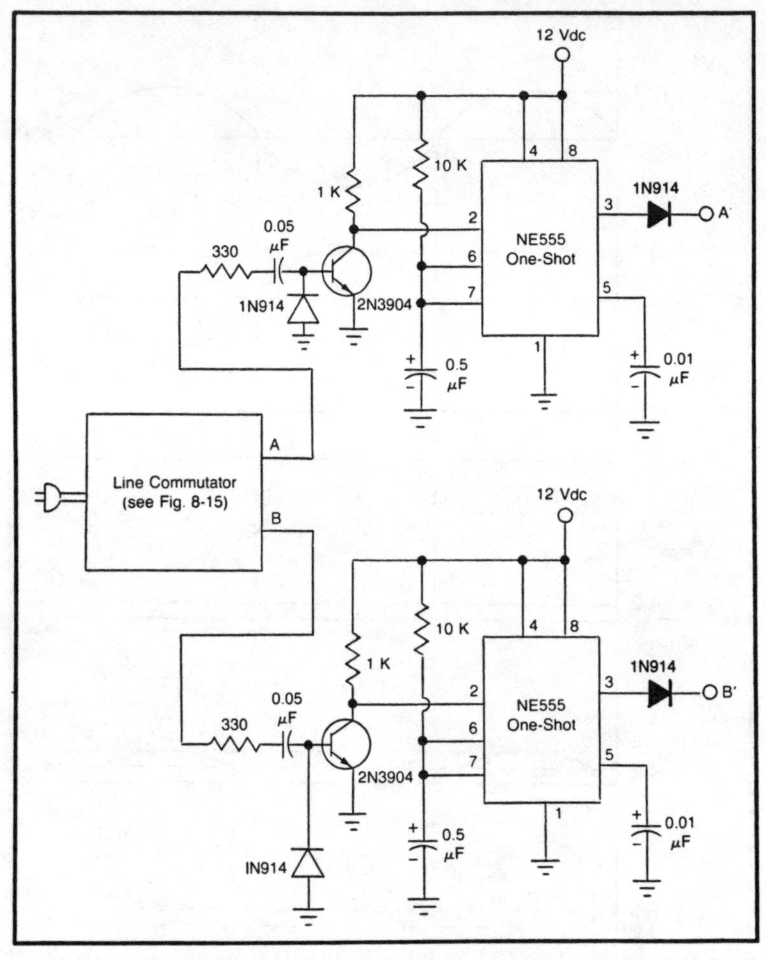

Fig. 8-17. Square-wave timer circuit for use with the line-commutator of Fig. 8-15.

could be used with minimum degradation of the sine wave; however, no converter should be connected to utility power lines without first consulting the power company. Safety equipment will have to be installed between the converter and the power lines to protect both the converter and power company employees that may be working on the lines.

HIGH-POWER SILICON CONTROLLED RECTIFIER CONVERTERS

Figure 8-18 shows a schematic of the power circuit section of a high-power converter that utilizes silicon controlled rectifiers (SCR) instead of transistors (Fitzgerald et al. 1971, p. 416). Silicon controlled rectifiers can carry more current and are more efficient and less expensive than transistors. The only disadvantage is that you must reverse the voltage across them in order to turn them off. The circuit of Fig. 8-18 utilizes a commutator capacitor (C3) to alternately reverse voltage across the two silicon controlled rectifiers.

The two silicon controlled rectifiers of Fig. 8-18 are alternately triggered by short pulses applied between pins E and F for SCR1 and pins G and H for SCR2. While SCR1 is conducting and SCR2 is off, capacitor C3 is building up a voltage equal to 2Vb. When SCR2 is triggered by its pulse, it applies voltage to the bottom half of the transformer primary and also applies the full capacitor voltage across SCR1 in the reverse direction. This reverse voltage causes SCR1 to turn off. This process is repeated in reverse after SCR2 develops −2Vb volts across C3. The peak no-load output voltage Vout(peak) is equal to the turns ratio of the transformer multiplied by 2Vb.

The size of capacitor C3 depends on the transformer used, the turn-off time for SCR1 and SCR2, and the size of the output load (see Fitzgerald et al. (1971, p. 416) for more information on selection of the commutating capacitor). An easy way to select C3 is to use trial and error while the output is connected to the lowest value of load resistance likely to be used.

Inductor L is placed between the battery and the silicon controlled rectifiers to buffer the supply current. A 10-mH coil should

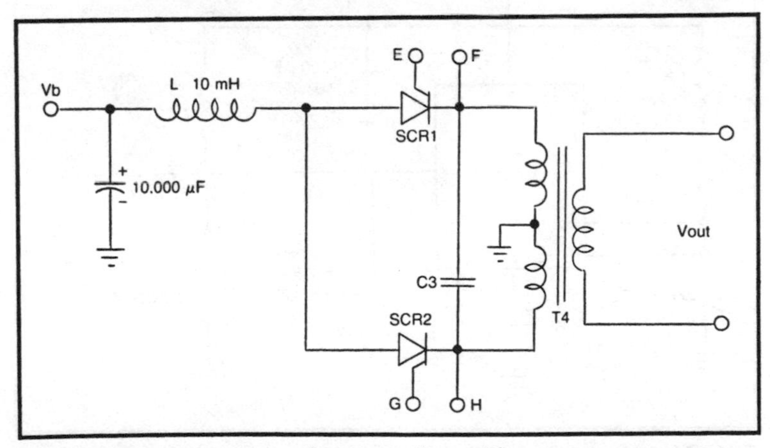

Fig. 8-18. Power circuit for the silicon controlled rectifier square-wave converter.

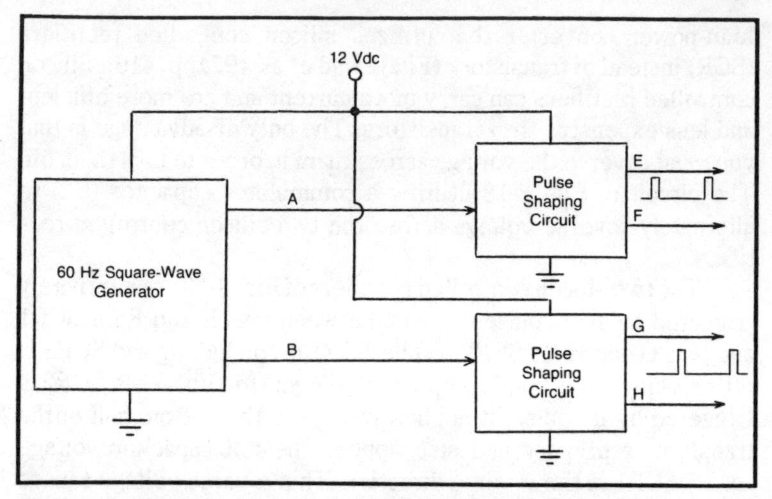

Fig. 8-19. Pulse-generating circuit used to trigger SCR1 and SCR2 in the SCR power circuit of Fig. 8-18.

suffice here. The 10,000-μF capacitor placed before inductor L helps filter out any high-frequency signals that might be fed back to the electronic pulse generator.

The selection of SCR1 and SCR2 depends on the size of the transformer used and the amount of power output desired. If Vb equals 60 volts and you want a power output of 5000 watts, then both

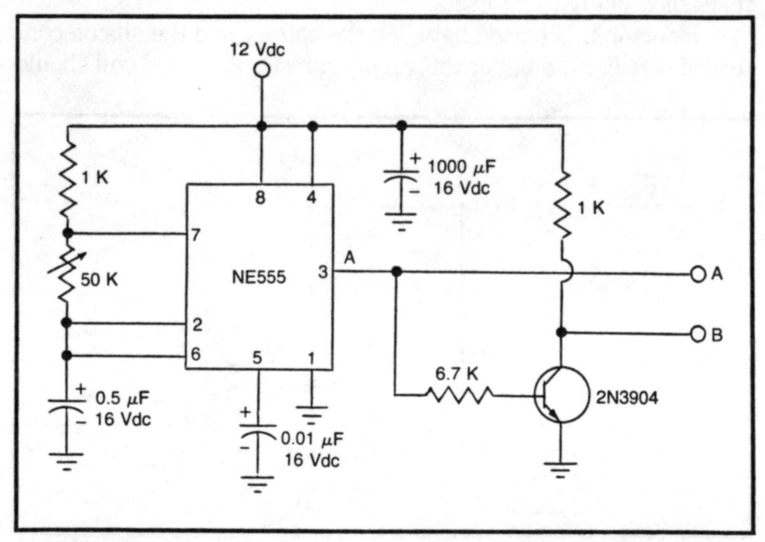

Fig. 8-20. 60-cycle square-wave generator for use with the SCR converter.

silicon controlled rectifiers should be able to handle at least 85 amps and have PIV ratings of at least 80 volts.

The transformer should also be chosen with power requirements in mind. If Vb equals 60 volts and you want an output of 120 Vac at 5000 watts, then the transformer needs a turns ratio of approximately 1—a current rating of at least 85 amps on the primary and 42 amps on the secondary. The current rating of inductor L should also match the current ratings for both the transformer and the silicon controlled rectifiers.

The two silicon controlled rectifiers are triggered by two short alternating pulses that are generated by the circuit in Fig. 8-19. This circuit utilizes a square-wave generator (see Fig. 8-20) to generate two square waves that are 180 degrees out of phase. These waveforms are then passed through pulse-shaping circuits (see Fig. 8-21), which produce short pulses on their outputs each time the input waveforms make a transition from negative to positive. These pulse-shaping circuits utilize small isolation transformers (T5) to apply the pulses to the silicon controlled rectifiers. The turns ratio of T5 should be approximately 1, but it is not necessary. Practically any small transformer will be sufficient here as long as it produces enough current to trigger the SCR.

As stated before, a disadvantage of the SCR converter is that the SCR can only be turned off by reversing the voltage across it. The commutating capacitor provides this voltage-reversing service, but it also causes some additional problems. If a commutating

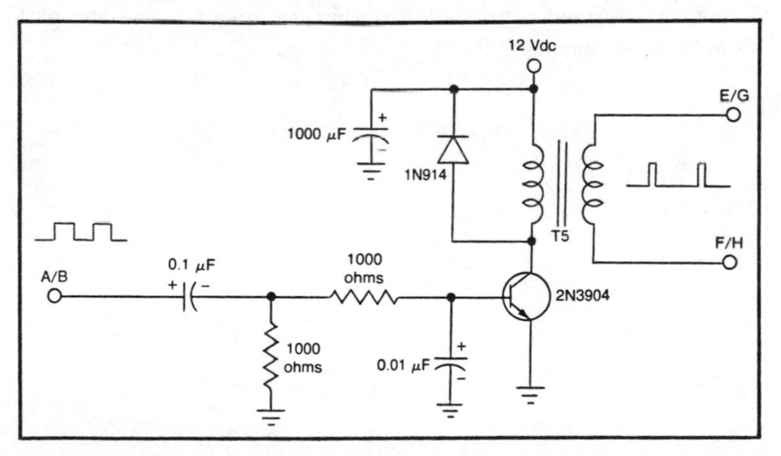

Fig. 8-21. Pulse-shaping circuit for producing alternating spikes from square-wave inputs to trigger the silicon controlled rectifier of Fig. 8-18.

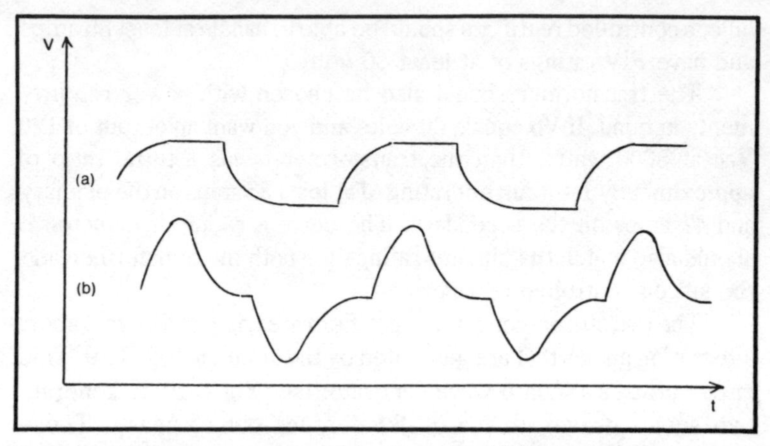

Fig. 8-22. (A) Proper output waveform of a loaded SCR converter. (B) Output waveform of the same converter with the load removed.

capacitor is chosen for a particular output load and then the output load is changed, the output waveform may change drastically. Figure 8-22A shows an acceptable output waveform of an SCR converter with the output loaded. Figure 8-22B shows the output of the same converter with no load on the output. These changes may or may not be acceptable, depending on the use of the converter.

This converter circuit is not recommended for the electronic novice. The power circuit requires much trial and error to find the correct size components that will mate together to produce the proper output waveform at the desired frequency and power level. The transistorized converter, however, is very reliable and part selections are not as critical.

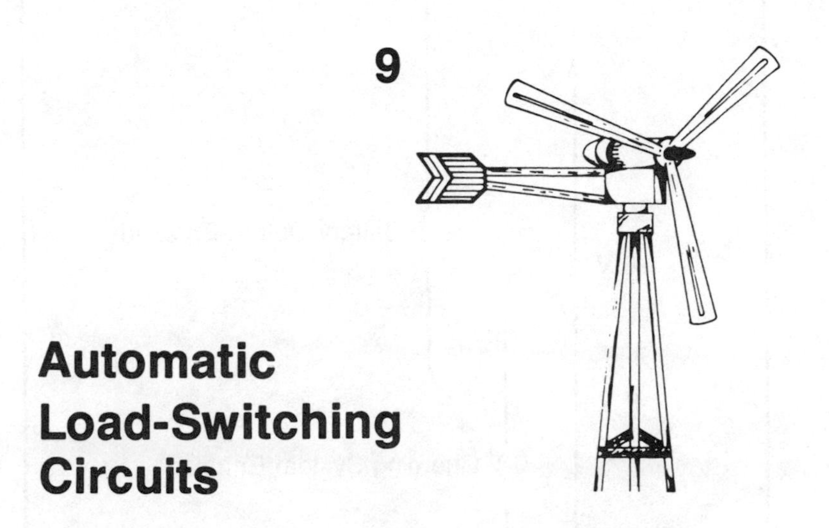

9 Automatic Load-Switching Circuits

Battery banks should always be protected from overcharging or undercharging. This can be done manually by periodically checking voltage levels on a daily basis or by implementing an automatic load-switching system. An automatic system can constantly monitor battery voltage and switch charging systems and loads in or out to maintain the battery-bank voltage at the desired level. This chapter describes the design of some very useful automatic load-switching circuits.

12-VOLT BATTERY BANKS

One possible load-switching scheme is shown in Fig. 9-1. Here the charging system is left on only for voltages below 16 volts, while the battery output is on for voltages greater than 9 volts. This load-switching scheme will effectively maintain battery-bank voltage between 9 volts and 16 volts. The only problem now is designing the switching circuits.

The first step in the design of a logical switching circuit is to draw a truth table. Figure 9-2 shows a truth table for the load-switching scheme of Fig. 9-1. In this truth table a "1" represents true, and a "0" represents false. For example, the first case in Fig. 9-2 shows a false for high voltage (greater than 16 volts) and a false for low voltage (less than 9 volts). This means the actual voltage is between 9 volts and 16 volts. The second case shows a false for high

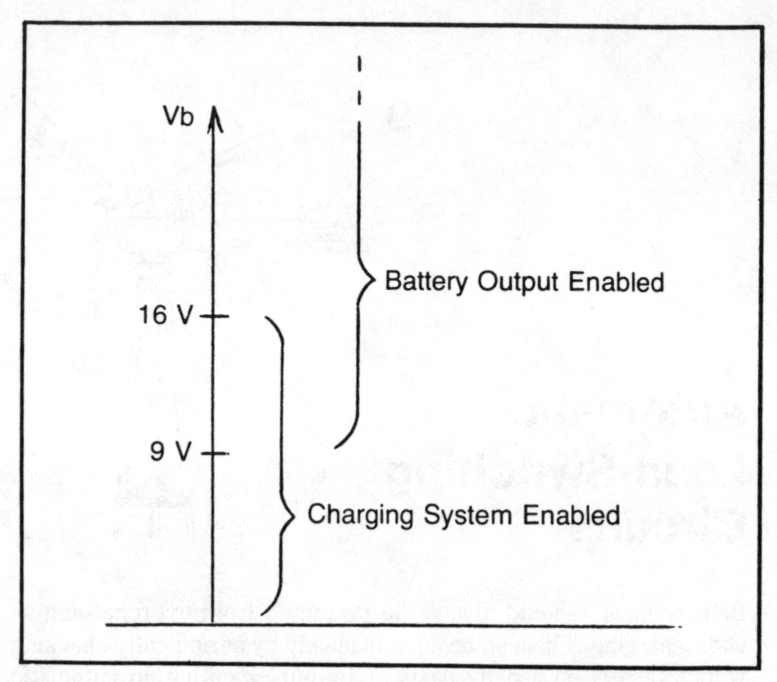

Fig. 9-1. Voltage levels for enabling or disabling battery output and charging systems.

voltage and a true for low voltage. In this case the actual voltage is less than 9 volts. The third case shows a true for high voltage and a false for low voltage. Here the actual voltage is greater than 16 volts.

From the truth table in Fig. 9-2 you can see that "charging system enabled" is exactly opposite "high voltage." In other words, the charging system is enabled only when the battery voltage is not above 16 volts. Also, "battery output enabled" is opposite "low

High Voltage (>16 V)	Low Voltage (<9 V)	Charging System Enabled	Battery Output Enabled
0	0	1	1
0	1	1	0
1	0	0	1

1 - True
0 - False

Fig. 9-2. Truth table for the graph in Fig. 9-1.

voltage." This means the battery output is enabled when it is not below 9 volts. If voltage detectors are now used to detect the low-voltage and high-voltage levels, their outputs can be inverted and used to control the charging system and the battery output. Figure 9-3 illustrates this electrical implementation of the truth table of Fig. 9-2.

In Fig. 9-3 the low-voltage detector outputs a high dc voltage (a dc voltage equal to the battery-bank voltage) when the battery voltage is below 9 volts. The inverter then changes this to a zero voltage level, which causes the battery output to be disabled. When the battery voltage is between 9 volts and 16 volts, the outputs of both detectors are zero, which means that the voltage is neither low nor high. The inverters then change these zero voltages to high dc voltages, which cause both the battery output and charging system to be enabled. If the battery voltage goes above 16 volts, the output of the high-voltage detector goes high, while the low-voltage detector remains low. In this case the charging system is turned off by the low output of the inverter, and the battery output remains on. The enabling circuits that use these high and low voltages to turn the systems on and off will be discussed later.

Figure 9-4 shows a more detailed electrical diagram of the same load-switching circuit. Two voltages (V1 and V2), both of which are proportional to battery-bank voltage (Vb), are compared to a 6.2-volt stable reference voltage to achieve the desired switching levels. This stable reference voltage is developed across the 1N4735 zener diode. Resistors R27 and R28 are chosen so that V1 is always equal to (6.2/9)Vb and V2 is always equal to (6.2/16)Vb. Thus, when Vb = 9 volts, V1 = 6.2 volts . When Vb goes

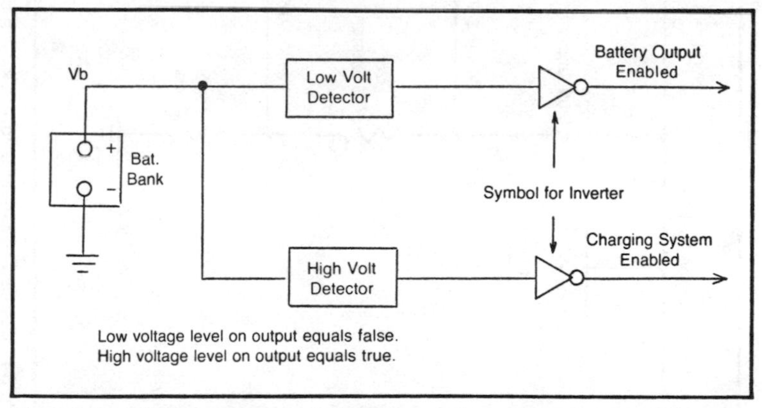

Fig. 9-3. Electrical diagram showing implementation of the truth table in Fig. 9-2.

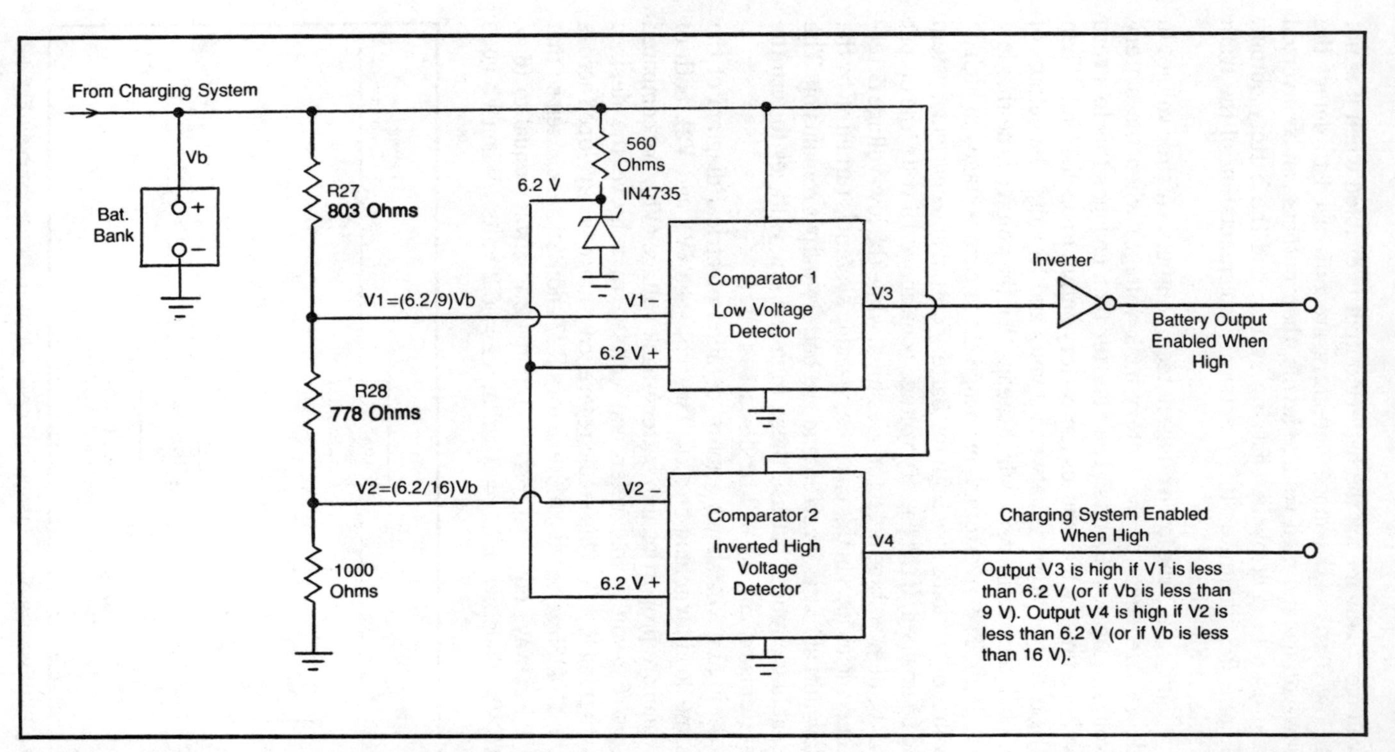

Fig. 9-4. Detailed electrical diagram for the circuit of Fig. 9-3.

below 9 volts, V1 goes below 6.2 volts and the output of comparator 1 goes high. The inverter output goes low at the same time and disables the battery output to prevent the battery voltage from going lower and damaging the batteries. When Vb = 16 volts, V2 = 6.2 volts. When Vb becomes greater than 16 volts, V2 becomes greater than 6.2 volts and the output of comparator 2 goes negative, disabling the battery charging system. This comparator does not require an inverter at the output because it is already operating in reverse. For V1 to be equal to (6.2/9)Vb and V2 to be equal to (6.2/16)Vb, resistor R27 must be 803 ohms and resistor R28 must be 778 ohms. Because these resistors are not standard, potentiometers must be used instead and adjusted to the proper resistance.

The voltage comparator used in the circuit of Fig. 9-4 is shown in Fig. 9-5. It is actually a high gain, low frequency, differential amplifier that takes over 5 seconds to respond to a voltage imbalance at the input. This slow response is necessary to prevent the load-switching circuit from reacting to fast, temporary changes in voltage levels.

The inverter circuit, shown in Fig. 9-6, is a simple circuit that changes a zero input voltage to a high-level output voltage and a high-level input voltage to a zero output voltage. This circuit was also used in Chapter 8.

Figure 9-7 shows a relay circuit for enabling battery charging systems or battery outputs. This circuit connects directly to the outputs of the load-switching circuit of Fig. 9-4. The sizes of the

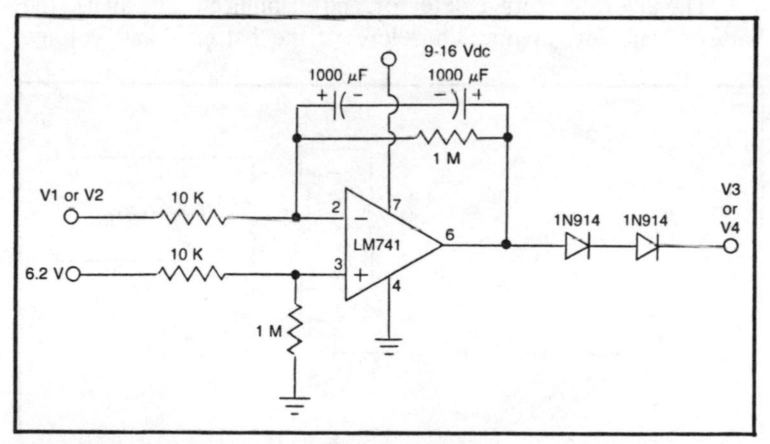

Fig. 9-5. Time-delay voltage comparator used in the circuit of Fig. 9-4 (also a high-gain, low-frequency, differential amplifier).

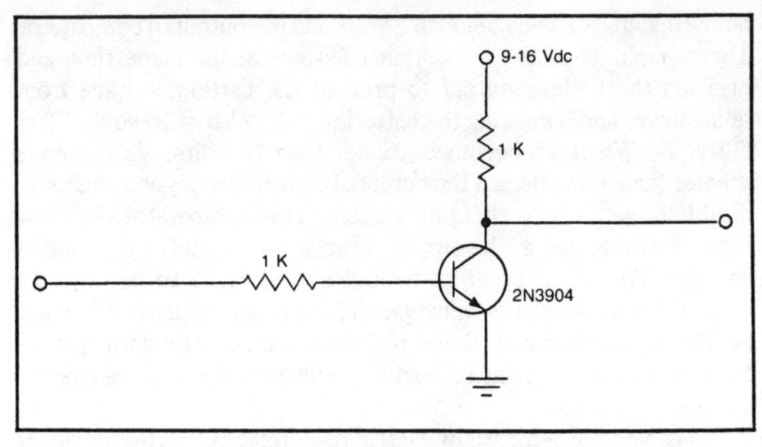

Fig. 9-6. Inverter used in the circuit of Fig. 9-4.

electronic components used in this circuit depend on the sizes of the relay or solenoid used. If the relay requires 1 amp to energize it, then transistor Q5 must be chosen to handle at least 1 amp. The value of resistor R29 (in ohms) can be calculated from the following equation:

$$R29 = \frac{(12 \text{ volts})(\text{current gain of Q5})}{2(\text{current required to energize relay})} \qquad \textbf{Eq. 9-1}$$

Because transistors usually have a current gain of about 100, a 1000-ohm resistor should be sufficient if the relay requires 1 amp.

The inverter, voltage detector, and enabling circuits all use the battery bank for power. Therefore, if the battery-bank voltage

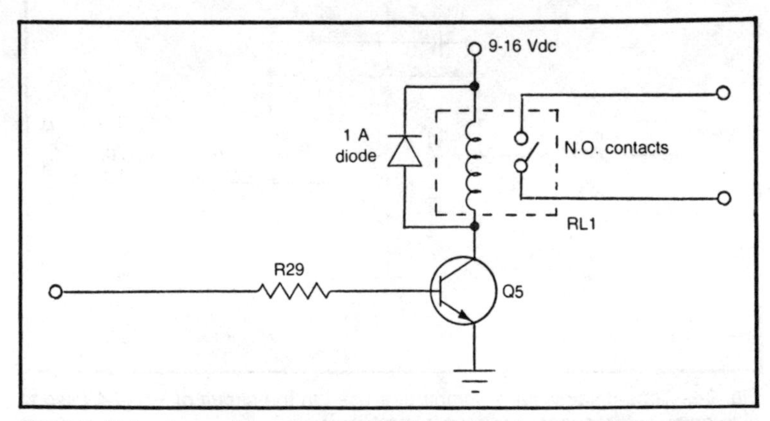

Fig. 9-7. Relay for enabling the battery charging system or battery outputs.

varies from 9 volts to 16 volts, the voltage used by these circuits does the same. This does not affect the detector circuits, because the 6.2-volt reference is regulated by the zener diode. The only effect this has on the inverters is that the high-level outputs will always be equal to the battery-bank voltage. This output voltage will still, however, always be large enough to turn on the transistors in the enabling circuits. If the battery voltage goes too low, it is possible that the relay in the enabling circuit will not energize. You should test this before the system is operational. If the relay does not energize at 9 volts, then use a relay with a lower voltage rating.

There is one problem with this load-switching circuit. Once the battery bank reaches 16 volts, the charging system is disabled. This has a tendency to waste energy that could be put to use elsewhere. Figure 9-8 shows a load-switching scheme that leaves the charging system on, but switches in an additional load when 16 volts is reached. This additional load must draw considerable power and should be something that is not an absolute necessity, because it will probably not be enabled very often. This could probably be a supplemental electric room heater or hot water preheater. It is still necessary, however, that all battery outputs be disconnected at some low-voltage level. If lead-acid batteries are used in the battery bank, this low-voltage level should be no less than 9 volts.

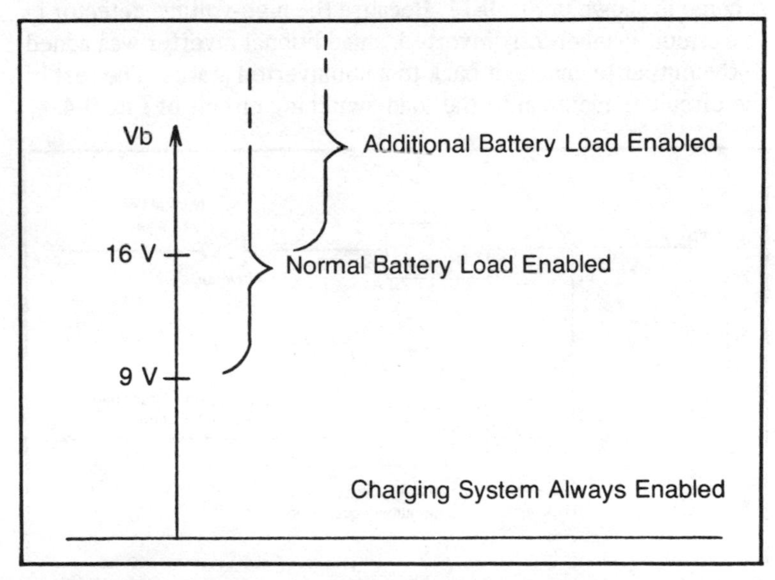

Fig. 9-8. Voltage levels for enabling or disabling the normal battery load and additional battery load.

High Voltage (>16 V)	Low Voltage (<9 V)	Normal Battery Load Enabled	Additional Load Enabled
0	0	1	0
0	1	0	0
1	0	1	1

1 - True
0 - False

Fig. 9-9. Truth table for the graph in Fig. 9-8.

Figure 9-9 shows the truth table for this load switching scheme, and Fig. 9-10 shows the electrical implementation. Notice that "normal battery load enabled" is opposite "low voltage," and "additional load enabled" is the same as "high voltage." Therefore, the output of the low-voltage detector should be inverted before being input to the enabling circuit, and the output of the high-voltage detector should not.

The more detailed load-switching circuit for this switching scheme is shown in Fig. 9-11. Because the high-voltage detector in this circuit is inherently inverted, an additional inverter was added to the output to change it back to a noninverted status. The rest of the circuit is identical to the load-switching circuit of Fig. 9-4.

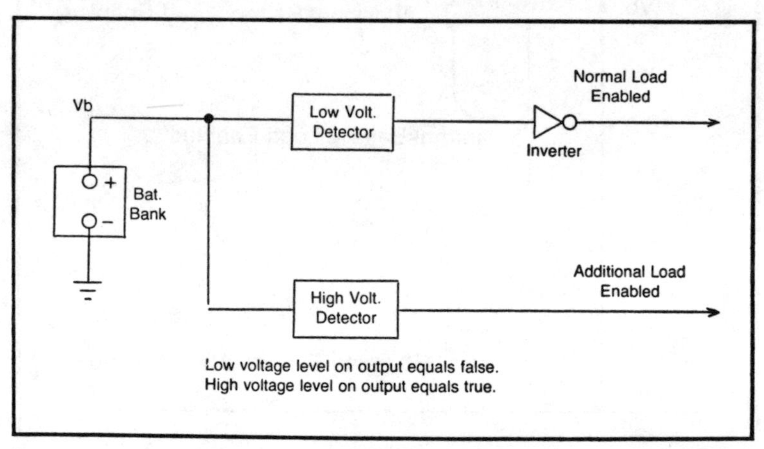

Fig. 9-10. Electrical diagram showing implementation of the truth table in Fig. 9-9.

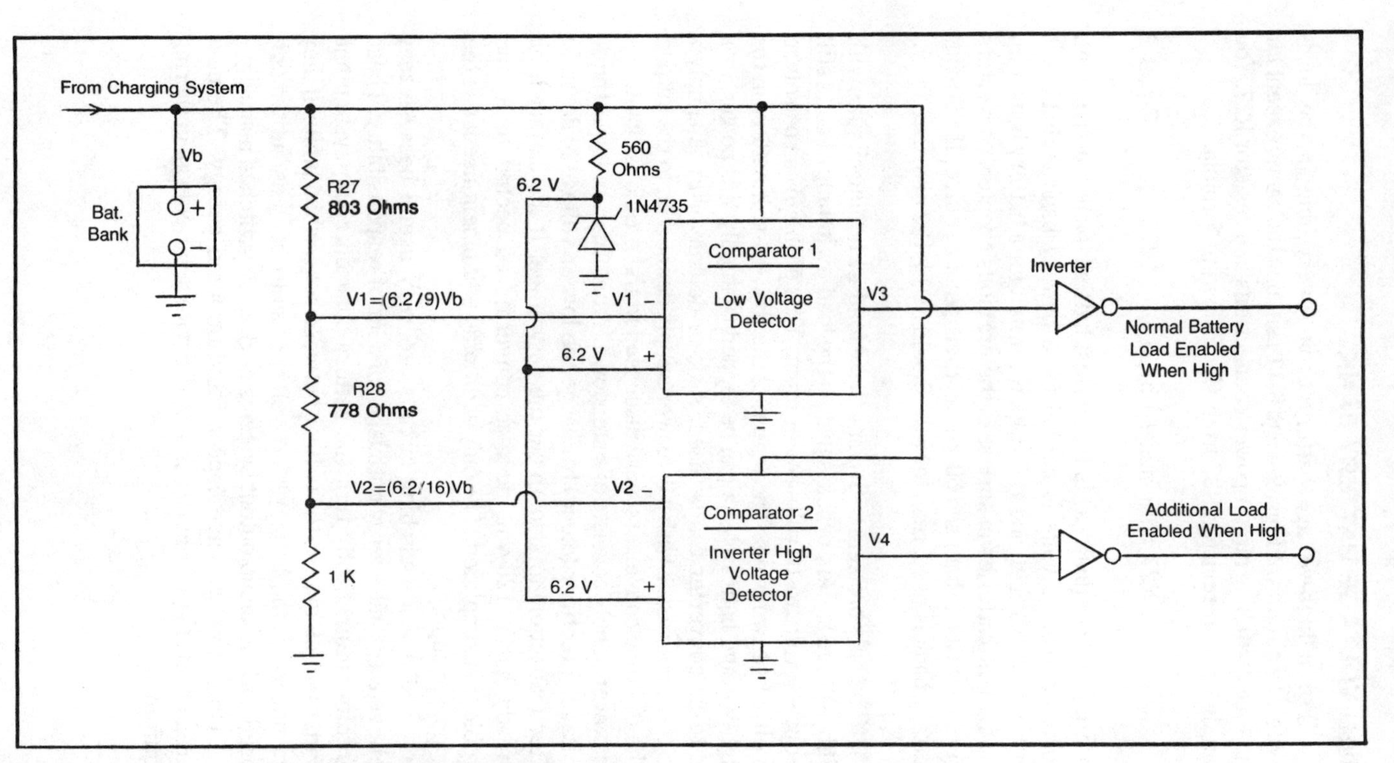

Fig. 9-11. Detailed electrical diagram for the circuit of Fig. 9-10.

HIGH-VOLTAGE BATTERY BANKS

For a high-voltage battery bank, everything in the load-switching circuits of Figs. 9-4 and 9-11 remains the same, except for the load switching circuit power connections and resistor R27. You can determine resistor R27 from the following formula:

$$R27 = (\text{nominal Vb})(215) - 1778 \qquad \textbf{Eq. 9-2}$$

In Eq. 9-2, "nominal Vb" is the normal battery-bank voltage level. For example, 12 volts is normal for a battery bank with 12-volt batteries in parallel, and 120 volts is normal for a battery bank with ten 12-volt batteries in series. In the previous chapter, converters were described that used 60 Vdc as their power source. If a 60-volt battery bank is used, resistor R27 should be 11,122 ohms, as shown by Eq. 9-2.

The only problem with using these load-switching circuits with high-voltage battery banks is that all the load-switching circuits still require 9 volts to 16 volts power input. This problem can be solved in either of two ways. The output of the first 12-volt battery in the high-voltage battery bank can be tapped to supply this power, or a dc-to-dc converter can be used. A dc-to-dc converter is simply a dc-to-ac converter with a dc power supply connected to the output. The dc-to-ac converters were discussed in the previous chapter. A dc power supply consists of a stepdown transformer to lower the ac voltage, a rectifier (diode) to change the low ac voltage to dc, and a filter (capacitor) to smooth out the dc voltage. If the battery bank already has a 120-volt dc-to-ac converter connected to it, the problem is simplified. The only additional equipment needed is the dc power supply.

These load-switching circuits are very useful because they relieve the wind-energy enthusiast of a lot of responsibility. These circuits automatically turn the battery-bank charging system and loads on or off to keep the voltage between prescribed limits. If they are not used, the battery-bank voltage has to be constantly monitored and the input/output systems have to be switched manually.

Until now we have concentrated on wind energy. The next three chapters present circuits for use with solar-energy gathering equipment.

10

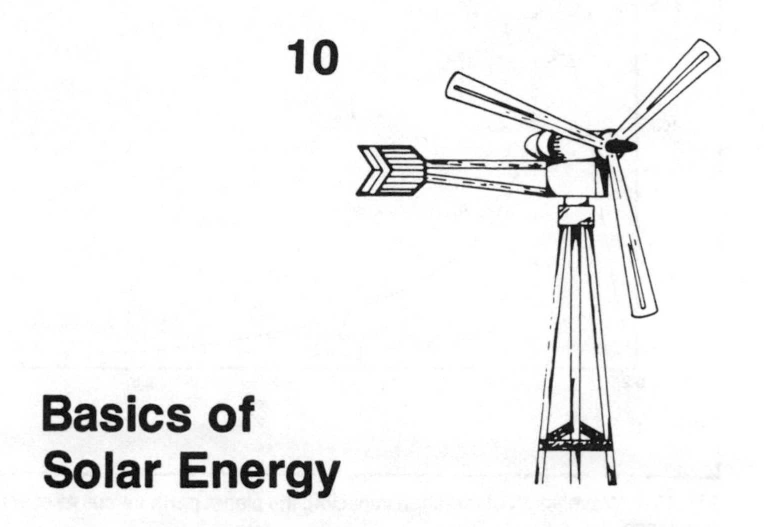

Basics of Solar Energy

Solar energy is our only truly renewable source of energy. The sun is 80 percent hydrogen, 18 percent helium, and 1.4×10^6 kilometers in diameter; it has a mass of 1.987×10^{30} kilograms and a temperature at the center of 20×10^6 degrees kelvin (Anderson 1983, p. 3). It transmits 83.3 million billion billion kilowatt hours of energy (8.33×10^{25} kWh) into space each day, 4.14×10^{15} kWh of which is received by the earth (Cheremisinoff and Regino 1978, p. 17). It is expected to continue radiating energy for billions of years.

AVAILABILITY OF SOLAR ENERGY

The intensity of the solar energy striking the earth is about 1353 watts per square meter (Anderson 1983, p. 3). The majority of that radiation has a wavelength between 0.2 μm and 2.6 μm (1×10^{-6} meters), as shown in Fig. 10-1. Visible light, which is between 0.4 μm and 0.7 μm, constitutes only a small portion of that energy (Cheremisinoff and Regino 1978, p. 18).

Even though this planet intercepts 1353 watts per square meter of this power (energy is measured in *watt-hours*), less than 80 percent of this actually reaches the earth's surface. The rest is absorbed in the atmosphere. The solar energy impacting this planet is used up in two ways: thermal radiation into space and photosynthesis, which is the process by which plants convert solar energy to chemical energy (Anderson 1983, p. 3). If only 20 percent of the

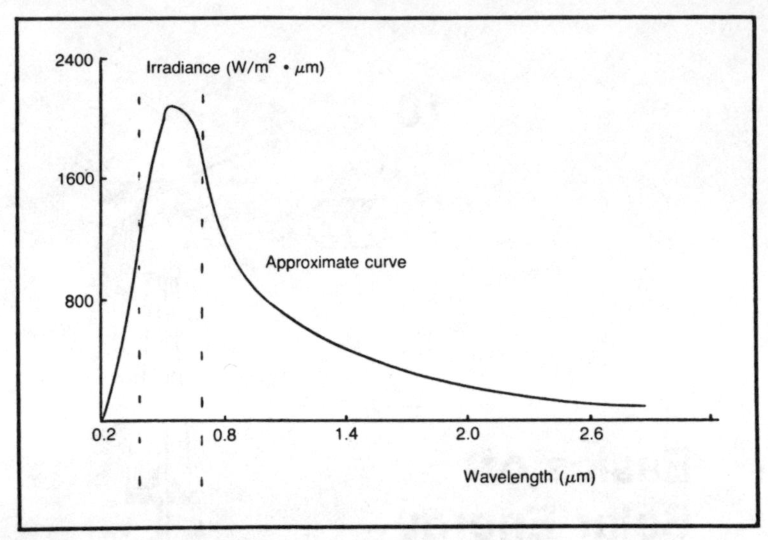

Fig. 10-1. Wavelength of radiation impacting the planet earth versus its spectral irradiance.

solar energy were absorbed by the atmosphere and all the solar energy striking the surface of the earth could be converted to usable energy, then a collector 50 meters square could supply more than enough energy for an average home.

Things, however, are never that easy. Solar energy is a low-level energy form. Large collector surfaces are required to produce usable levels of energy, and the solar radiation must be concentrated if high temperatures are desired. Also, solar energy is not always available. Some type of medium must be used to store available energy and release it at night or in bad weather. This means that a larger collector must be used to simultaneously provide enough energy for storage and immediate use. The energy from the sun is free, but converting it to a reliable and usable form is not.

SOLAR INCLINATION

To have a thorough knowledge of solar energy you must understand the geometrical relationship between the earth and the sun and how that relationship changes through the year.

The earth rotates daily in an equatorial plane about its polar axis. It also rotates yearly about the sun in the earth orbital plane. These relationships are shown in Fig. 10-2. The angle between the polar axis and a line perpendicular to the earth orbital plane is the

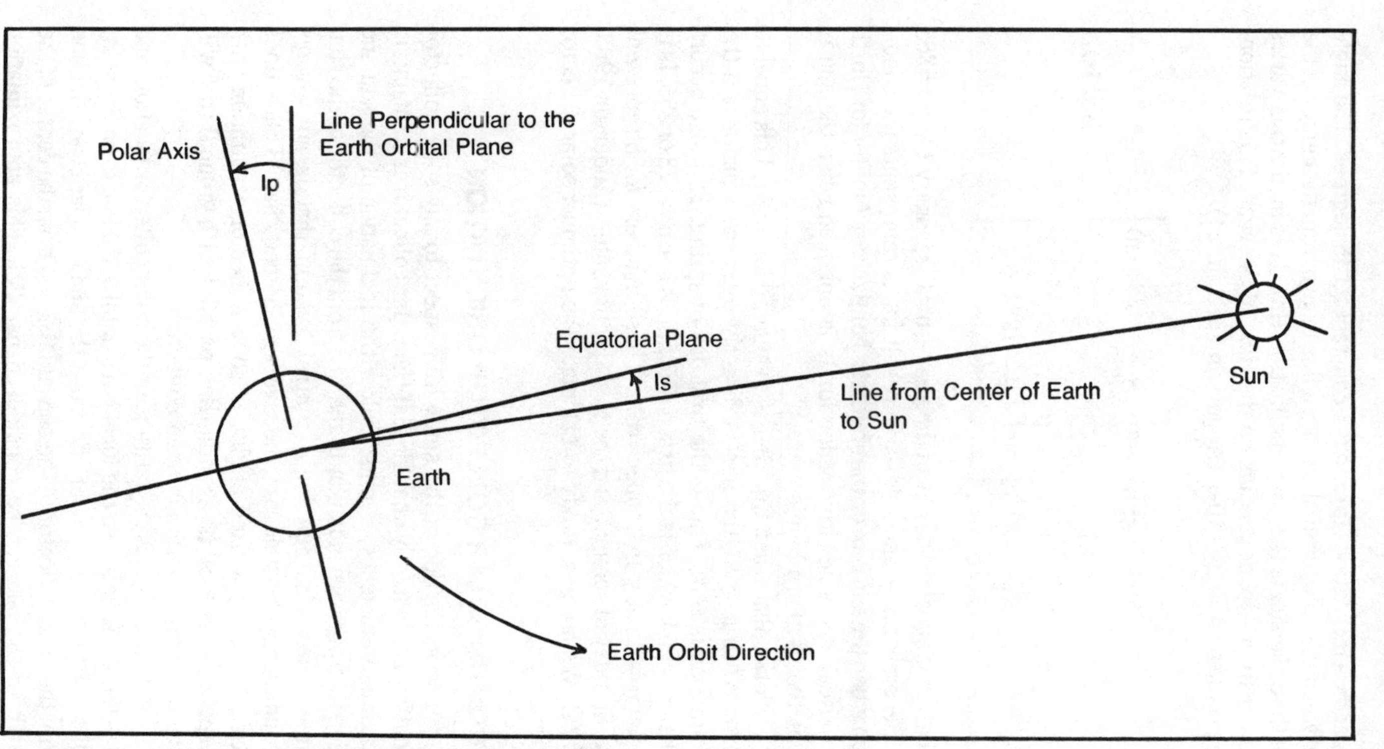

Fig. 10-2. Geometrical relationship between the sun and earth.

polar axis inclination (Ip) and is a constant 23.45 degrees. The angle between the equatorial plane and a line drawn from the center of the earth to the sun is the solar inclination (Is). This inclination varies from +Ip to −Ip, depending on the time of the year. This relationship is shown in Eq. 10-1 (Anderson 1983, p. 24).

$$Is = -Ip \sin\left[\frac{360}{365}(284+n)\right] \qquad \textbf{Eq. 10-1}$$

$$= -23.45 \sin\left[\frac{360}{365}(284+n)\right]$$

where n is equal to the day of the year—n=1 is January 1 and n=365 is December 31—and a positive result for Is represents the condition when the sun is over the southern hemisphere (winter for North America), and a negative value for Is is obtained when the sun is over the northern hemisphere.

Notice that when Eq. 10-1 is evaluated for n=81, Is is equal to zero. When n = 81 the sun is directly over the equator and the inclination is zero. This is the first day of spring. This also occurs when n = 264, which is the first day of fall. When n = 173 or 356, Is is equal to −Ip or +Ip, respectively. These numbers for n represent the first day of summer and the first day of winter. (Anderson 1983, p. 24). We are now ready to calculate the optimum solar collector orientation.

OPTIMUM SOLAR COLLECTOR ORIENTATION

The inclination of the sun with respect to any solar collector location is constantly changing. If the solar collector is to maintain maximum efficiency, it must follow the inclination of the sun and keep its front perpendicular to the line from the collector to the sun. This requires an automatic control system that is not always economically feasible for the average homeowner. There does exist, however, a fixed angle for solar collector orientation with respect to the local horizontal (line parallel with ground) that will give optimum results for fixed collectors.

In Fig. 10-3, L is the latitude of the solar collector location, and Ih is the angle between the local horizontal and a line from the solar collector to the sun at noon. Because the sun is so far away, the line from the solar collector to the sun and the line from the center of the earth to the sun can be considered parallel. This approximation

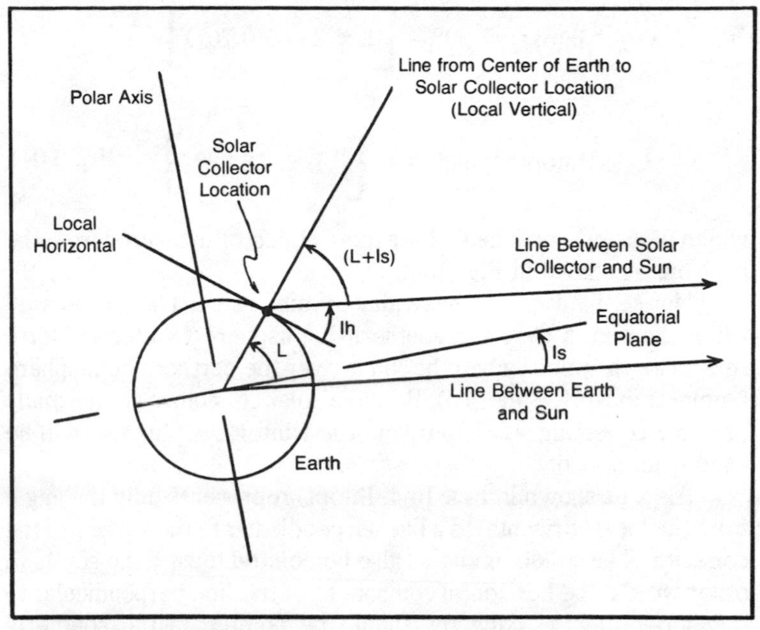

Fig. 10-3. Geometrical diagram for calculating the angle (Ih) between the sun and the local horizontal at noon.

allows the angle between the local vertical and the line from the collector to the sun to be approximated as (L + Is). Because the local horizontal and the local vertical are 90 degrees apart, we can find Ih as follows:

$$Ih = 90° - (L + Is). \qquad \textbf{Eq. 10-2}$$

If Eq. 10-1 is substituted for Is in Eq. 10-2, we obtain

$$Ih = 90° - \left\{L - 23.45 \quad \sin\left[\frac{360}{365}(284 + n)\right]\right\}. \qquad \textbf{Eq. 10-3}$$

Equation 10-3 gives the inclination angle from the horizon to the sun (at noon) for any day of the year and any location in the northern hemisphere.

The only part of Eq. 10-3 that is not constant is the sine term. If the rms value of the sine term is used instead, a constant value can be found. This is similar to finding the dc equivalent (in terms of energy) of an ac signal. The rms value of the sine function is 0.707. Once this is substituted, the equation becomes

$$Ih(opt) = 90° - \left[L \pm 23.45(0.707) \right]$$

or

$$Ih(opt) = 90° - \left[L \pm 16.58° \right] \quad \textbf{Eq. 10-4}$$

which is the optimum angle for a fixed collector measured from the horizon, as shown in Fig. 10-4.

Notice that Eq. 10-4 has a plus or minus sign. The plus is used when the sun is over the southern hemisphere (winter in North America); the minus when the sun is over the northern hemisphere (summer in North America). Because solar collectors are normally used for collecting solar energy in the winter, the plus sign will be used from now on.

Also, as shown in Fig. 10-4, Ih(opt) represents only the angle from the local horizontal to a line perpendicular to the surface of the collector. The collector should also be pointed toward the south. In other words, the horizontal component of the line perpendicular to the surface of the collector should be pointed in the southerly direction.

As an example of the usefulness of Eq. 10-4, suppose that a solar collector is located at 38.49 degrees latitude (Colorado Springs, Colorado). Using Eq. 10-3 first, we see that Ih varies from 28.06 degrees on the first day of winter to 51.51 degrees on the first days of spring and fall. We naturally assume that a fixed collector should be mounted at some angle between these two values if it is to take maximum advantage of the energy from the sun during this half of the year. Now using Eq. 10-4, we calculate Ih(opt) to be 34.93 degrees. This is the collector angle that gives best results for a fixed collector located at 38.49 degrees latitude. Table 10-1 gives a list of

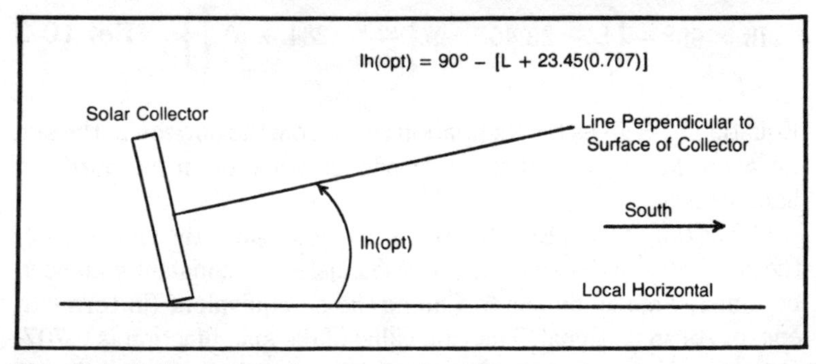

Fig. 10-4. Optimum angle for a fixed solar collector.

Table 10-1. Latitude Angles for Several Major Cities in the United States. (Continues through page 106.)

State	City	Latitude in degrees
AK	Homer	59.38
AL	Birmingham	33.34
AL	Mobile	30.41
AL	Montgomery	32.18
AR	Fort Smith	35.20
AR	Little Rock	34.44
AZ	Phoenix	33.26
AZ	Prescott	34.39
AZ	Tucson	32.07
AZ	Winslow	35.01
AZ	Yuma	32.40
CA	Bakersfield	35.25
CA	China Lake	35.41
CA	Daggett	34.52
CA	Fresno	36.46
CA	Long Beach	33.49
CA	Los Angeles	33.56
CA	Mount Shasta	41.19
CA	Needles	34.46
CA	Oakland	37.44
CA	Red Bluff	40.09
CA	Sacramento	38.31
CA	San Diego	32.44
CA	San Francisco	37.37
CA	Santa Maria	34.54
CA	Sunnyvale	37.25
CO	Colorado Springs	38.49
CO	Denver	39.45
CO	Grand Junction	39.07
CO	Pueblo	38.17
CT	Hartford	41.56
DC	Washington-Sterling	38.57
DE	Wilmington	39.40
FL	Apalachicola	29.44
FL	Daytona Beach	29.11
FL	Jacksonville	30.30
FL	Miami	25.48
FL	Orlando	28.33
FL	Tallahassee	30.23
FL	Tampa	27.58
GA	Atlanta	33.39
GA	Augusta	33.22
GA	Macon	32.42
GA	Savannah	32.08
HI	Hilo	19.43
HI	Honolulu	21.20
HI	Lihue	21.59
IA	Burlington	40.47
IA	Des Moines	41.32
IA	Mason City	43.09

State	City	Latitude in degrees
IA	Sioux City	42.24
ID	Boise	43.34
ID	Lewiston	46.23
ID	Pocatello	42.55
IL	Chicago	41.47
IL	Moline	41.27
IL	Springfield	39.50
IN	Evansville	38.03
IN	Fort Wayne	41.00
IN	Indianapolis	39.44
IN	South Bend	41.42
KS	Dodge City	37.46
KS	Goodland	39.22
KS	Topeka	39.04
KS	Wichita	37.39
KY	Lexington	38.02
KY	Louisville	38.11
LA	Baton Rouge	30.32
LA	Lake Charles	30.07
LA	New Orleans	29.59
LA	Shreveport	32.38
MA	Boston	42.22
MD	Baltimore	39.11
MD	Patuxent River	38.17
ME	Caribou	46.52
ME	Portland	43.39
MI	Alpena	45.04
MI	Detroit	42.25
MI	Flint	42.58
MI	Grand Rapids	42.53
MI	Houghton	47.10
MI	Traverse City	44.44
MN	Duluth	46.50
MN	International Falls	48.34
MN	Minneapolis/St. Paul	44.53
MN	Rochester	43.55
MO	Columbia	38.49
MO	Kansas City	39.18
MO	Springfield	37.14
MO	St. Louis	38.45
MS	Jackson	32.19
MS	Meridian	32.20
MT	Billings	45.48
MT	Cut Bank	48.36
MT	Clasgow	48.13
MT	Great Falls	47.29
MT	Helena	46.36
MT	Lewistown	47.03
MT	Missoula	46.55
NC	Ashville	35.26
NC	Cape Hatteras	35.16
NC	Charoltte	35.13

State	City	Latitude in degrees
NC	Greensboro	36.05
NC	Raleigh-Durham	35.52
ND	Bismarck	46.46
ND	Fargo	46.54
ND	Minot	48.16
NE	Grand Island	40.58
NE	North Omaha	41.22
NE	North Platte	41.08
NE	Scottsbluff	41.52
NH	Concord	43.12
NJ	Lakehurst	40.02
NJ	Newark	40.42
NM	Albuquerque	35.03
NM	Clayton	36.27
NM	Farmington	36.45
NM	Roswell	33.24
NM	Tucumcari	35.11
NM	Zuni	35.06
NV	Elko	40.50
NV	Ely	39.17
NV	Las Vegas	36.05
NV	Reno	39.30
NV	Tonopah	38.04
NV	Winnemucca	40.54
NY	Albany	42.45
NY	Binghamton	42.13
NY	Buffalo	42.56
NY	Massena	44.56
NY	New York City (Central Park)	40.47
NY	New York City (La Guardia)	40.46
NY	Rochester	43.07
NY	Syracuse	43.07
OH	Akron-Canton	40.55
OH	Cincinnati	39.04
OH	Cleveland	41.24
OH	Columbus	40.00
OH	Dayton	39.54
OH	Toledo	41.36
OH	Youngstown	41.16
OK	Oklahoma City	35.24
OK	Tulsa	36.12
OR	Astoria	46.09
OR	Burns	43.35
OR	Medford	42.22
OR	North Bend	43.25
OR	Pendleton	45.41
OR	Portland	45.36
OR	Salem	44.55
PA	Allentown	40.39
PA	Erie	42.05
PA	Harrisburg	40.13

State	City	Latitude in degrees
PA	Philadelphia	39.53
PA	Pittsburgh	40.30
PA	Wilkes-Barre-Scranton	41.20
RI	Providence	41.44
SC	Charleston	32.54
SC	Columbia	33.57
SC	Greenville-Spartanburg	34.54
SD	Huron	44.23
SD	Pierre	44.23
SD	Rapid City	44.03
SD	Sioux Falls	43.34
TN	Chattanooga	35.02
TN	Knoxville	35.49
TN	Memphis	35.03
TN	Nashville	36.07
TX	Amarillo	35.14
TX	Austin	30.18
TX	Brownsville	25.54
TX	Corpus Christi	27.46
TX	Dallas	32.51
TX	El Paso	31.48
TX	Fort Worth	32.50
TX	Houston	29.59
TX	Laredo	27.32
TX	Lubbock	33.39
TX	Midland-Odessa	31.56
TX	Port Arthur	29.57
TX	San Antonio	29.32
TX	Waco	31.37
UT	Bryce Canyon	37.42
UT	Cedar City	37.42
UT	Salt Lake City	40.46
VA	Norfolk	36.54
VA	Richmond	37.30
VA	Roanoke	37.19
VT	Burlington	44.28
WA	Olympia	46.58
WA	Seattle-Tacoma	47.27
WA	Spokane	47.38
WA	Yakima	46.34
WI	Eau Claire	44.52
WI	Green Bay	44.29
WI	La Crosse	43.52
WI	Madison	43.08
WI	Milwaukee	42.57
WV	Charleston	38.22
WV	Huntington	38.22
WY	Casper	42.55
WY	Cheyenne	41.09
WY	Rock Springs	41.36
WY	Sheridan	44.46

latitude angles for major cities in the United States (Anderson 1983, p. 485). These can be used in Eq. 10-4 to calculate the optimum solar collector orientation angle.

CONVERTING SOLAR ENERGY TO HEAT ENERGY

Heat in a material is a function of the amount of molecular motion in that material. When light strikes an object, some of the light is reflected and some is absorbed. The molecules that absorb the light do this by converting the energy in the photons (smallest measure of light) to motion. This molecular motion causes the temperature of the object to rise (Cheremisinoff and Regino 1978, p. 34).

To increase the efficiency of this process, you must reduce the reflected light. You can do this by painting the surface of the object with a high-temperature, flat, black paint. Because black is absence of color, most of the light is absorbed, and very little is reflected.

Once the surface of the collector has been heated by solar radiation, it can lose that heat by *radiation* or by *convection.* Heat radiation is similar to light radiation but has a much shorter wavelength. Convection—the transfer of heat from one material to another through contact—is a useful property that provides a means by which the heat generated in the collector can be transferred to a more useful location. This transfer is usually accomplished with flowing water or air. Radiation, however, is not a useful property and must be reduced. This is usually done by placing insulation behind the collector and a glass cover on the front. The glass allows visible radiation to enter the collector while preventing most of the heat radiation from exiting.

The next chapter describes in more detail different solar collectors and the circuits for controlling their operation.

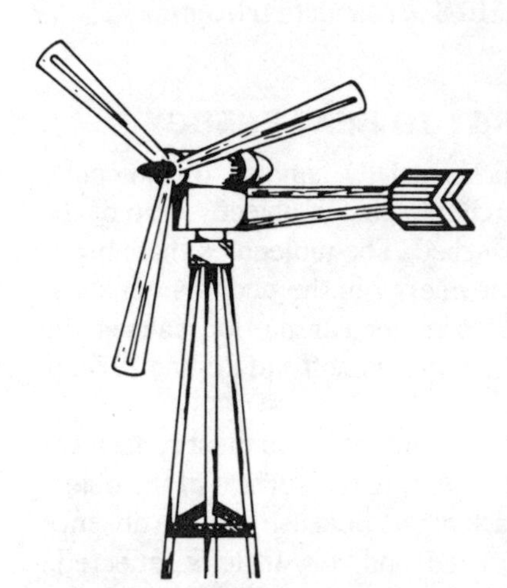

11

Circuits for Automatic Control of Solar-Energy Gathering Devices

There are two basic types of solar heat-energy gathering devices: the *parabolic-reflecting surface type* and the low-heat, nonconcentrating *solar panels*. Because of their peculiar shape and highly polished surface, the parabolic reflectors reflect and focus sunlight onto a small location, called the *focal point* or *focus*. Figure 11-1 shows a graph of a parabolic surface along with the location of its focal point.

The two-dimensional equation for the parabolic surface is

$$X = \sqrt{4pY} \quad \textbf{Eq. 11-1}$$

or

$$Y = X^2/4p \quad \textbf{Eq. 11-2}$$

where X is the distance in one direction, Y is the distance in a direction perpendicular to X, and p is the location of the focus on the Y-axis equal to ½. If the parabola is built on a large scale where all the dimensions are measured in feet, then p is equal to ½ foot or 6 inches. If it is built on a small scale where dimensions are measured in inches, then p is equal to ½ inch. If the parabolic surface is built exactly in accord with Eqs. 11-1 or 11-2, all the sunlight will be focused on the focal point when the Y-axis of the parabolic reflector is pointed toward the sun.

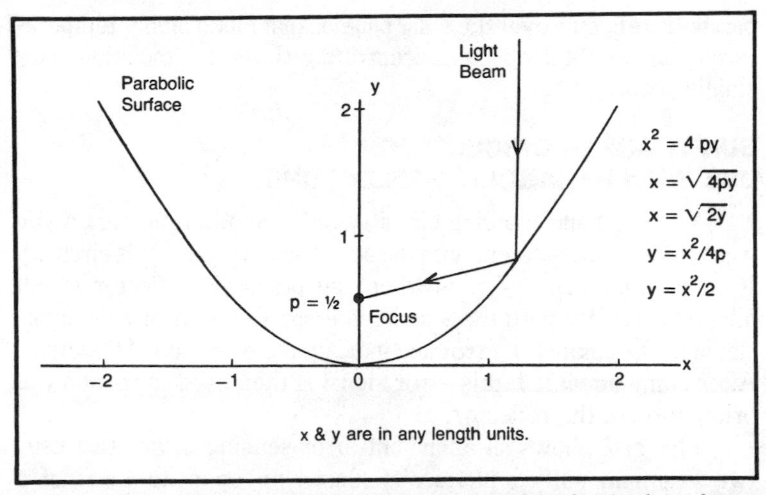

Fig. 11-1. Graph of a parabola showing the location of the focal point or focus.

Figure 11-2 shows two different parabolic reflectors, one of which requires precise alignment in one direction (*unidirectional*) while the other requires precise alignment in two directions (*bidirectional*). The focus of the unidirectional parabolic reflector is a line extending the length of the reflector. The focus of the bidirectional parabolic reflector is a single point.

Nonconcentrating solar panels do not require precise alignment and are, therefore, much less expensive and more common. While the parabolic reflector requires sun-tracking circuitry, mechanical direction control, and a heat transfer system, the solar panel requires only a heat transfer system. The advantage of the

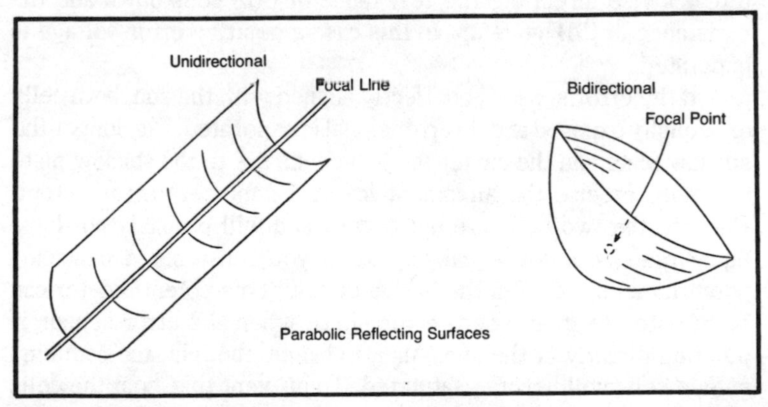

Fig. 11-2. Two types of parabolic reflectors.

parabolic reflector over the solar panel is that much higher temperatures can be obtained by concentrating the solar radiation on a smaller spot.

SUN-TRACKER CIRCUITS FOR USE WITH PARABOLIC REFLECTORS

You need sun-tracking circuitry on parabolic reflectors if you want automatic alignment with the sun. The purpose of this circuitry is to produce error signals when the parabolic reflector is not aligned correctly with the sun. This error signal must be proportional to the amount of error and indicate the direction of the error. After being amplified, this error signal is then used to correct the orientation of the reflector.

Fig. 11-3 shows an alignment-error sensing device that uses two cadmium sulfide photocells placed on opposite sides of a shadow plate. When this device is placed on a parabolic reflector with the shadow plate pointed in the Y-axis direction, it produces an error signal for misalignment in one direction.

As shown in Fig. 11-3, when the error-sensing device is misaligned with the sun in the clockwise direction, cadmium sulfide cell number 1 (Cd1) is exposed to light, while cell number 2 (Cd2) is in the shadow of the shadow plate. The exposure of Cd1 causes its resistance to go down. The lack of exposure on Cd2 causes its resistance to rise. Because these cells are part of a resistor bridge with 24 volts placed across it, an error voltage is developed across the bridge when the resistances of the cells are unbalanced. For this case a negative error signal is generated.

If the error sensor is misaligned with the sun in the counterclockwise direction, the resistance of Cd2 goes down and the resistance of Cd1 goes up. In this case a positive error voltage is generated.

If the error sensor is perfectly aligned with the sun, both cells are equally exposed and no error signal is generated. The longer the shadow plate and the closer the photocells are to the shadow plate the more precise the alignment has to be for zero error output. Because the two cells are not perfect and will probably not have equal resistance for equal exposure, you must put a balancing potentiometer (R30) in the bridge circuit. This potentiometer can be adjusted to give a zero error signal when the error sensor is pointing directly at the sun. Also, in bright sunlight, the cadmium sulfide cells will become saturated. To prevent this, coat the cells with a thin layer of opaque paint until the resistance of the cells is

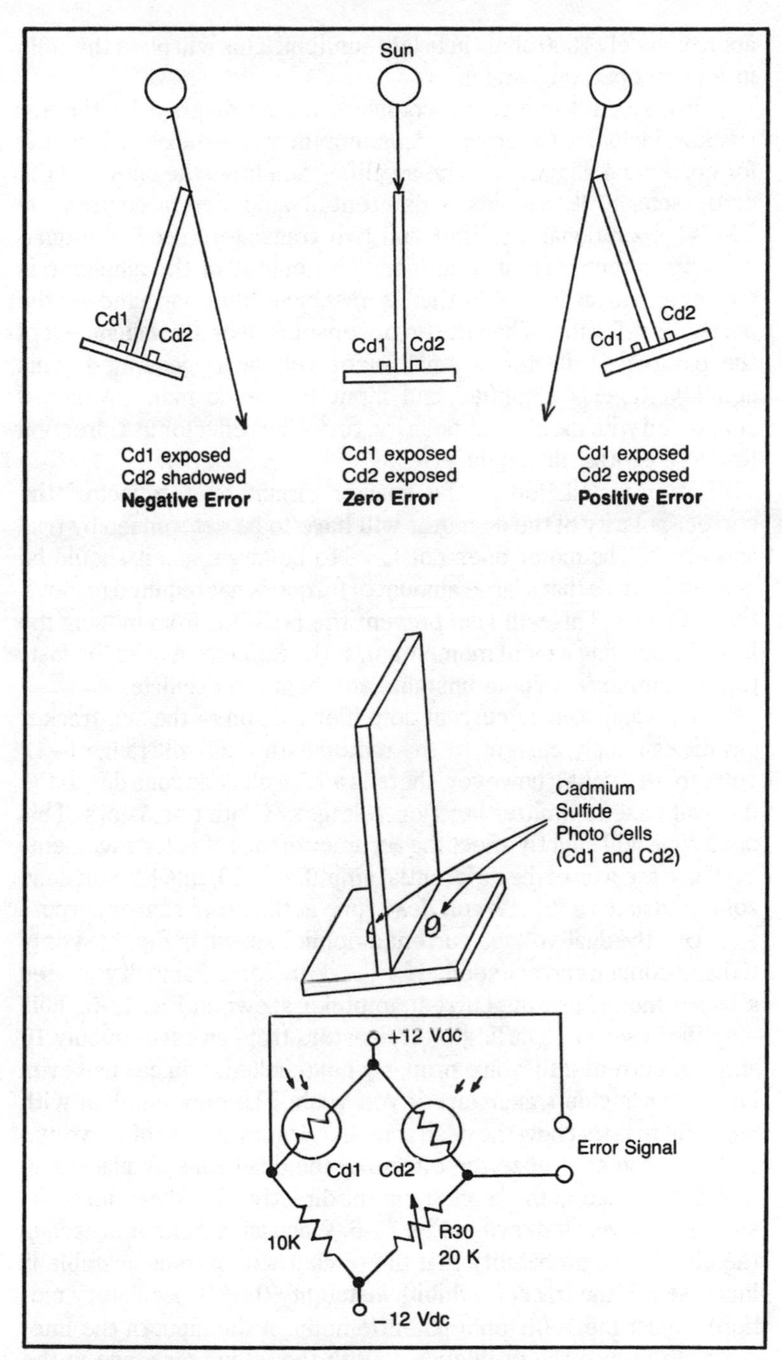

Fig. 11-3. Solar alignment error sensor.

approximately 7000 ohms in bright sunlight. This will place the cells in a good operating region.

Figure 11-4 contains a complete circuit diagram for the sun tracker, including the error sensor, amplifier, and the electric motor for correcting alignment. The amplifier amplifies the output of the error sensor. It contains a differential amplifier made from an LM741 operational amplifier and two transistors used as a dual-polarity output current amplifier. The output of the amplifier is connected to a dc motor that is mechanically connected to the parabolic reflector. When the error sensor senses a misalignment of the parabolic reflector, a small error voltage is generated. This small voltage is amplified and input to the dc motor, which if connected with the correct polarity, turns the reflector in a direction to decrease the alignment error.

After installation of the tracker circuit on a reflector, the correct polarity of the dc motor will have to be determined by trial and error. The motor does not have to be large, and it should be geared down so that a large amount of torque is not required to move the reflector. This will also prevent the reflector from moving too fast and building a lot of momentum. If the reflector moves too fast, the system may become unstable and begin to oscillate.

The dual-voltage current amplifier section of the sun-tracker circuit can apply current to the motor with a 24-volt range (−12 volts to +12 volts); however, there is a 1.2-volt dead zone due to the 0.6-volt base-to-emitter junction voltages of both transistors. This dead zone will slightly affect the accuracy of the reflector alignment. Because the gain of the differential amplifier is 10, this 1.2-volt dead zone translates to a 0.12-volt dead zone at the error-sensor output.

Use the dual-voltage current amplifier shown in Fig. 11-4 only if the maximum current needed for the dc motor is 1 amp. If you need a larger motor, use the current amplifier shown in Fig. 11-5. This amplifier uses two Darlington transistors that can easily supply 10 amps of current if they are properly heat-sinked. You can use even larger Darlington transistors if you wish. The only problem with using them is that now the dead zone is 2.4 volts instead of 1.2 volts.

You can neutralize the effects of the dead zone by placing an integrator stage in the tracker circuit directly after the differential amplifier stage, as shown in Fig. 11-6. When an integrator is used in the circuit, the probability that the tracker will become unstable is increased. If the tracker exhibits instability (highly oscillatory motion), adjust the 100K-ohm potentiometer at the input of the integrator to reduce the oscillations. With the integrator stage in the

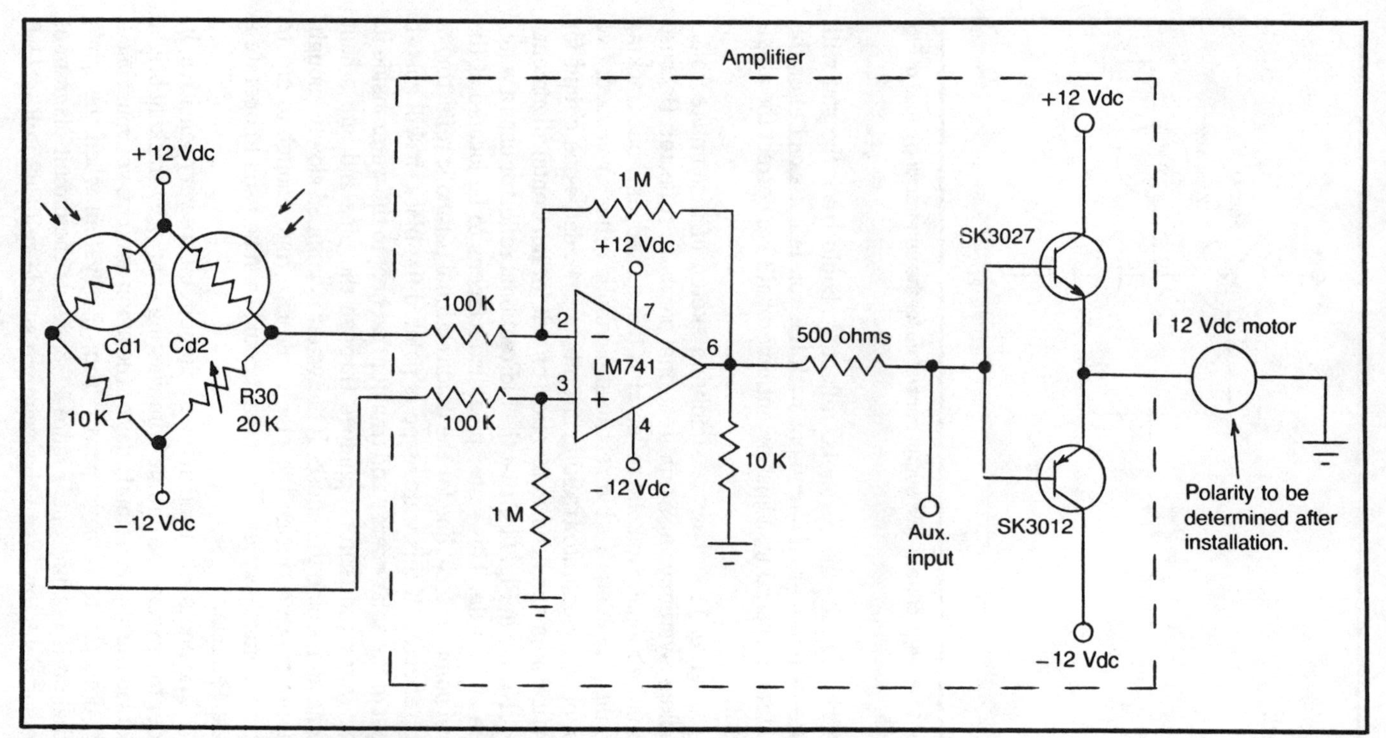

Fig. 11-4. Complete sun-tracker circuit for use with parabolic reflectors.

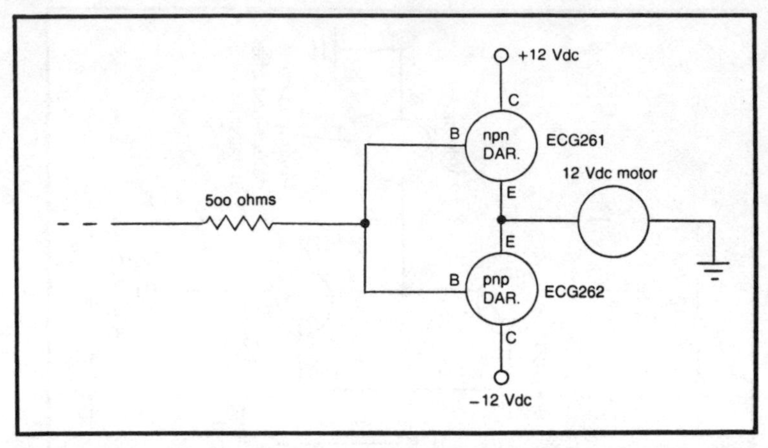

Fig. 11-5. High current-gain output transistors for the sun-tracking circuit of Fig. 11-4.

tracker circuit, the parabolic reflector should track the sun with little or no error. If alignment with the sun is not exact after the system is operating, adjust potentiometer R30 to correct the alignment.

Figure 11-7 shows a simplified sketch of a complete solar-tracking system using a unidirectional parabolic reflector. Because a unidirectional parabolic reflector is being used, only one tracking circuit is needed. If a bidirectional parabolic reflector is used, two completely separate tracking circuits are needed—one circuit for vertical alignment and an identical circuit for horizontal alignment.

Notice in Fig. 11-7 that the bidirectional reflector uses a water pipe as its axle. This water pipe also happens to be located at the focal point (or focal line for the unidirectional parabolic reflector) of the reflector. If this water pipe is painted flat black, it will absorb most of the solar energy and transfer that heat to the water inside it. The water can then be pumped through the pipe and into a heat exchanger inside the house. If the water is pumped slowly enough, the temperature inside the pipe could rise high enough to create steam from the water. This steam could then be used to operate a steam engine.

The photocells used in the tracker circuit are very sensitive. In order to prevent the circuit from wasting energy by tracking bright spots in the sky on a cloudy day or the moon at night, you must use a sun detector to remove power from the system when the sun's radiation is not present. Figure 11-8 shows the circuit diagram of the sun detector. When the cadmium sulfide cell Cd3 in Fig. 11-8

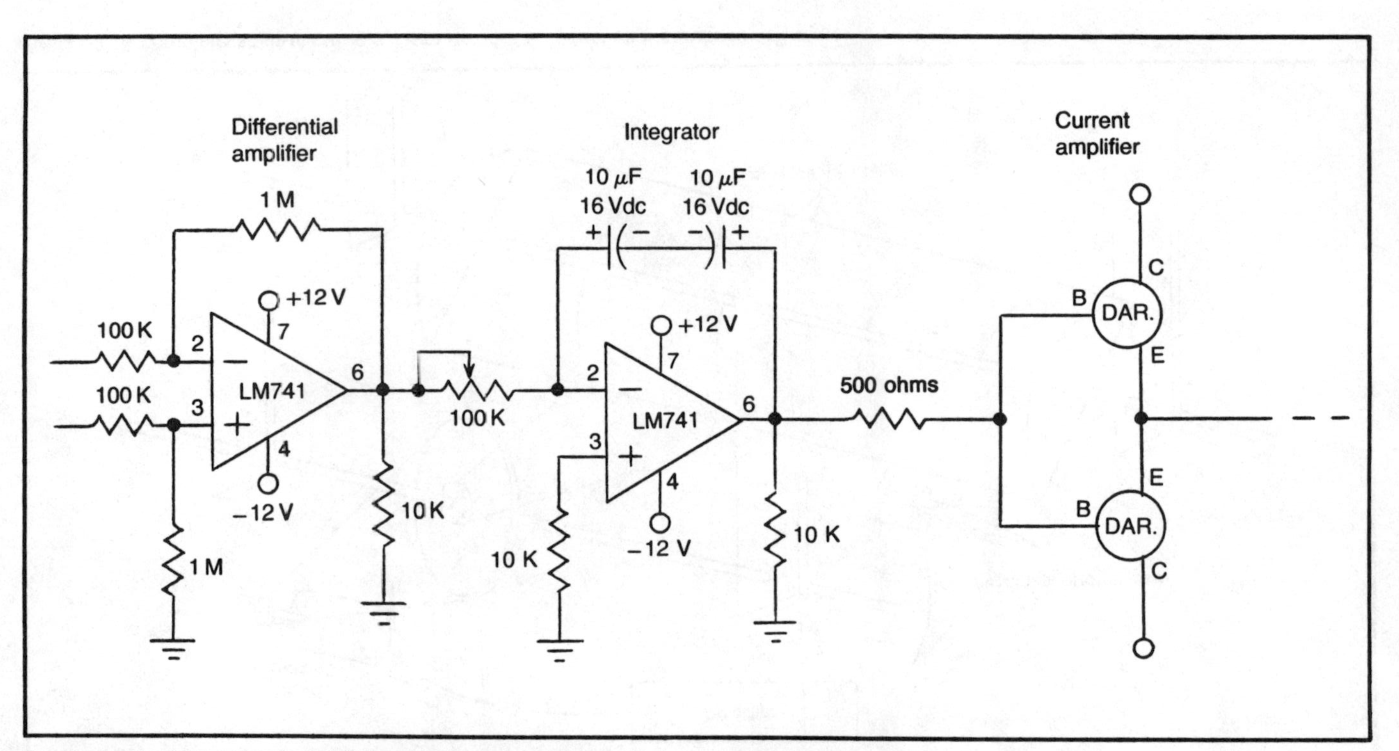

Fig. 11-6. Circuit showing the addition of an integrator stage to the solar-tracking circuit.

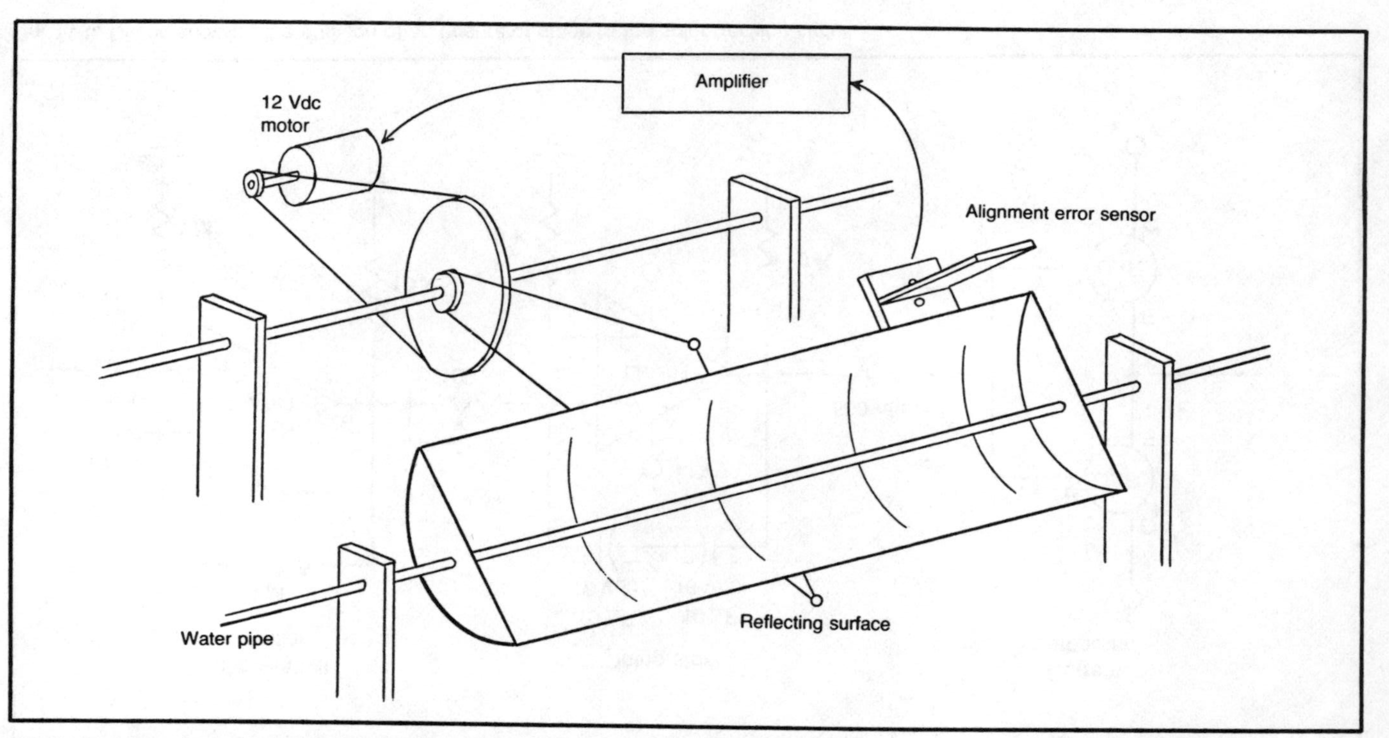

Fig. 11-7. A complete solar-tracking system using a unidirectional parabolic reflector.

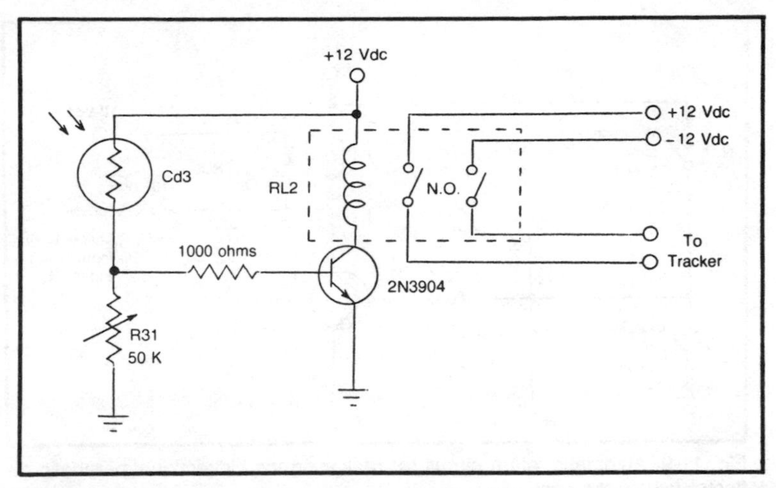

Fig. 11-8. Sun detector for removing power from tracking circuits when solar radiation is absent.

detects sunlight, relay RL2 is energized and power is applied to the tracker. The intensity of sunlight needed to energize the relay can be adjusted by varying potentiometer R31.

The unidirectional parabolic reflector is by far the simpler of the two reflectors in terms of alignment and sun tracking. If positioned horizontally, such as the one in Fig. 11-7, the unidirectional reflector requires only one tracking circuit and one sun detector for operation. The bidirectional reflector requires two tracking circuits, one sun detector, and also an automatic realignment circuit for returning the reflector back to the early morning position to await the rising sun. If an automatic realignment circuit is not used, the tracker will track the sun into the west at sundown and then be unable to detect the rising sun in the east the next morning. Figure 11-9 contains a circuit diagram of this automatic return circuit.

The automatic return circuit uses two limit switches—one for detecting maximum westerly position and one for detecting maximum easterly position. The westside limit switch is placed on the reflector base so that when the reflector reaches the maximum westerly direction, the switch closes. When this switch closes, capacitor C4 in Fig. 11-9 is discharged and relay RL3 is energized by the transistor. This relay applies power to the horizontal alignment motor through the auxiliary input on the tracker circuit of Figure 11-4. Relay RL3 remains energized for no more than 20 seconds, and the actual time limit can be varied by adjusting potentiometer

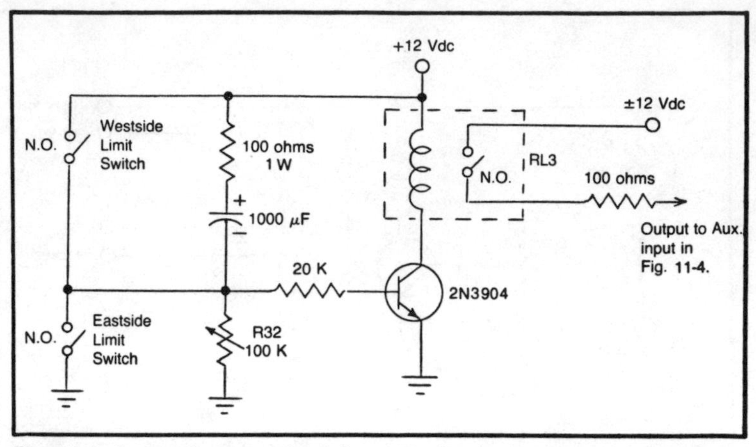

Fig. 11-9. Automatic return circuit for realigning the bidirectional parabolic reflector toward the east.

R32. The voltage polarity (positive or negative 12 volts) to make the motor move the reflector in an easterly direction depends on the particular mechanical construction. This voltage is applied through relay RL3 to the output transistor section of the tracker circuit (see Fig. 11-4) and overrides the signal from the alignment error sensor.

When you use this automatic return circuit, place the photocell for the sun detector somewhere on the reflector frame. In addition, the sun-detector circuit must only remove power from the alignment error sensor and the operational amplifier section of the amplifier. This way, as soon as the reflector begins to move back to the easterly direction, the sun detector will sense the loss of sunlight and remove power from the alignment error sensor and operational amplifier but not from the output transistors.

You should put the eastside limit switch on the reflector base so that it closes when the reflector reaches its maximum easterly direction. When this switch closes, it recharges capacitor C4 and deenergizes relay RL3, regardless of whether or not the time limit has been reached. This removes power from the motor and stops the reflector. The reflector will now sit and wait for the sun to rise and will dissipate only a very small amount of electrical power during the night.

The relay used for RL3 in Fig. 11-9 does not have to be large. Actually, for this particular circuit, the relay should be small enough to be energized by 60 milliamps. A small reed relay would work well. If a larger relay is used, the 20K-ohm resistor connected to the base of the transistor must be exchanged for one with a small value. Also, to keep the time delay constant, the 1000-μF capacitor must be exchanged for a larger one. The product of this capacitance and resistance of the resistor in the base circuit should remain the same. This product determines the time delay.

CIRCUITS FOR TEMPERATURE CONTROL OF SOLAR HEATING SYSTEMS

As previously stated, all solar heating systems require a system for transferring heat from the panel to the place or object to be heated. These systems usually employ water or air as the heat transfer medium. Because the sun is not always out, some type of controller must be used to disable the system when the temperature of the panel or collector approaches or goes below the temperature of the item being heated. Figure 11-10 shows the circuit diagram of a differential thermostat that can be used for this purpose.

The thermostat circuit of Fig. 11-10 uses two thermistors (TM1 and TM2) as temperature sensing elements. TM1 is located inside the solar panel or collector, and TM2 is located on the item or in the room to be heated. When the temperature of TM1 is higher than that of TM2 by a set amount, relay RL4 is energized and power is applied to the water pump or air blower. The air or water then transfers the heat from the solar collector to the item to be heated. Potentiometer R33 is used to set the desired temperature differential for system activation. Once the temperature of TM2 rises enough to reduce the temperature differential between it and TM1 below the set value, relay RL4 is deenergized and the air or water flow is stopped.

If it is possible for the temperature in the room or on the item being heated by the solar panel to become too hot, you must add a second circuit to remove power from the heat transfer system, transfer the heat to a separate room or item, or transfer the heat to a heat storage system. The circuit in Fig. 11-11 can do these things.

When the temperature of thermistor TM3 in Fig. 11-11 reaches a preset value (set by adjusting potentiometer R34), relay RL5 is energized. This relay can have normally open (N.O.) contacts or normally closed (N.C.) contacts. Normally open contacts can be used to apply power to a bypass value when relay RL5 is energized. This changes the flow path of the water or air. If desired, the N.C. contacts can be used to remove power from the water pump or blower when relay RL5 is energized, but, removing power from

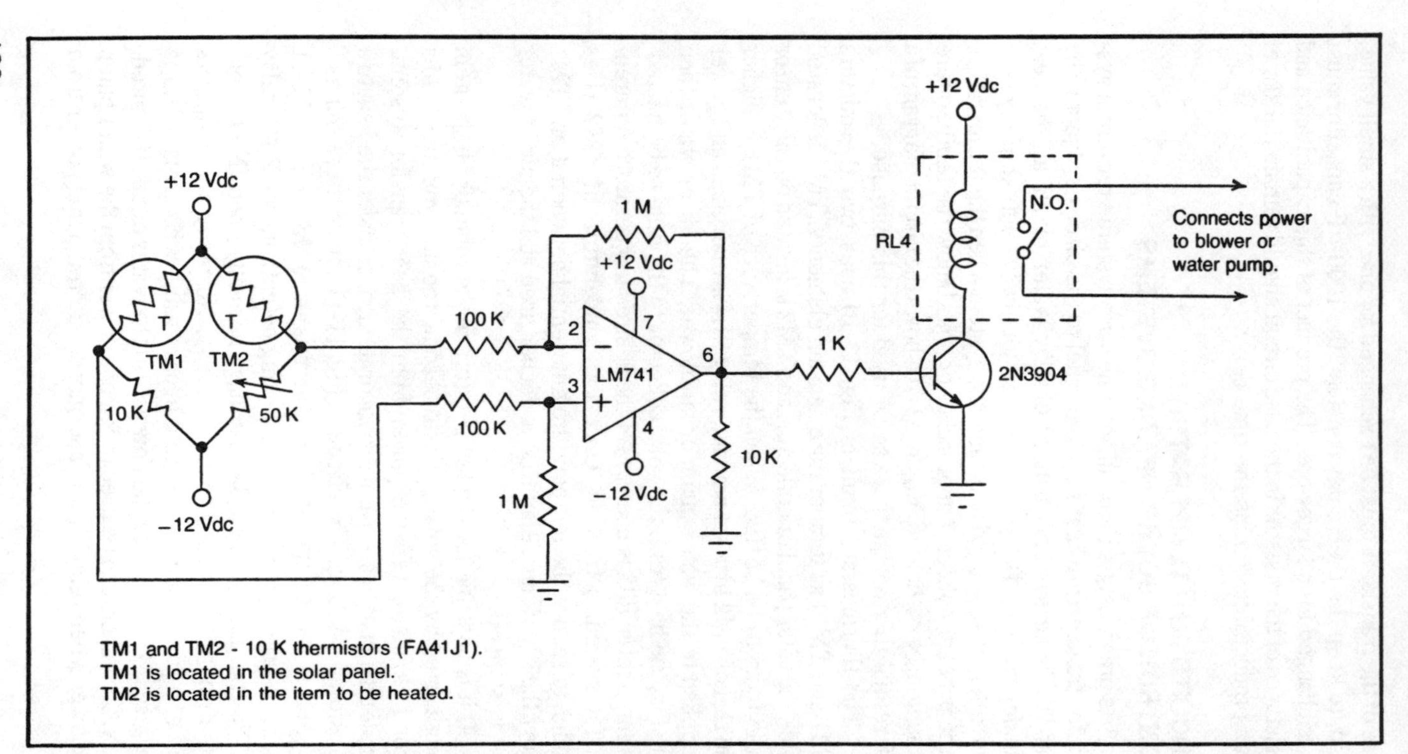

Fig. 11-10. Differential thermostat for use on solar-collector heat-transfer systems.

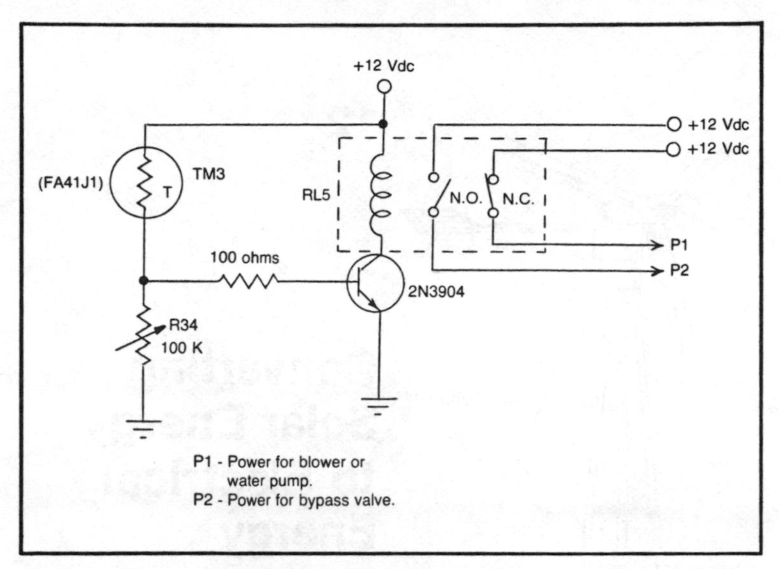

Fig. 11-11. Thermostat for limiting high temperatures in a solar heating system.

the heat transfer medium circulation system can be dangerous to the solar collector because it may allow heat in the collector to reach destructive levels.

Like wind power, solar power is free once the initial cost and the small operating costs are overlooked. It requires only a small amount of technical knowledge and a moderate amount of work to tap these free sources of power.

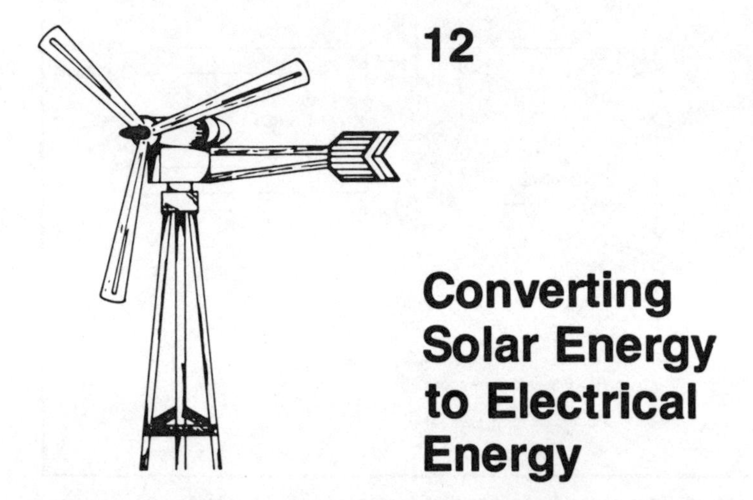

12

Converting Solar Energy to Electrical Energy

There are several ways of converting solar energy to electrical energy. The first is a mechanical process that uses the heat produced by sunlight to power a boiler that, in turn, powers a steam engine for generating electricity. Parabolic mirrors can be used in this mechanical process to concentrate the sunlight on the boiler so higher temperatures can be achieved. Sunlight can be converted to heat very efficiently, but the mechanical losses after the boiler can reduce the efficiency drastically.

Another mechanical process for converting solar energy to electrical energy uses thin polyethylene strips called *solar muscles* (Ray 1980, p. 126). When these strips are stretched, sunlight striking them causes them to contract. This translational motion can then be converted to rotational motion to run a generator.

The third method is purely electrical and uses *photovoltaic cells* to convert solar energy to electrical energy. This method is very simple but also very expensive. The 1982 retail cost of these photovoltaic cells (solar cells) is about $20 per watt. A house requiring an average of 2,000 watts of power would need $40,000 worth of cells to power it. This price does not include the cost of a dc-to-ac converter large enough to convert this 2,000 watts of dc power to ac power. Because the sun shines only about one-third of the time, a battery bank is needed for energy storage and three times as many cells are needed to store enough energy to produce this constant 2,000-watt average. The cost is now $120,000, not including the battery bank and converter. Assuming the cells can be purchased at a discount for large volumes, the entire system would probably cost between $100,000 and $150,000.

Obviously, at present prices, this method of power production is not recommended for the average homeowner; however, much research is now being done to improve the efficiency and lower the cost of photovoltaic cells. Perhaps within 10 years the cost of these cells will be competitive with other forms of power production.

PHOTOVOLTAIC CELL CHARACTERISTICS

Photovoltaic cells are made from a silicon wafer containing impurities that produce negative charges. The front of this wafer is coated with a thin layer of silicon containing impurities that produce positive charges. On top of this thin layer are conductive strips that connect to form the negative terminal. The back of the wafer, coated with nickel, is used as the positive terminal. When sunlight strikes the front of the cell, hole-electron pairs are formed in the junction between the two layers and contribute to current flow to the front and back connections (Crouch, Christie, and Malmstadt 1981, p. 69).

Photovoltaic cells are typically about 10 percent efficient. At noon on a clear day the sun strikes the earth with approximately 1000 watts per square meter. A 1-square meter solar panel containing photovoltaic cells can, then, convert 100 watts of the power to electrical power.

The output power of a photovoltaic cell is affected by three things: the strength of the sunlight, the load resistance connected to its output, and the temperature of the cell. At full sunlight with no load connected, a typical solar cell will produce 0.45 volts (open circuit voltage or Voc), zero current, and zero power. At full sunlight with the output shorted, the cell will produce a current equivalent to its short circuit current (Isc) and, theoretically, zero voltage and zero power. A graph of a typical voltage-current curve is shown in Fig. 12-1. If a load resistance other than zero (short circuit) or infinity (open circuit) is placed on the cell, power is produced depending on the intensity of the sunlight striking the cell (Pierson 1978, p. 88).

Open-circuit voltage is typically 0.45 volts and short-circuit current depends on the size of the cell. Isc can be 0.05 amps to 2 amps per cell. The voltage-current relationship for a photovoltaic cell can be approximately described by the following equation:

$$I = Isc \times \exp[-4(V/Voc)^k] \qquad \textbf{Eq. 12-1}$$

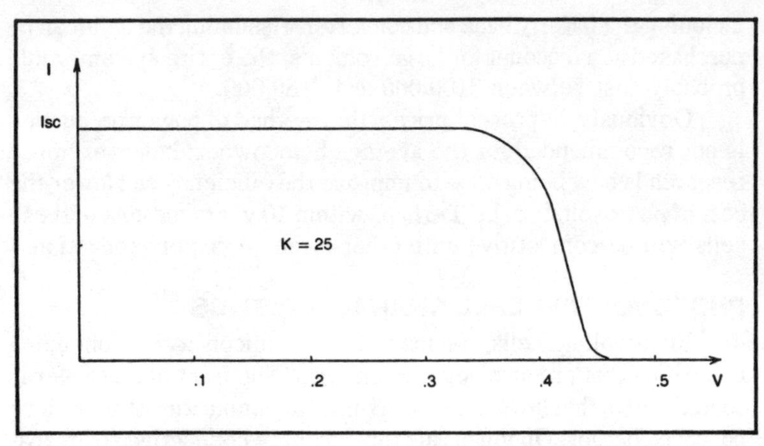

Fig. 12-1. Current-voltage curve for a photovoltaic cell.

where I is the actual output current of the cell, V is the actual output voltage, and k is a constant that may be different, depending on which cell is used.

Output power is equal to output voltage multiplied by current, and voltage is equal to output current multiplied by the load resistance. These relationships can be combined as follows to show that output power is dependent on load resistance:

$$P = VI \qquad \textbf{Eq. 12-2}$$
$$= RI^2$$
$$= R\{Isc \times \exp[-4(RI/Voc)^k]\}^2 \qquad \textbf{Eq. 12-3}$$

From Eqs. 12-1 and 12-2 we see that the output voltage (Vmp) for maximum output power is

$$Vmp = Voc(1/4k)^{1/k} \qquad \textbf{Eq. 12-4}$$

If Voc = 0.45 volts and k = 25, then Vmp = 0.374 volts. From Eq. 12-1 this voltage output occurs when the current output is 96 percent of Isc or 0.96 Isc. Because the output power is equal to volts multiplied by current, the maximum output power (Pm) is 0.96Isc×Vmp when k is 25. The actual equation for Pm is

$$Pm = Voc \times Isc(1/4k)^{1/k} \exp(-1/k) \qquad \textbf{Eq. 12-5}$$

A graph of the output power versus voltage for Isc=1 amp and Voc=0.45 volts is shown in Fig. 12-2.

Temperature inversely affects the efficiency of a photovoltaic cell. As the temperature goes up, the output current goes down for a constant sunlight intensity. The cells must, therefore, have some way to dissipate heat.

One good technique for dissipating the heat from the cell, and at the same time using the wasted heat, is to use the solar cells as part of the heat-gathering backing of a forced-air solar heating panel. The electrical power generated by the solar cells can be used to power a fan to pull air through the solar panel and into the house. At the same time the excess heat from the photovoltaic cells is carried along with the rest of the heat to warm the house.

POWER SUPPLY FOR RESISTIVE LOADS

As previously discussed, output power is a function of load resistance. Figure 12-2 shows that for photovoltaic cells with k=25, Voc=0.45 volts, and Isc=1 amp, the maximum output power is 0.359 watts and Vmp is 0.374 volts. Because power is also equal to V^2/R, the load resistance for maximum power output (Rmp) is

$$\text{Rmp} = (\text{Vmp})^2/\text{Pm} \qquad \textbf{Eq. 12-6}$$
$$= \text{Voc}(1/4k)^{1/k}/[\text{Isc}\exp(-1/k)] \qquad \textbf{Eq. 12-7}$$
$$= 0.389 \text{ ohms}$$

Therefore, to get maximum power from a photovoltaic solar cell with Isc=1 amp, Voc=0.45 volts, and k=25, it must have a load resistance of 0.389 ohms.

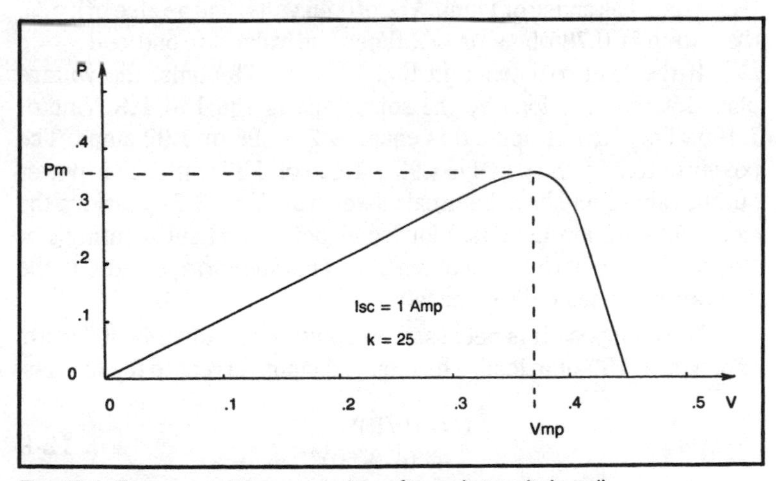

Fig. 12-2. Power output versus voltage for a photovoltaic cell.

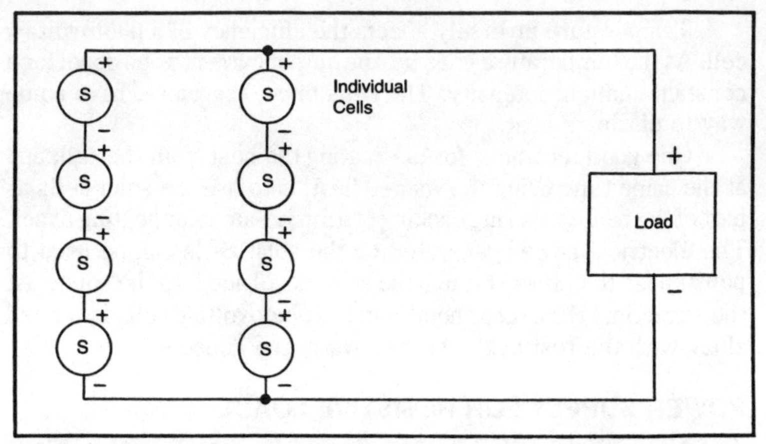

Fig. 12-3. Two parallel arrays of four series cells connected to a load.

If several cells are connected in series and then several series arrays are connected in parallel, as shown in Fig. 12-3, the equation for Rmp changes slightly. Because voltage adds in series, the value of Voc must be the value of Voc for one cell multiplied by the number of cells in series. Current adds in parallel, so the value of Isc must be the value of Isc for one cell multiplied by the number of parallel arrays. For example, in Fig. 12-3 there are two parallel arrays of four series cells. For this solar panel, Eq. 12-7 now becomes

$$Rmp = 4Voc(1/4k)^{1/k}/[2Isc\ \exp(-1/k)] \qquad \textbf{Eq. 12-8}$$

If each cell has an Isc of 1 amp, Voc of 0.45 volts, and a value of k=25, then Rmp is 0.78 ohms or 4/2 times the value for one cell.

If the load resistance in Fig. 12-3 is 0.78 ohms, the voltage placed across the load by the solar cells is equal to 4 × Vmp or 1.496. The current applied is equal to 2 × .96 or 1.92 amps. The power output by the panel is 1.92 × 1.496 or 2.87 watts. This power output value could have been calculated from Eq. 12-5 by making the same substitution (2 × Isc) for Isc as before and substituting 4 × Voc for Voc, where 2 is the number of parallel arrays and 4 is the number of series cells in each array.

Now suppose it is necessary to apply approximately 400 watts of power to a 50-ohm load. The required input current to the load is

$$\begin{aligned} I &= (P/R)^{1/2} \\ &= (400/50)^{1/2} \qquad \textbf{Eq. 12-9} \\ &= 2.828 \text{ amps} \end{aligned}$$

If you use 1-amp cells, you need three parallel arrays to supply this current. From Eq. 12-1 (with V=Vmp=.374 volts, Isc=1 amp, and k=25), I = 0.96 amps per array. If three arrays are used, 2.88 amps can be applied to the load if the correct number of series cells is used.

By substituting 3 × Isc for Isc, 50 ohms for Rmp, and N × Voc for Voc in Eq. 12-7, we can calculate the value for N. This value will be the number of series cells to be placed in each of the three arrays. The new equation is

$$N = \frac{(Rmp)(3Isc)\exp(-1/k)}{Voc(1/4k)^{1/k}}$$
$$= 385$$

Each array will have 385 cells. The total number of cells in all three parallel arrays is 1156. The total power supplied, using Eq. 12-5, is

$$\begin{aligned} Pm &= VocIsc(1/4k)^{1/k}\exp(-1/k) \\ &= (385)(0.45)(3)(1)(1/k)^{1/k}\exp(-1/k) \\ &= 415 \text{ watts} \end{aligned}$$

Notice that 385Voc was used in place of Voc, and 3Isc was used in place of Isc.

These equations cannot be used without a value for k. Because manufacturers do not use such a parameter, you must determine the particular value you need. You can do this by drawing several graphs

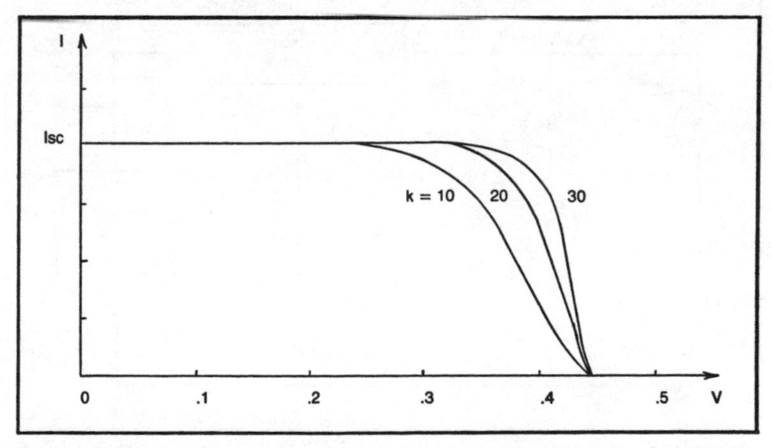

Fig. 12-4. Current-voltage curves for photovoltaic cells with K equal to 10, 20, and 30.

using different values for k and comparing them to the current-voltage curve supplied by the manufacturer. Several of these curves are shown in Fig. 12-4.

The equations may take a little time and effort, but the time will be well spent. The values for N, the number of series cells needed, calculated with these equations will be very close to the optimum number of cells to be used for maximum output and minimum cost. An hour spent calculating can mean a savings of hundreds of dollars.

POWER SUPPLY FOR ELECTRONIC DEVICES

Using solar cells to power electronic devices is very similar to using solar cells to power resistive loads, except for two things: the equivalent load resistance of the electronic device is usually not constant; and electronic devices usually require a constant voltage. Obviously, these two characteristics of electronic devices are not compatible when using solar cells as the power source.

There are two ways to maintain a constant voltage on the electronic load. You can use an electronic voltage regulator, as shown in Fig. 12-5, or a battery, as shown in Fig. 12-6. The battery has an advantage over the electronic regulator because of its energy storage capability; however, the battery voltage tends to vary slightly with the amount of charge it contains. If voltage regulation is very critical, both methods can be combined to give excellent voltage regulation and constant operation when there is no sun.

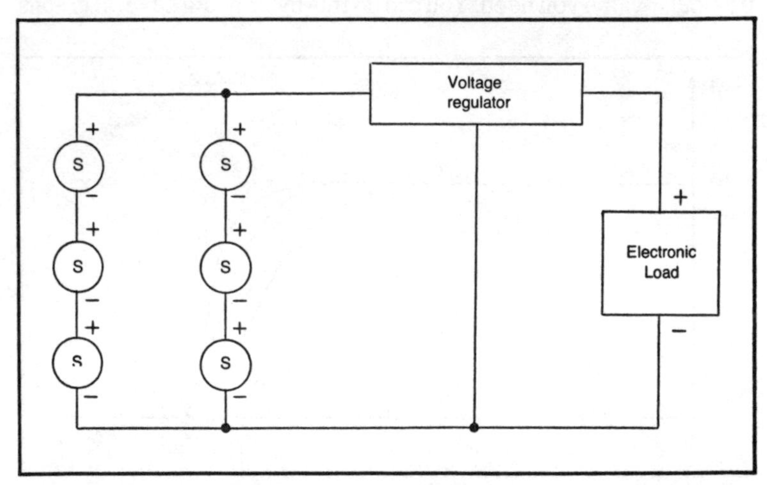

Fig. 12-5. Electronic device powered by solar cells and using an electronic voltage regulator to maintain a constant voltage.

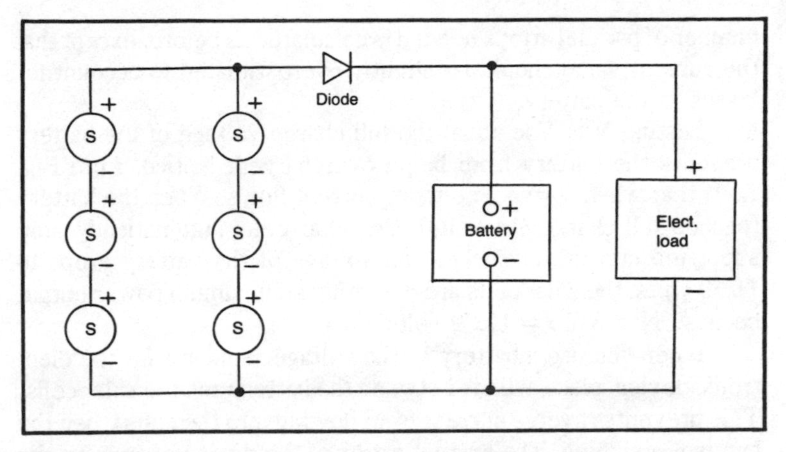

Fig. 12-6. Electronic device powered by solar cells and using a battery as a voltage regulator.

When powering electronic devices with solar cells, you can determine the number of cells needed in a way similar to the method described in the last section. You first calculate the equivalent resistive load of the electronic device and then use the procedure for resistive loads. You can estimate the equivalent resistive load (Req) from the formula

$$\text{Req} = \frac{\text{Required Input Voltage}}{\text{Average Current Draw}} \qquad \text{Eq. 12-10}$$

After you obtain Req, increase it by about 10 percent if you are using an electronic voltage regulator.

For example, if the electronic device requires 12 volts and 1.3 amps, the equivalent resistance is Req=9.23 ohms. Adding 10 percent because of the voltage regulator gives 10.15 ohms. Since the current required is 1.3 amps, three parallel arrays of cells with an Isc of 0.5 amps can be used. Putting these values (Rmp=10.15 ohms, Isc=0.5 amps, Voc=0.45 volts) in Eq. 12-7 gives a value for N, the number of cells in each series array, of 39. The total number of cells needed is 3 × 39 or 117.

If you use a battery to regulate voltage, the process is slightly different. The full charge voltage of the battery should be equal to Voc multiplied by the number of series cells in each array. For example, a 12-volt lead-acid battery has a full charge voltage of about 14 volts. With Voc=0.45 volts, you need 31 series cells. The number of parallel arrays needed is calculated as before, except that the current value should be slightly overestimated to account for losses in the battery.

Letting N × Voc equal the full charge voltage of the battery prevents the battery from being overcharged. Notice, from Fig. 12-1, that when V=Voc, no more current flows. When the battery reaches full charge (14 volts), the solar cells automatically stop supplying current, and when the voltage of the battery drops to 11.59 volts, the solar cells are operating at maximum power output because N × Vmp = 11.59 volts.

When you use a battery as the voltage regulator for the electronic device, place a diode between the battery and the solar cells. This prevents reverse current from flowing into the cells when the sun is not shining. The current rating of the diode depends on the amount of current being supplied by the solar panel and should equal at least 1.3 times that current value.

OPERATING DC MOTORS

A dc motor can also be considered a resistive load; however, its equivalent resistance is inversely related to the mechanical load applied to the motor. If the mechanical load is constant, the equivalent resistance remains constant, but if the mechanical load increases, the equivalent resistive load decreases.

If the equivalent resistive load value used for Rmp in Eq. 12-7 is too high, the dc motor will not respond well to overloads. When the motor is overloaded, its equivalent resistance decreases. If this resistance falls below the resistance for maximum output power (Rmp), the motor slows down even more because less power is being supplied by the solar cells.

The value used for Rmp when solving for N should be slightly less than the average equivalent operating resistance. If this is done the motor will respond well to overloads. As the motor becomes overloaded, its equivalent resistance decreases and approaches the resistance for maximum output power. The solar cells then apply more power to the motor and try to prevent it from slowing down further.

It is also good to slightly overestimate the average current draw of the motor, because the solar cells will not be operating at maximum power output for normal loads. It turns out that overestimating the average current draw is synonymous with underestimating the equivalent resistance. For example, suppose the motor to be powered by the solar cells requires 6 volts and an average current of 2 amps. First, overestimate the current by about 10 percent. This gives an average current draw of 2.2 amps. Before overestimating the current, the equivalent resistance was 3 ohms; after overestimating the current, it is 2.73 ohms.

Using 2.73 ohms for Rmp in Eq 12-7 and assuming that five parallel arrays of solar cells, with Isc=0.5 amps, Voc=0.45 volts, and k=25 are to be used, you can obtain a value for N:

$$N = \frac{(Rmp)(5Isc)\exp(-1/k)}{Voc(1/4k)^{1/k}}$$
$$= 18$$

Each of the five parallel arrays contains 18 cells, for a total of 90 cells.

When connecting photovoltaic solar cells to dc motors, no special circuitry is needed between the cells and the motor, as shown in Fig. 12-7.

CHARGING BATTERIES

Much of the information in this section has already been covered in the previous section on using solar cells to power electronic devices. The rules are simple. Always use a diode between the cells and the battery to prevent reverse current leakage, and choose a value of N such that N × Voc is equal to the full charge voltage of the battery.

The number of parallel arrays depends on the charge rate

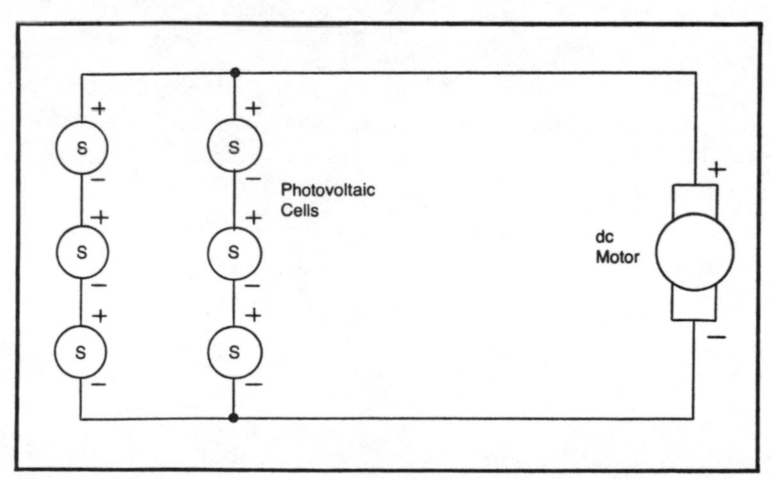

Fig. 12-7. Circuitry for using photovoltaic cells to operate dc motors.

desired. If you use cells with an Isc of 1 amp, the maximum charge current will be about 0.96 times the number of parallel arrays.

Even though photovoltaic cells are not an economical means of producing power today, they have the potential for being the power source of the future. They are extremely simple to use and require hardly any maintenance once installed. As current research progresses they should become competitive with other power sources in the near future.

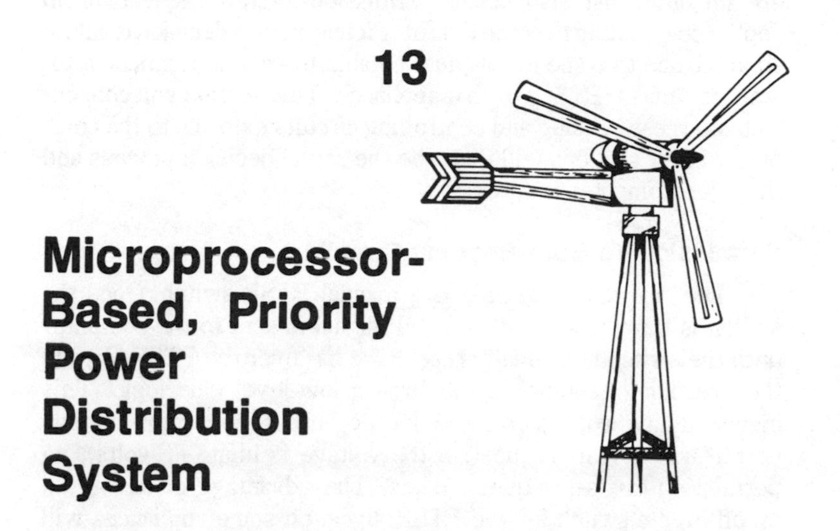

13

Microprocessor-Based, Priority Power Distribution System

It is possible, although the probability is low, for every electrical appliance in a home to be on at the same time. This means the power system which supplies electricity to the house must be capable of providing enough current to meet this maximum demand. An alternate-energy electrical system large enough to provide this kind of power would be beyond the budget of most homeowners. If some kind of *priority power distribution system* could be employed to control the number of appliances allowed to come on at any specific time, a significant savings would be made in the cost of the electrical power system.

Such a priority power distribution system could utilize a microcomputer built around a microprocessor to receive *permission-to-start requests* (PTSR) from the appliances and to send *permission-to-start commands* (PTSC) back, based on its knowledge of priorities and the number of appliances already on. The credit for the idea behind this type of system should be given to John A. Kuecken. He presented the concept along with a few sensing and control circuits in his book *How to Make Home Electricity from Wind, Water and Sunshine* (Kuecken 1979, p. 85). Some of the terms used in this chapter are borrowed from this book.

SENSING AND CONTROL CIRCUITS

If the computer is to make decisions on which appliances are allowed to come on, it must first be able to sense which appliances are on and must also sense permission-to-start requests from appliances wanting to come on. After it has made a decision to allow an appliance to come on, it must be able to send a permission-to-start command (PTSC) to that appliance. This section will concentrate on these sensing and controlling circuits external to the computer. Later sections will describe the actual decision process and the microcomputer hardware.

Permission-To-Start Request Circuits

The circuit of Fig. 13-1 is a manual PTSR switch. Once the switch is thrown, a constant $\overline{\text{PTSR}}$ signal is sent to the computer until the switch is manually reset. The bar over the PTSR term in the preceding sentence symbolizes a low-level true logic. This means a low voltage on the PTSR output line represents a permission-to-start request. If the voltage is high (+5 volts), no permission-to-start request exists. The advantage of having an on-off toggle switch for the PTSR signal on some appliances will become more apparent after reading the next section concerning the decision-making process.

If a momentary, manual $\overline{\text{PTSR}}$ signal is desired, the circuit of Fig. 13-2 can be used. When using this circuit, you turn the appliance on, push the PTSR button, and wait for the computer to apply power to the appliance. If no PTSC is sent by the computer to apply power to the appliance, you must wait and try again later. Once the appliance is on, it can be turned off at any time by the computer if a higher-priority appliance requests permission to come

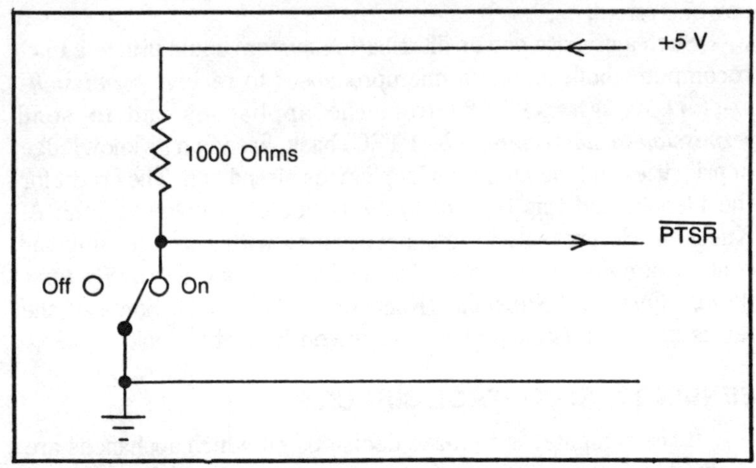

Fig. 13-1. On-off toggle permission-to-start request switch.

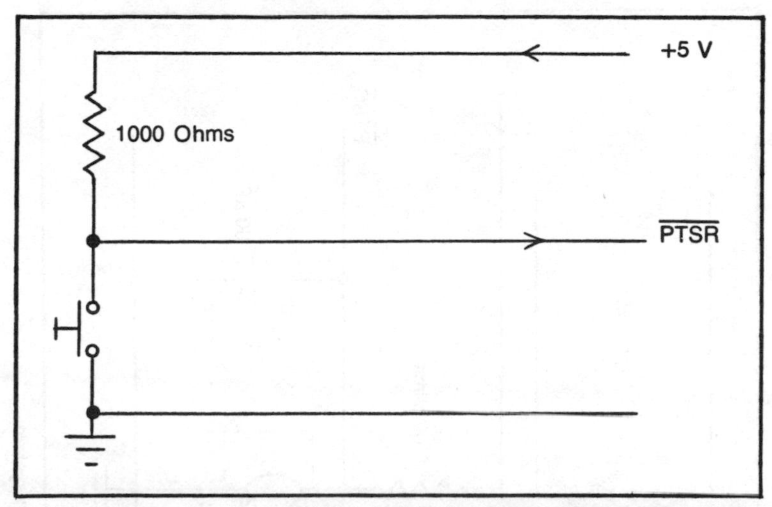

Fig. 13-2. Temporary closure pushbutton permission-to-start request switch.

on. On the other hand, if the on-off toggle switch of Fig. 13-1 is used, you can turn the appliance on and then flip the PTSR switch on and leave. The computer will apply power to the appliance with a PTSC signal as soon as possible, depending on its priority and the number of other appliances on. Once the appliance is on, however, the computer will not cancel the PTSC until the switch is manually reset. This toggle-switch PTSR circuit would be ideal for use with appliances such as washing machines. Once the washing machine starts, you should not turn it off until it has completed its cycle, at which time it will turn itself off anyway. The circuit of Fig. 13-2 (momentary switch) could be used on such things as shop tools. Because an operator must be with this equipment when in use, a cancellation of the PTSC to the equipment would cause no harm, just a delay in completion of the work.

There are some appliances in the house that operate automatically. If these are also to be controlled by the computer, an automatic PTSR signal generator circuit must be attached to the appliance. Kuecken describes several of these circuits in his book.

One such appliance requiring an automatic PTSR signal is a thermostat-controlled heating system. The thermostat is actually a temperature-controlled switch that closes when the temperature drops below a set level. This switch applies 28 Vac to the gas solenoid valve, which, in turn, allows gas to flow to the burner where it is lit by the pilot light. When the heat exchanger reaches a set temperature, the blower comes on. Figure 13-3 shows a circuit

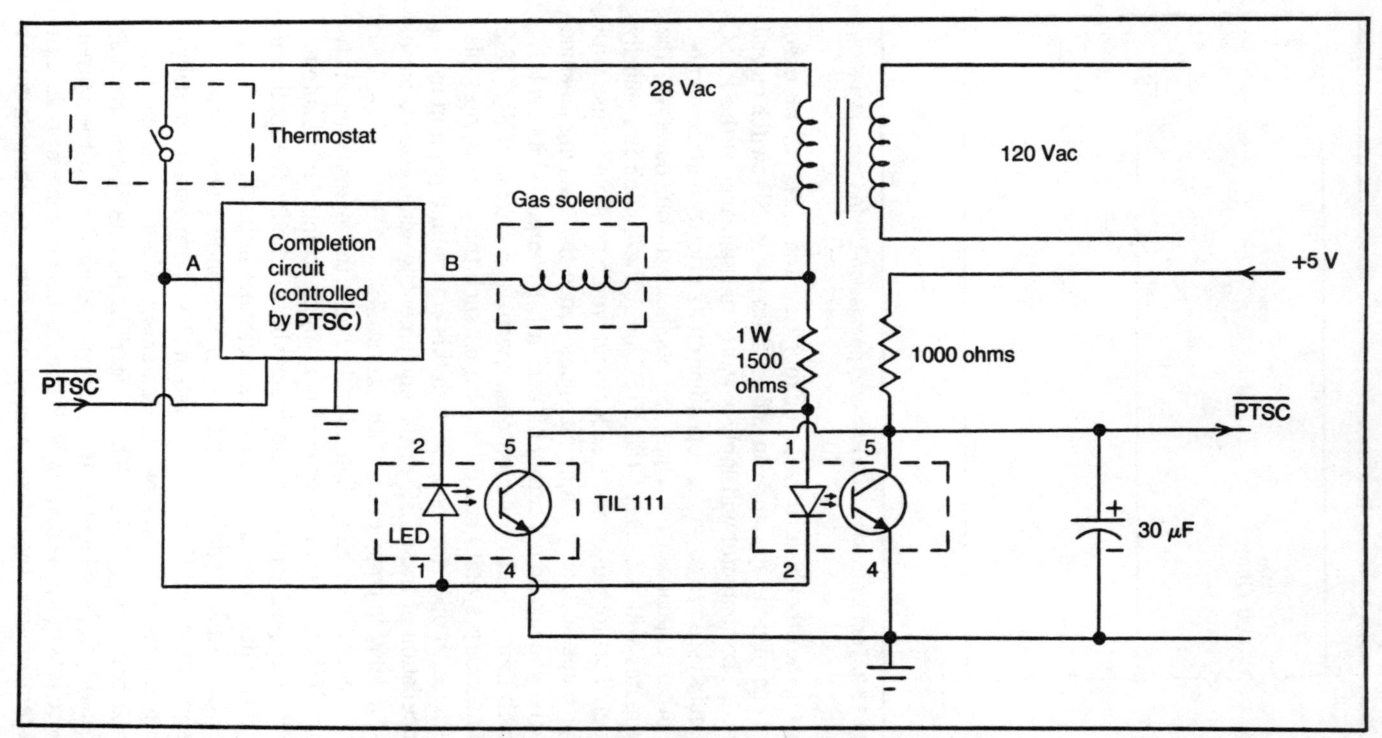

Fig. 13-3. Automatic permission-to-start request signal generator for a gas, home-heating system.

that can be attached across the solenoid valve to send an automatic PTSR signal. Notice also the PTSC completion circuits in series with the gas solenoid. This circuit allows the solenoid to operate only after receiving a PTSC signal from the computer. This completion circuit is shown in Fig. 13-4.

When the thermostat switch closes, 28 Vac is applied to the two optocoupler circuits (TIL 111) that are wired in parallel. One TIL 111 is wired so that the LED is forward biased with a positive voltage, while the other TIL 111 is wired so that its LED is forward biased with a negative voltage. With the two circuits wired this way, one output transistor will always be conducting when an alternating current is applied. When at least one transistor is conducting, a low-level signal (PTSR) is sent to the computer. The 30-μF capacitor ensures a smooth output signal.

Each of the PTSR circuits shown have required +5 volts for operation. These 5-volt power sources can all be from one central power supply. This simplifies the system by requiring extra wire instead of extra power supplies.

Permission-To-Start Command Circuits

The PTSC completion circuit for the heating system (see Fig. 13-4) utilizes an MOC 3010 optocoupler with a bilateral switch output. The bilateral switch output is connected to the input of an RS 276-1001 triac. When 5 volts is applied to the LED input of the MOC 3010 through a 270-ohm resistor, the bilateral switch begins to conduct alternating current to the gate of the triac. When the

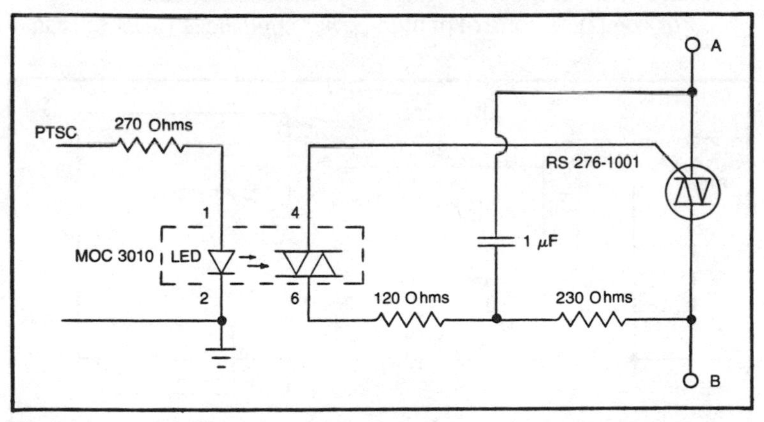

Fig. 13-4. Completion circuit which is controlled by a PTSC signal from the computer and is used to apply power to the gas solenoid of a home-heating system.

alternating current is applied to its gate, the triac turns on and allows current to flow to the gas solenoid. Because the PTSR signal is on as long as the thermostat switch is closed, the computer will not cancel the PTSC signal until the temperature has reached a value high enough to reset the thermostat switch.

The circuit of Fig. 13-5 can be used by the computer to turn on resistive-load appliances that draw less than 6 amps of current. Such appliances might include incandescent reading lamps, sun lamps, soldering irons, and clothes irons. This circuit also utilizes the MOC 3010 optocoupler connected to an RS 276-1001 triac. For this circuit, however, the extra capacitor and resistor are not used between the bilateral switch and triac. They are only used if the triac is operating an inductive load. Because current lags voltage in an inductive load, you must add a low-pass filter (capacitor and resistor) at the gate input to cause the gate voltage to lag also. The 270-ohm resistor is always used in series with the MOC 3010 input to limit current through the LED, otherwise the LED would be destroyed by the high current forced through it by the 5 volts.

A circuit that allows the computer to control low-current (less than 6 amps) inductive-load appliances is shown in Fig. 13-6, which, except for the capacitor and resistor values, is very similar to Fig. 13-4. It requires larger resistors because the input voltage is 115 Vac rather than 28 volts, as in Fig. 13-4. The capacitor value is different because the time constant (RC product) must remain the same. This circuit can be used to control appliances such as televisions, radios, blenders, electric fans, and any other low-power appliance with an electric motor or a transformer input.

For computer control of high-power appliances (greater than 6

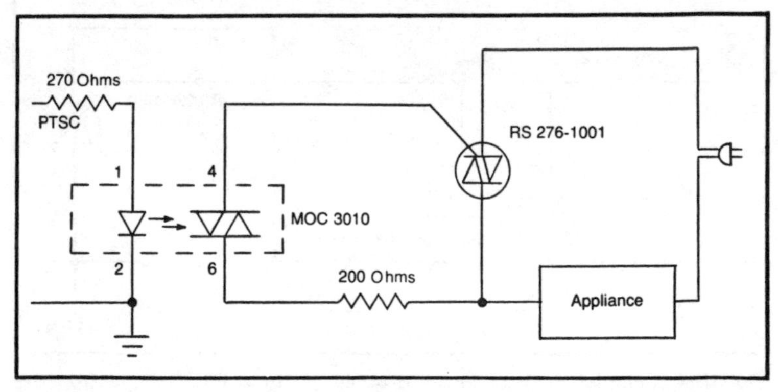

Fig. 13-5. Permission-to-start command circuit for low-current (less than 6 amps) resistive-load appliances.

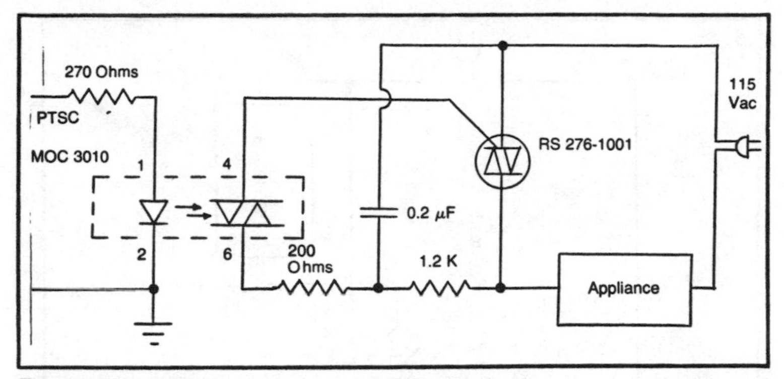

Fig. 13-6. Permission-to-start command circuit for low-current inductive-load appliances.

amps) you can use the circuit of Fig. 13-7, which also uses the MOC 3010 optocoupler and the RS 276-1001 triac, but the triac now energizes a relay with high-current contacts. When a 5-volt signal is applied to the input of the optocoupler through the 270-ohm resistor, the triac energizes the relay, which closes the circuit and applies power to the appliance. You can use this circuit to control washing machines, clothes dryers, shop tools.

A refrigerator should not be controlled by the computer unless it is given the highest priority, because refrigeration systems require some time after shutdown for the pressure in the reservoir to decrease. If the computer shuts the refrigerator off for a higher-priority device and then turns it back on without sufficient time between, the motor could burn out trying to overcome the back pressure which has not bled off (Kuecken 1979, p. 90).

The same type of situation is encountered with a room air conditioner; however, the on-off PTSR switch can be used with this appliance and the computer cannot cancel the PTSC until the toggle switch is manually reset. You must still be careful to turn off the air conditioner before resetting the PTSR switch in case there are some automatic shutdown procedures built into the appliance.

Current-Draw Sensor Circuits

So far, circuits for receiving permission-to-start requests and sending permission-to-start commands have been presented. The computer must also know how many appliances are on before it can make logical decisions as to whether or not to allow another appliance to come on. It must also be able to sense whether or not an appliance is on so it can cancel the permission-to-start command.

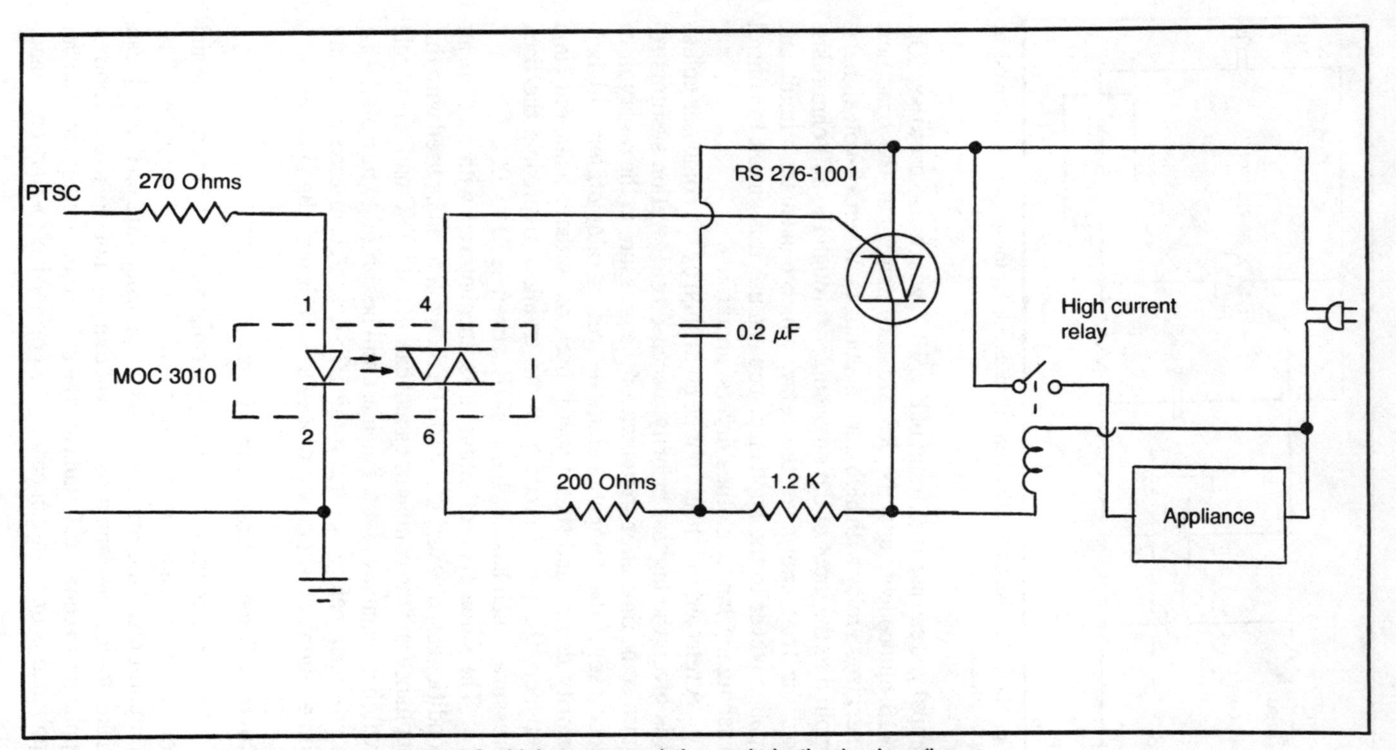

Fig. 13-7. Permission-to-start command circuit for high-current resistive- or inductive-load appliances.

The current-sensor circuits shown in Fig. 13-8 and 13-9 can be used by the computer to sense the on-off condition of appliances.

The circuit in Fig. 13-8 is a *current-draw* (CD) *sensor* for low-current appliances (less than 6 amps). It uses a transformer with its primary winding wired in series with the appliance. The primary of the transformer should be rated for more than 6 amps at 115 Vac. The secondary should be rated for no less than 1 milliamp at 6 Vac. When the appliance draws ac current through the primary of the transformer, an ac voltage is applied to diode D8. This ac voltage is rectified by D8 and filtered by the 30-μF capacitor. The filtered voltage is applied to the 1K-ohm resistor, through which it forces current into the base of the transistor. This base current causes the transistor to conduct and lower the voltage on the CD output line from +5 volts to near 0. The low-voltage level is sensed by the computer as an indication that the appliance is drawing current.

The circuit of Fig. 13-9 is a current-draw sensor for high-current appliances. It is basically the same circuit as that of Fig. 13-8 except it uses two transistors in a Darlington configuration for higher gain, a germanium diode for less voltage drop, and an inductive-loop pickup instead of a transformer. The number of loops needed and the length (L) of the inductive loop depend on the amount of current the appliance draws. Therefore, it will be a trial and error process to get the sensitivity high enough to produce a low-level CD signal when the appliance is drawing current. One side of the multiple-wire loop should be taped lengthwise to one power conductor of the appliance. The other side of the loop should be placed at least 6 inches away from the power conductor. The reason for using an inductive loop instead of a transformer is that high-current transformers are very expensive. This method is much less expensive and should work reasonably well.

Now that sensing and control circuits have been discussed, we need to describe how the computer uses the sensor data to make decisions. The next section outlines the decision-making process with the use of a program flowchart.

PROGRAM FLOWCHART

This section does not contain a detailed listing of the *Priority Power Distribution System* (PPDS) computer program. It does, however, contain a logical flowchart that can easily be adapted to a particular homeowner's needs. Once a working flowchart has been devised, it can be interpreted into machine language and stored in

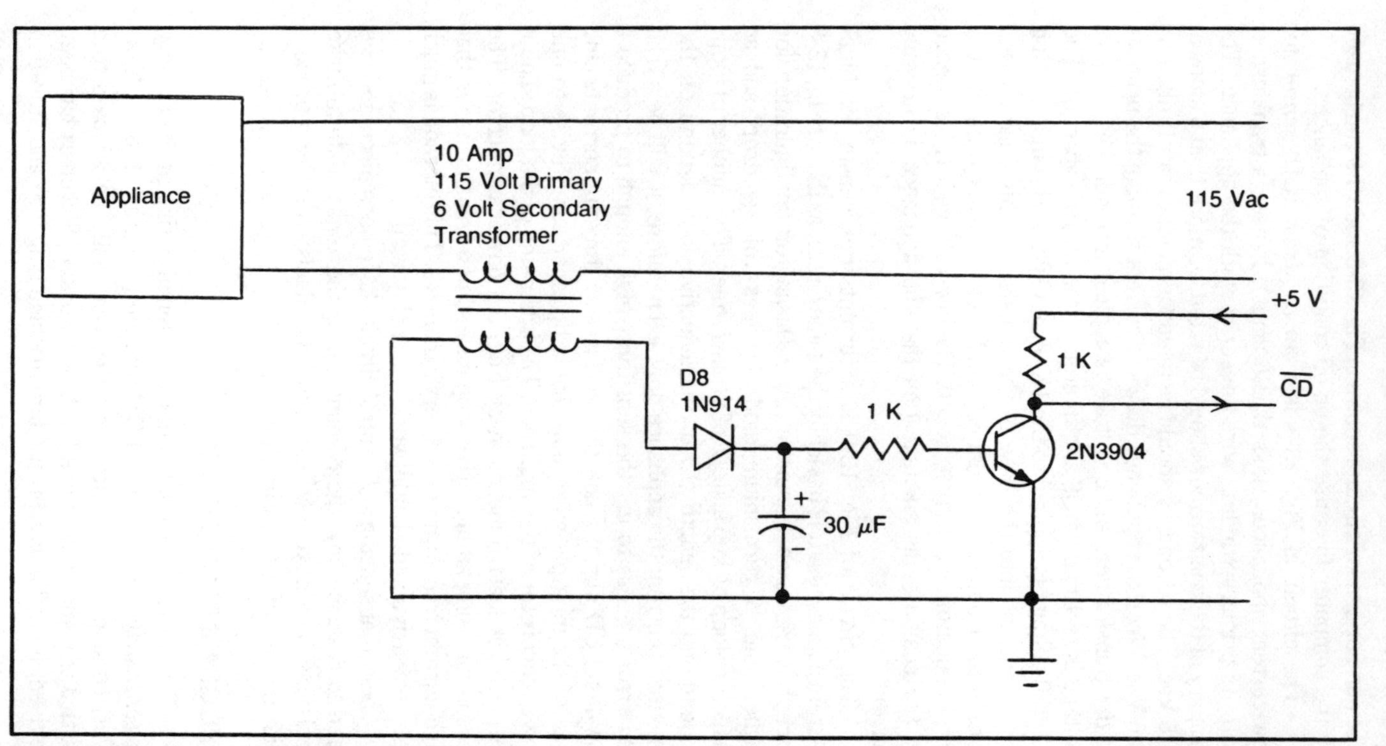

Fig. 13-8. Current sensor for low-current appliances.

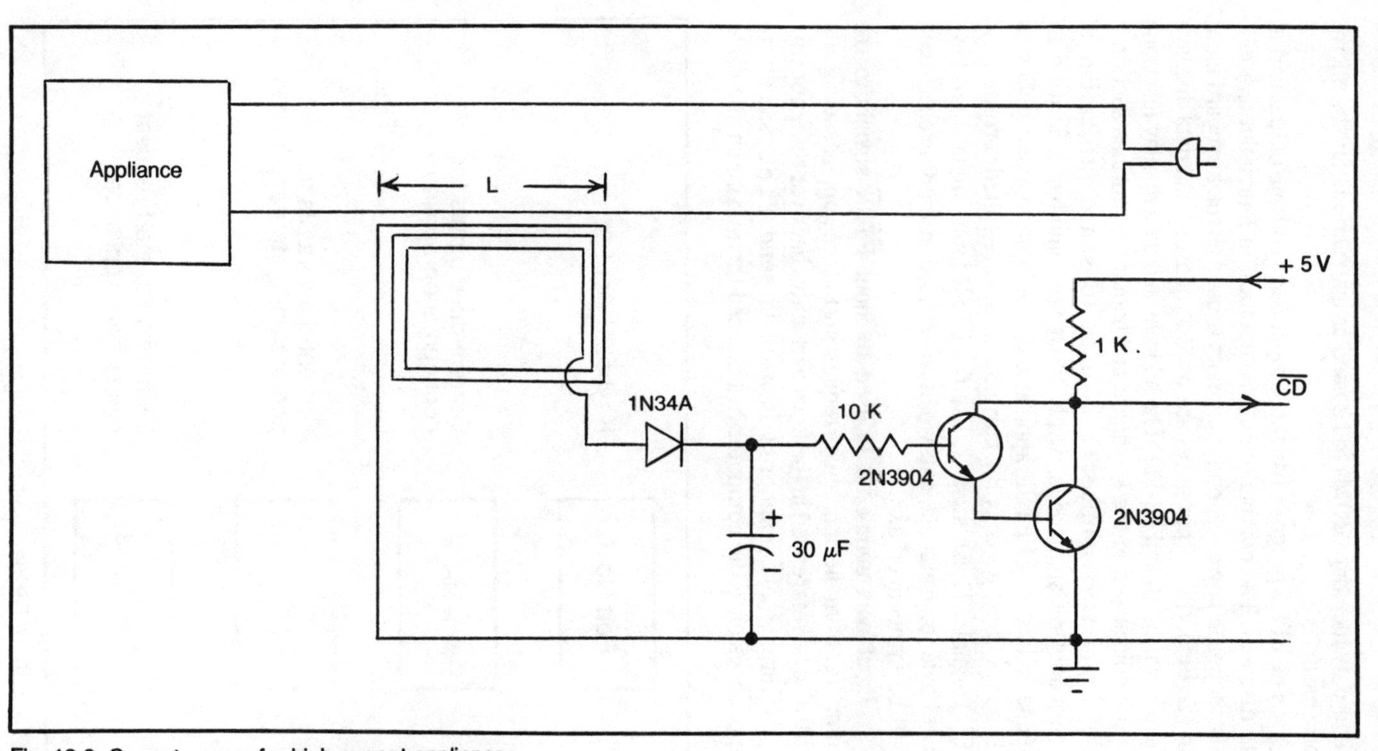

Fig. 13-9. Current sensor for high-current appliances.

read-only memory (ROM) to be used as instructions for the microprocessor.

The PPDS program has three input ports and one output port at its disposal. The computer will have a total of four input and four output ports; however, one input port is used as the keyboard input and three of the output ports are used as address and data displays (see Figs. 13-10 and 13-11). The address and data display functions will be discussed in the keyboard and bootstrap routines section.

Input port number 1 is used to sense utilization of continuous-on appliances, while input port number 2 is used to sense utilization of PTSR appliances. The input signals applied to these two input ports by the current sensors are called current-draw (CD) signals. These signals, like the PTSR input signals, are low-level true logic signals, and when referenced in the flowchart they will be symbolized as $\overline{CD}$.

Input port number 3 is used to input $\overline{PTSR}$ signals to the computer from the appliances requesting to come on. These signals will also be referenced in the flowchart with a bar over the acronym.

Output port number 3 is used to send PTSC signals to appliances once the computer had decided the maximum allowable

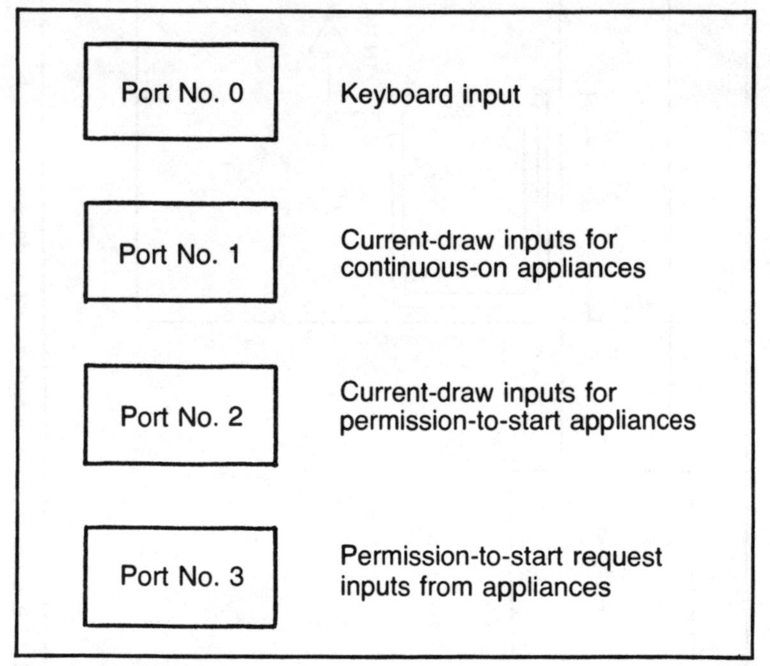

Fig. 13-10. Input ports.

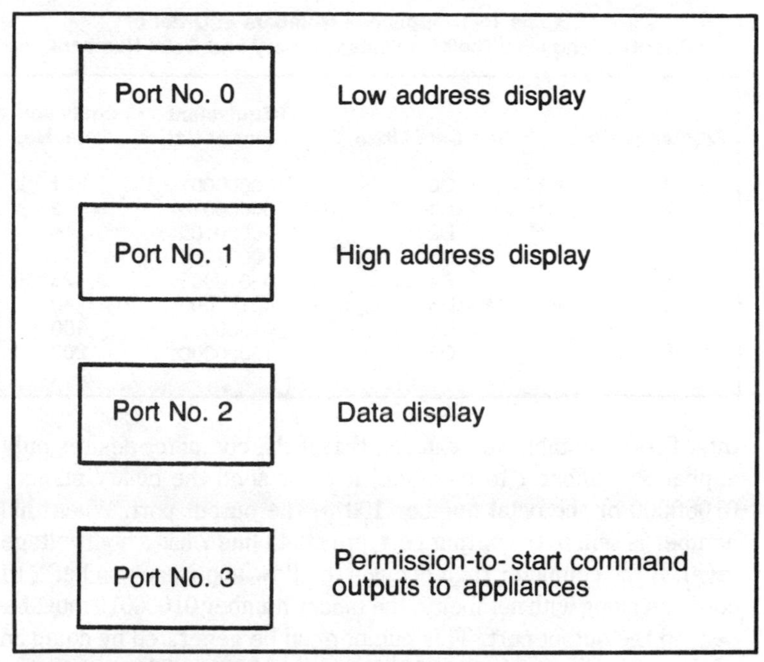

Fig. 13-11. Output ports.

current will not be exceeded. These signals are high-level true logic, and their acronym does not have the bar over it. A high voltage level (+5 volts) on one of these output lines causes the appliance associated with that output line to come on.

Each input port contains eight individual input lines and therefore is able to sense signals from eight separate appliances. Likewise, each output port contains eight output lines and is capable of controlling eight different appliances. The computer design presented in this chapter has enough input and output ports to service eight PTSR appliances and another input port to sense utilization of eight continuous-on appliances.

Each of the eight lines on the output port is dedicated to a particular appliance. If the computer desires certain appliances to come on, it places high voltage levels on their corresponding data lines. If only one appliance is allowed to come on, its data line on the output port will be high while the others are all low. If all appliances are allowed to come on, all data lines will be high. Table 13-1 gives a list of appliance numbers in order of priority along with the high data lines required to select these appliances. The table also shows the equivalent binary and octal numbers represented by these high data

Table 13-1. Appliance Numbers in Order of Priority Along with Their Equivalent Binary and Octal Numbers.

Appliance No.	High Data Lines	Equivalent Binary No.	Equivalent Octal No.
1	D0	00000001	1
2	D1	00000010	2
3	D2	00000100	4
4	D3	00001000	10
5	D4	00010000	20
6	D5	00100000	40
7	D6	01000000	100
8	D7	10000000	200

bits. From the table you can see that if the computer desires only appliance number 7 to come on, it must send the binary number 01000000 or the octal number 100 to the output port. When this number is sent to the output port, only data line 7 has a high voltage level. If the computer then decides to allow appliance number 2 to come on along with number 7, the binary number 01000010 must be sent to the output port. This number can be generated by doing an AND operation with 01000000 and 00000010. The computer, therefore, must always remember the last number sent to the output port.

Now suppose that appliances 7, 2, and 3 are on. The computer then decides to turn off only appliance number 2. The last command or number sent to the output port was 01000110, and the number now desired is 01000100. The last number can be generated by inverting the binary number of the appliance to be turned off, then doing an AND operation with it and the first number that the computer must recall from memory. For example, the number for inverted appliance 2 is 11111101. When this is ANDed with the old command (01000110), the result is 01000100. This new command is sent to the output port and only appliances 7 and 3 are allowed to come on.

Table 13-2 provides a list of the reserved memory locations in the random-access memory other than the program locations. Because the computer must keep track of the current drawn by all appliances, there must be one current code (CC) stored in memory for each appliance, whether PTSR or continuous-on. This CC value can be octal equivalent of the maximum current the appliance will draw. The current codes of the appliances being operated can be added by the computer to produce a current sum (CS) value. This CS value can then be compared to a maximum current draw value (MC) to determine if additional appliances can be turned on. Both the CS and MC values also require reserved memory space.

Each continuous-on appliance also requires one reserved memory space for a current added flag ($\overline{CA}$). If the CA flag is set to logic 0 for a particular appliance, the current code for that appliance has already been added to the current sum value. If the CA flag is set to logic 1, the current code for that appliance has not been added to the current sum.

Each PTSR appliance requires another reserved memory location for an appliance operating flag ($\overline{AO}$). If the AO flag is set to logic 0, the appliance was operating when last observed, and its CC value is still added into the CS value. If the AO flag is set to logic 1, the appliance is not operating, and its CC value is not added into the CS value.

As mentioned earlier, the computer must also remember the last 8-bit PTSC signal sent to output port number 3. This value also requires one reserved memory space.

The condensed program flowchart for the PPDS is shown in Fig. 13-12. The first block in the program is the continuous-on appliance current summation block. This part of the computer program is used to sense which continuous-on appliances are actually operating at the time and to add the current codes to the current sum CS for the ones that are operating. A more detailed breakdown of

Table 13-2. Reserved Memory In RAM (Random-Access Memory) Other Than Program Locations.

	Total Reserved Memory Locations
Each continuous-on appliance	
One location per appliance for current code CC.	
One location per appliance for current added flag CA.	n × 2 (n = number of continuous-on appliances)
Each PTSR appliance	
One location per appliance for current code CC.	
One location per appliance for appliance operating flag AO.	i × 2 (i = number of PTSR appliances)
One memory location for current sun CS.	1
One memory location for maximum current draw value MC.	1
One memory location for copy of last 8-bit PTSC signal.	1

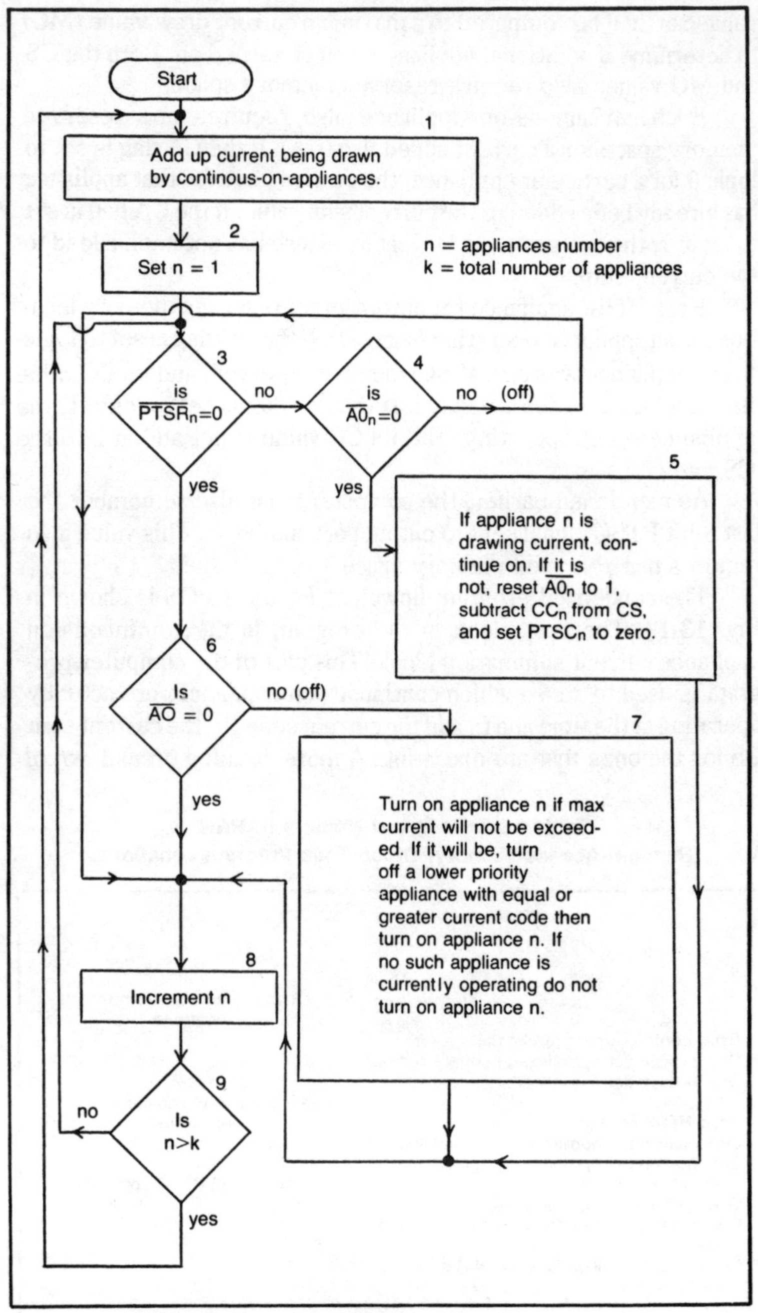

Fig. 13-12. Computer program flowchart for the priority power distribution system.

the continuous-on appliance current summation block is shown in Fig. 13-13.

Block 2 is used to initialize an appliance number counter. It is first set to 1. The program completes one cycle with all its attention focused on appliance number 1; then the counter is incremented to 2, and the process is repeated for appliance number 2. After the last appliance has been analyzed for permission-to-start request and current-draw, the program recycles to the beginning of the continuous-on appliance current summation block.

Block number 3 is simply a PTSR signal tester. If the PTSR signal for appliance number n is at logic 0, the appliance is requesting to come on. If the PTSR signal is at logic 1, the appliance is not requesting to come on. This PTSR signal is tested by inputting the 8-bit data word from input port number 3 and testing the corresponding data bit for high or low logic.

Block number 4 is an AO flag tester. If AO_n is set to logic 1, the appliance is not on. Because it is also not requesting permission to come on (known from block number 3), no action need be taken. If AO_n is set to logic 0, the computer knows that a PTSC has recently been sent to this appliance; however, it still does not know if the appliance is still operating.

If the AO_n flag is set to logic 0 from block 4, the next step is to check the current-draw inputs to see if appliance number n is still operating. This is done in block 5. If it is still drawing current, no action is taken. If it is not drawing current, AO_n is reset to logic 1, current code CC_n is subtracted from the current CS, and the PTSC for appliance n is cancelled. A more detailed flowchart of block 5 is given in Fig. 13-14.

If appliance n is requesting permission to start ($PTSR_n$=0 in block 3), the next step is again to check the appliance operating flag. This time, however, different actions are taken at the output of the block. If AO_n is set to logic 0, the appliance that is requesting permission to start is already on; therefore, no action is required. If AO_n is set to logic 1, the appliance is not on. In this case, the next step is to go to block 7.

Block 7 is used to turn on appliance n if the maximum current will not be exceeded. If maximum current will be exceeded, a lower-priority appliance with an equal or greater current code is turned off and appliance n is then turned on. If no appliance with lower priority and equal or greater current code is currently operating, no action is taken. If appliance n is turned on, AO_n is set to logic 0, and its current code CC_n is added to the current sum CS. The

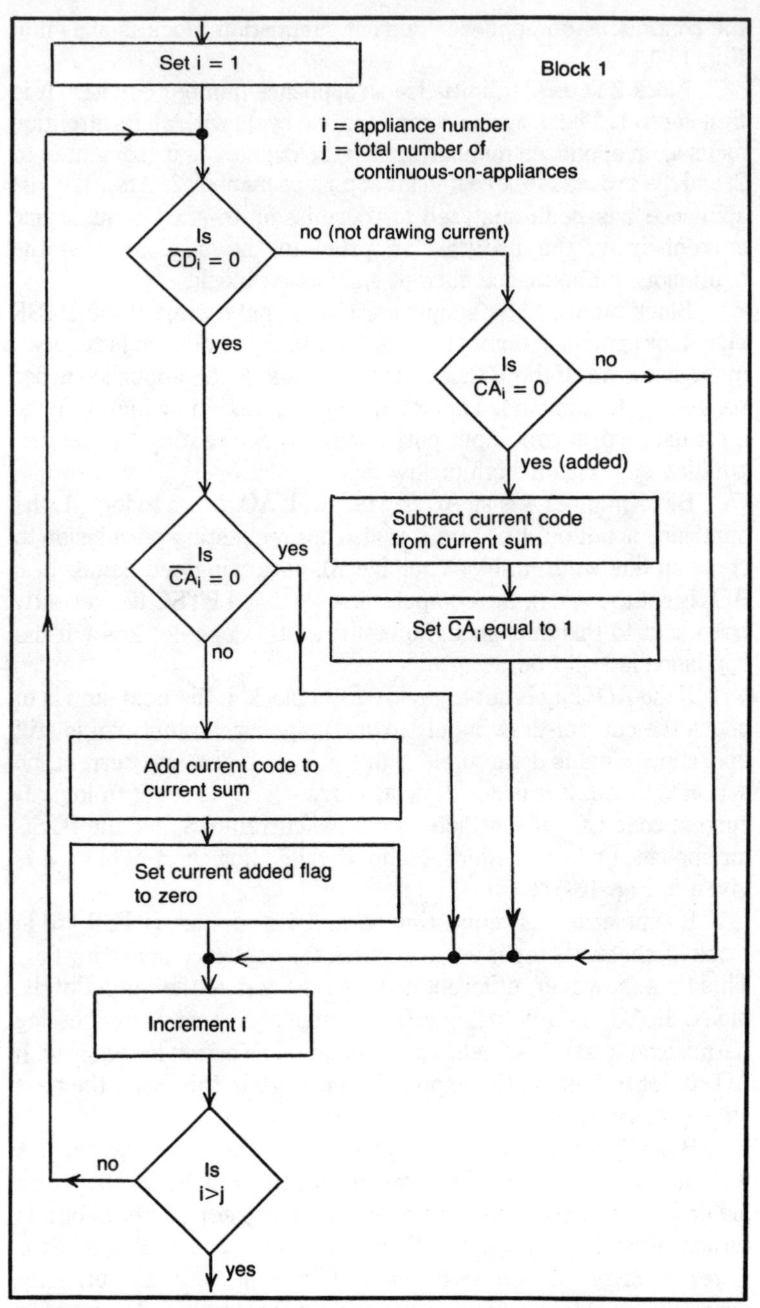

Fig. 13-13. Continuous-on appliance current summation block for the PPDS computer program.

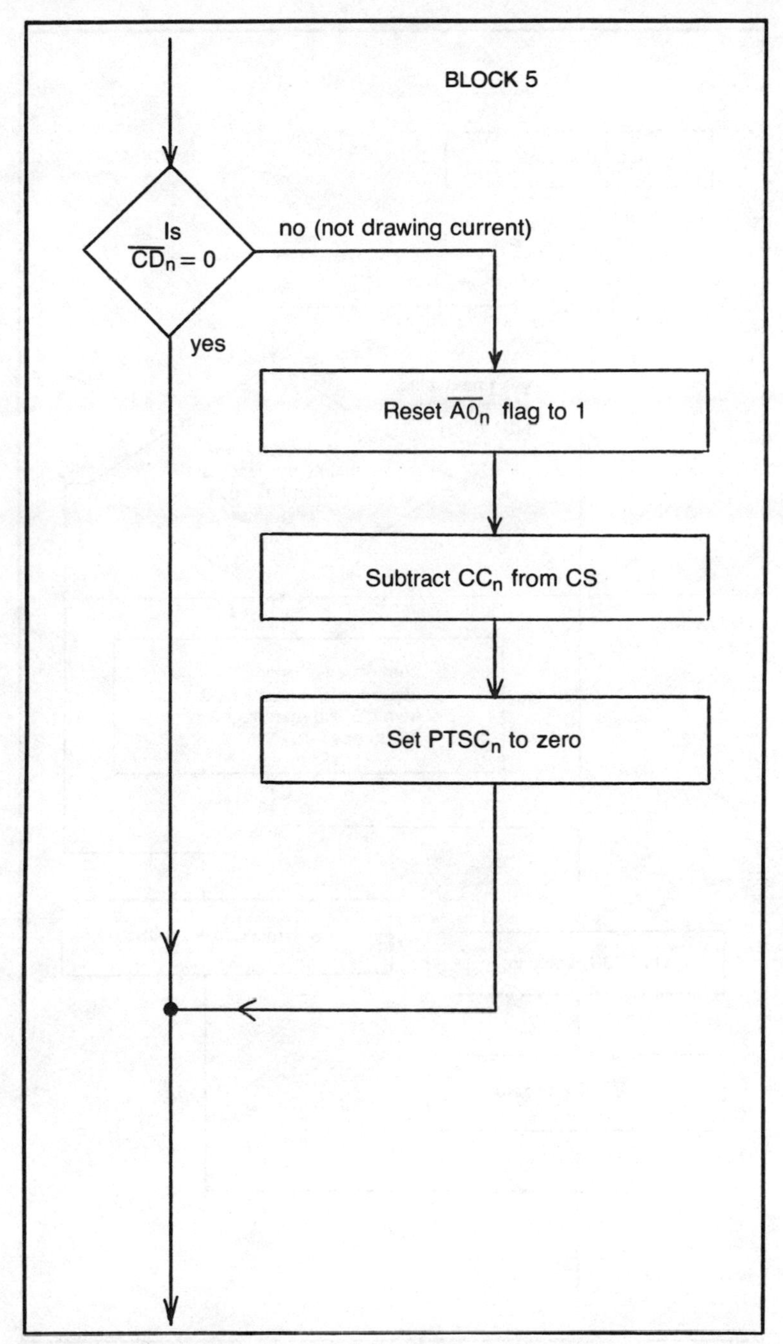

Fig. 13-14. Flowchart for block 5 of the PPDS computer program.

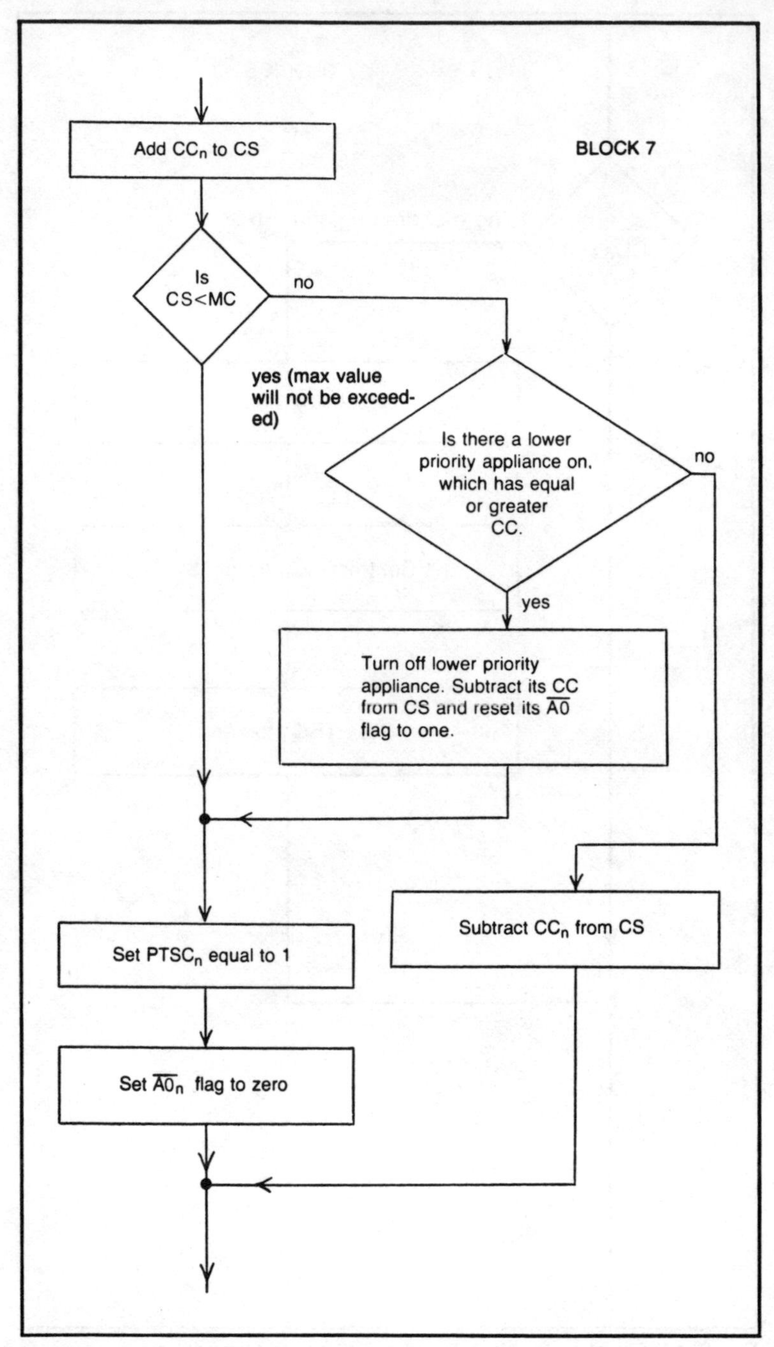

Fig. 13-15. Flowchart for block 7 of the PPDS computer program.

flowchart for block 7 is shown in Fig. 13-15.

Block 8 is used to increment appliance n. If this appliance number is not greater than the total number of PTSR appliances (k), the process is repeated at block 3 with the next appliance. If n is greater than k (last appliance checked), the program is reinitiated. The decision of whether or not n is greater than or less than k is made in block 9.

We have described a logical decision-making process for the priority power distribution system. We must now interpret this flowchart into machine language and store it in random-access memory so the microprocessor can execute the program. Until now, no mention of a particular microprocessor has been made. There are several good microprocessor chips on the market, but from here on the Intel 8080A is referred to. Appendix B contains a listing of the 8080A Instruction Set that can be used to write a machine language program based on the previous flowcharts (Rony 1981, p. 127).

Each 3-digit octal code instruction in Appendix B is an 8-bit data word sometimes called a byte. Some instructions require 1 byte, while others require 2 or 3 bytes. Byte 2 is symbolized as B2 and byte 3 as B3. With some instructions such as LXI B, the two additional bytes, B2 and B3, are simply data to be loaded into registers B and C. These two additional bytes are supplied by the programer and can be any value he chooses. A single byte of data can be manipulated in the same way with an instruction such as MVI B.

The two additional bytes can also be used as pointers to other memory locations. The JNZ command is a conditional jump instruction that uses B2 and B3 as the address of the memory location to which it is jumping. B2 is the least significant byte of the address, and B3 is the most significant byte. All addresses consist of 16 bits or 2 bytes.

The IN and OUT instructions use one additional byte of data to address input and output devices. Because only 8 bits are used in these addresses, no more than 256 input or output devices can be connected to this microprocessor; however, this usually proves to be plenty.

This section provides the necessary tools to write the machine language computer program (software) for the priority power distribution system. The next section describes the circuitry necessary for the hardware portion of the system.

MICROCOMPUTER CIRCUITS

There are many different ways to design an operating *mi-*

crocomputer. Some are designed for minimum cost, and others are designed for minimum size. Lhomond Jones of Ogden, Utah and I collaborated on the design presented in this section. It was designed not for minimum cost or size, but for parts availability. All microcomputers do have, however, three things in common: some way to input data, some program for manipulating the data, and some way to output the results of that manipulation.

A block diagram of this microcomputer design is shown in Fig. 13-16. The RAM block contains the random-access memory in which the power distribution program is stored. The EPROM block contains the electrically programable read-only memory (EPROM), which, in turn, contains the keyboard and bootstrap routines. The keyboard and bootstrap programs allow the user to enter and execute programs through the keyboard. This program, plus an EPROM programer circuit, is presented in later sections.

The CPU (*central processing unit*) is all contained in one microprocessor chip. All the arithmetic and logical operations are conducted in this block. The Intel 8080A microprocessor chip used in this design has seven 8-bit registers to aid in the manipulation of data.

The system controller circuit, also shown in Fig. 13-16, is used to decode instructions from the CPU into memory-write, memory-read, data-out, and data-in instructions. These instructions are then used to aid in the control of the memory and input/output (I/O) devices. These require two commands before they can be enabled. One command, mentioned above, must come from the system controller. The other must come from the memory chip selector. These selector circuits are used to decode the memory or I/O device address that is sent out on the 16-bit address line.

The clock circuit shown in Fig. 13-16 provides precise timing pulses to coordinate the activities of the CPU and system controller.

The following seven subsections describe these various circuits in more detail.

CPU, Clock, and Controller Circuits

The CPU, clock, and system controller circuits make up the heart of the microcomputer. They are shown in Fig. 13-17. As previously stated, all the decision-making logic operations take place in the CPU. For this design an Intel 8080A microprocessor integrated circuit chip is used for the CPU. Many technicians and engineers agree that this is the best microprocessor currently on the market. It is readily available, and there are plenty of other mi-

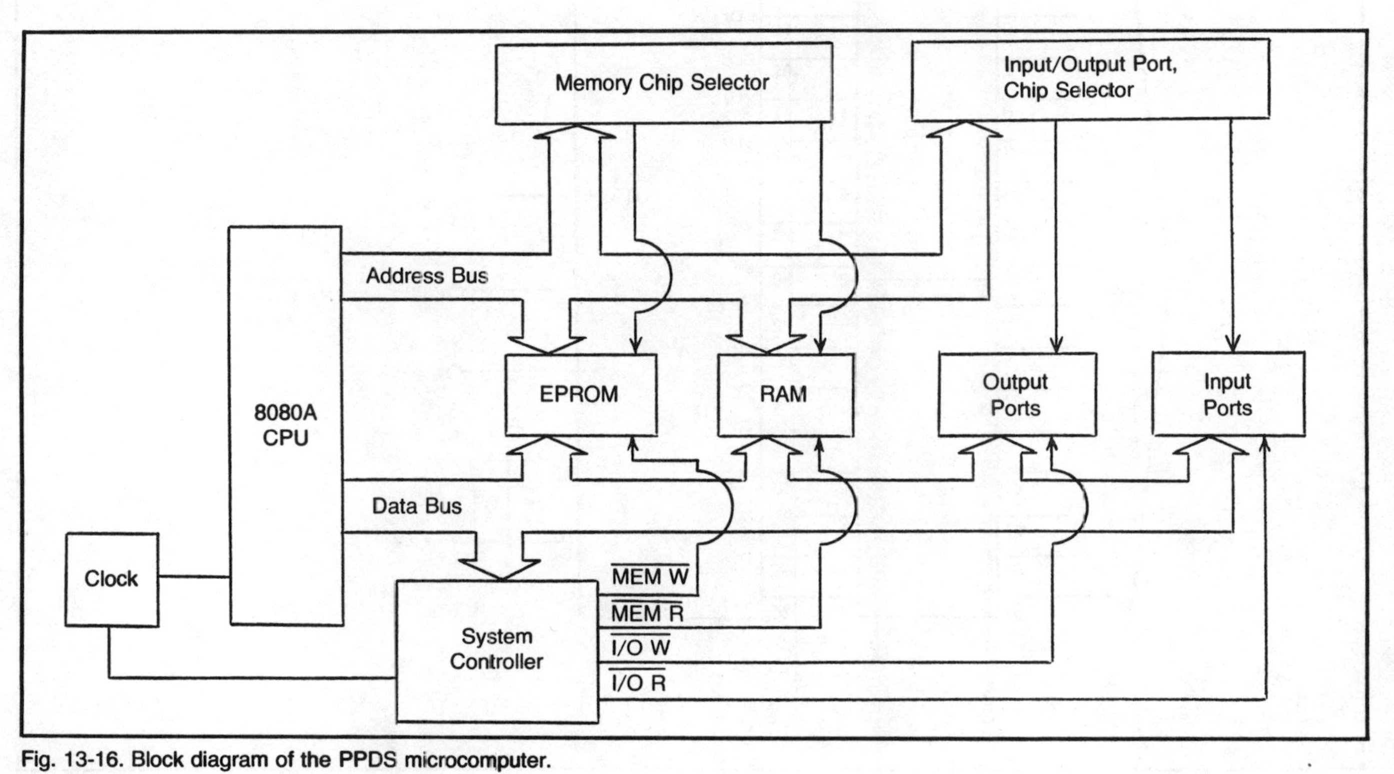

Fig. 13-16. Block diagram of the PPDS microcomputer.

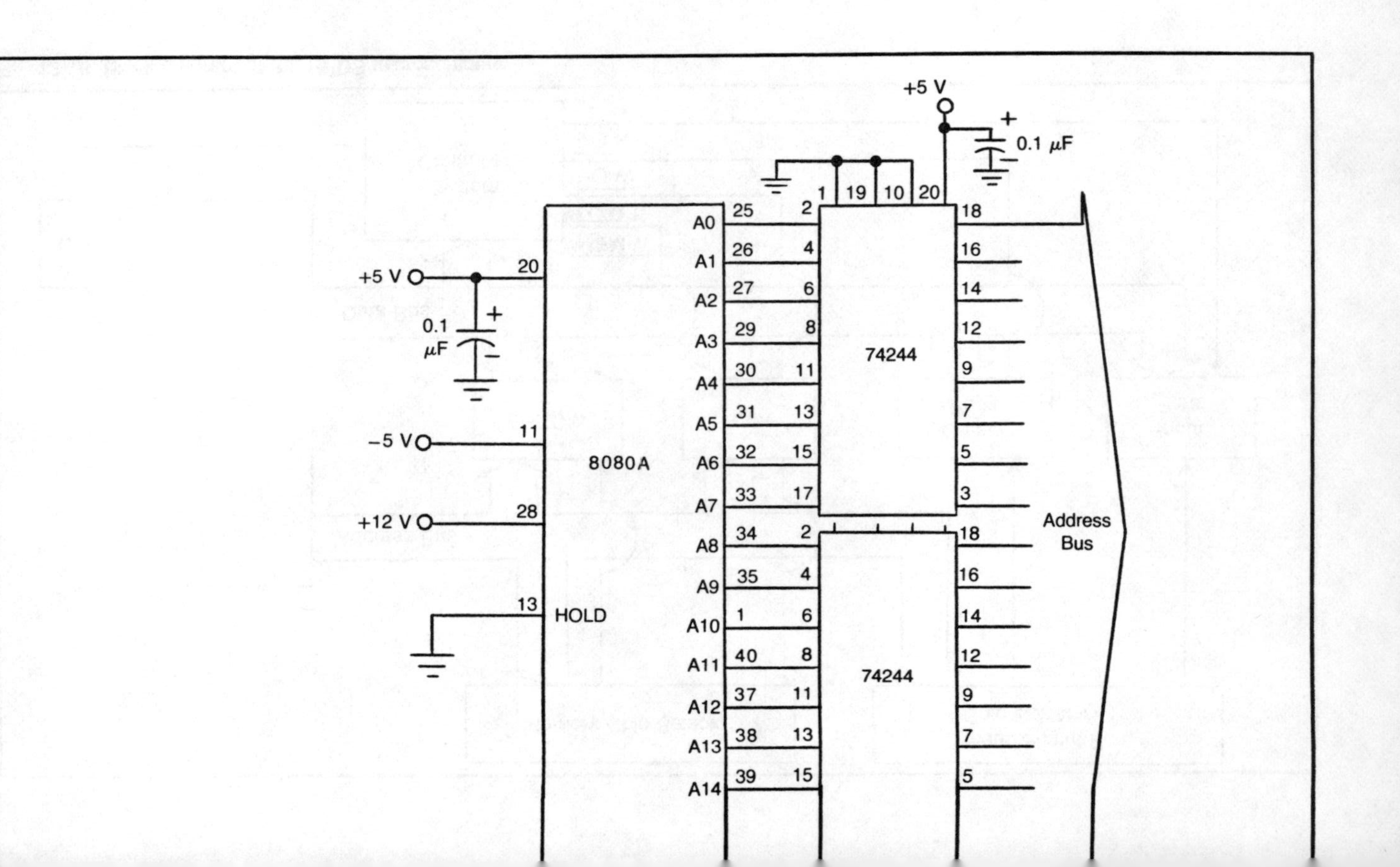
+5 V
0.1 µF
+5 V
0.1 µF
−5 V
+12 V
8080A
HOLD
A0
A1
A2
A3
A4
A5
A6
A7
A8
A9
A10
A11
A12
A13
A14
74244
74244
Address Bus

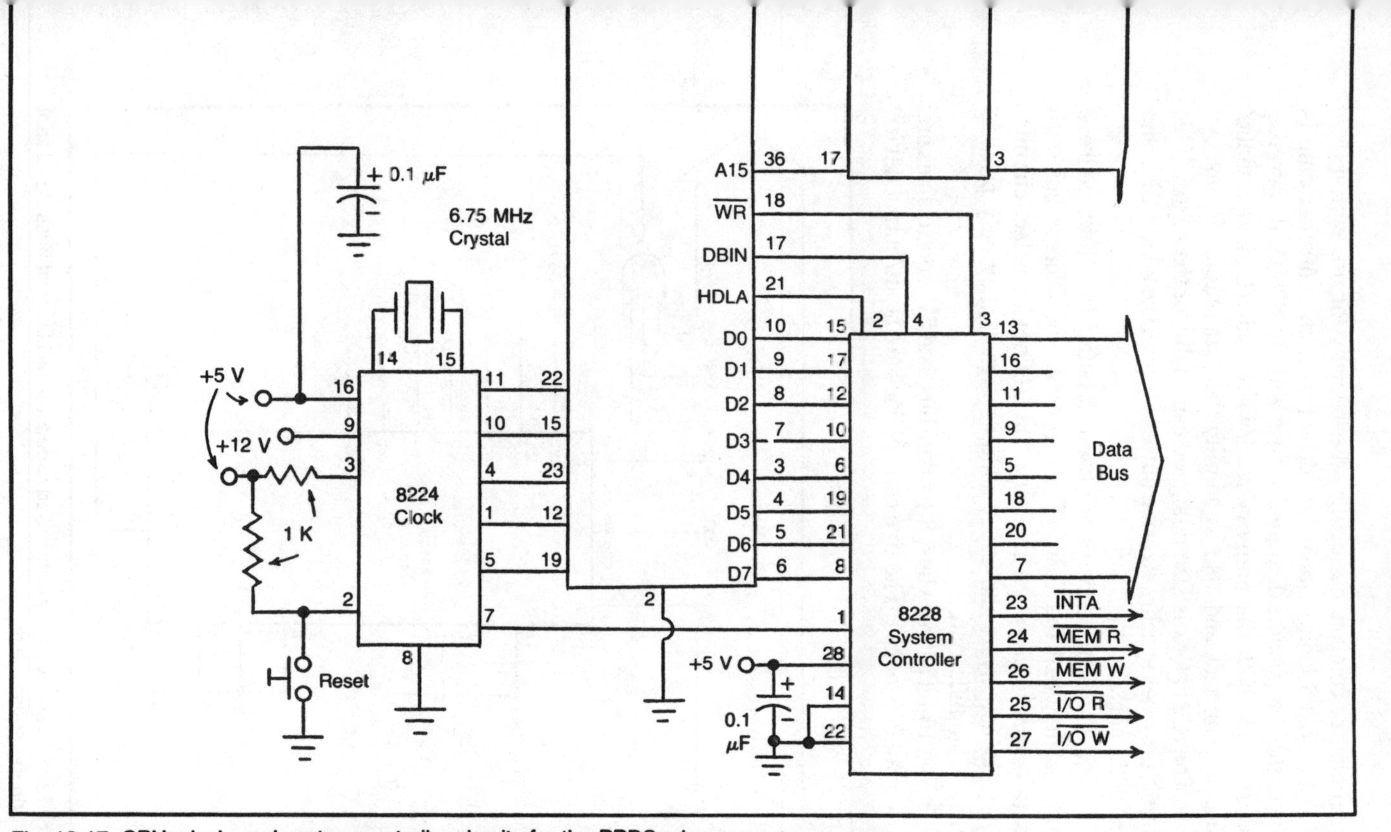

Fig. 13-17. CPU, clock, and system controller circuits for the PPDS microcomputer.

crocomputer components designed especially for use with it.

The 8228 bidirectional bus driver/system controller chip is one of the components designed for use with the 8080A. Its purpose is to decode both the memory-read/write and I/O device input/output commands and also to amplify the data output signals.

The 8224 clock is also designed especially for the 8080A. This circuit provides synchronization pulses for both the 8080A CPU and the 8228 chip.

The 74244 bus drivers are used to buffer the 16-bit address lines as they exit the CPU. These provide the address bus with extra power to drive the memory and I/O device select circuits. They do not invert the input address signal; they only provide current amplification.

The 16-bit address bus, 8-bit data bus, and five control lines are now available for use. The five control lines are input/output device

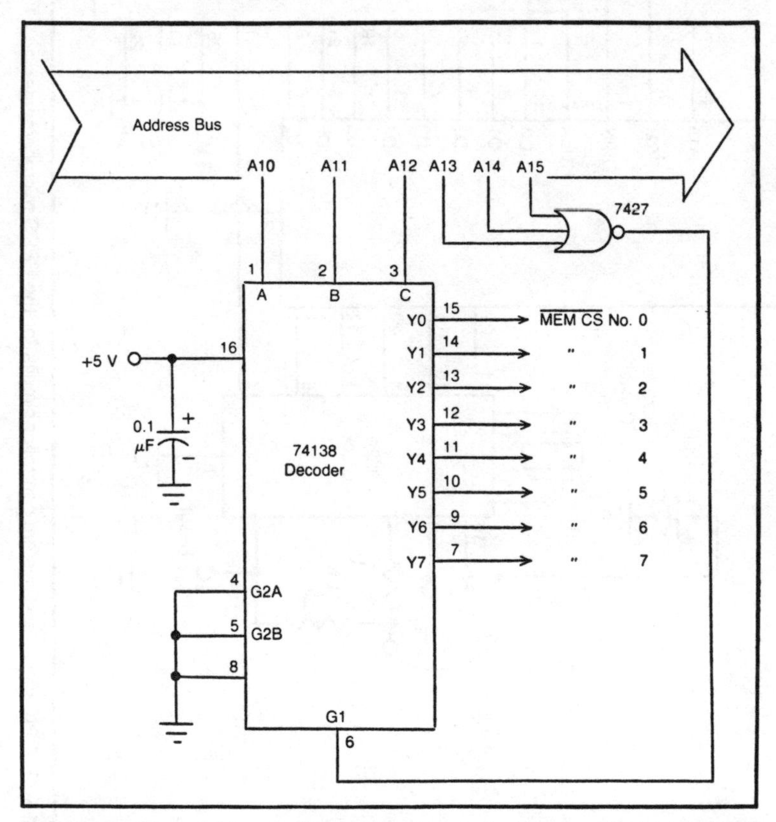

Fig. 13-18. Memory chip selector used to decode the binary address signal sent out on the address bus.

Table 13-3. Function Table for the Instruments 1976, p. 7-135). 74138 Decoder (Engineer Staff Texas

Inputs					Outputs							
G1	G2A and G2B	C	B	A	Y0	Y1	Y2	Y3	Y4	Y5	Y6	Y7
L	X	X	X	X	H	H	H	H	H	H	H	H
H	L	L	L	L	L	H	H	H	H	H	H	H
H	L	L	L	H	H	L	H	H	H	H	H	H
H	L	L	H	L	H	H	L	H	H	H	H	H
H	L	L	H	H	H	H	H	L	H	H	H	H
H	L	H	L	L	H	H	H	H	L	H	H	H
H	L	H	L	H	H	H	H	H	H	L	H	H
H	L	H	H	L	H	H	H	H	H	H	L	H
H	L	H	H	H	H	H	H	H	H	H	H	L

L = Low level, H = High level, X = Irrelevant

write (I/O W), input/output device read (I/O R), memory-write (MEM W), memory-read (MEM R), and interrupt-enable (INT A). the interrupt-enable signal will not be used in this design, but is available for experimentation. The bar over the mnemonic expressions symbolizes a negative true logic. In other words, when I/O W is at logic 0, the computer wants to write data into the output device.

Memory Chip Selection

The first 10 address lines (AO-A9) are sent directly to the memory circuits to select one of the 1024 memory locations in the chips. The last six address lines (A10-A15), however, must be sent to the decoder device to select a particular memory chip. This memory chip selector circuit is shown in Fig. 13-18.

A 74138 decoder chip is used as the binary code decoder in Fig. 13-18. This chip will accept a 3-digit binary input and convert it to a one-of-eight output. Each of these eight outputs can be connected to the chip select (CS) input of a 1024-byte memory chip, giving a 8192-byte (8K) memory capacity. A function table showing the relationship between the inputs and outputs of the 74138 is shown in Table 13-3.

Because there are actually six address lines that can be used for chip selection, this computer has the capability to address 64K or 65,536 bytes of memory; however, this design will only use 2K of memory; 1K of read-only memory and 1K of random-access memory.

In the circuit of Fig. 13-18 the additional address lines are tied to pin 6 of the 74138 decoder through a 7427 triple input NOR gate.

When wired this way the 74138 decoder is disabled if any of the last three address lines are high and no memory chip is selected at all. A pinout diagram of the 7427 is shown in Fig. 13-19.

Even though this design uses only 2K of memory, it is capable of addressing 8K. If you decide to expand the system, you need only add more memory citcuits.

Electrically Programable Read-Only Memory (EPROM)

Figure 13-20 shows the wiring of a 2708 EPROM. The first 10 address lines (AO-A9) are connected directly to the chip to select any one of the 1024 bytes of stored memory. The $\overline{\text{MEM R}}$ and memory chip select number 0 ($\overline{\text{MEM CS}}$ No. 0) command lines are connected to the chip select ($\overline{\text{CS}}$) input of the 2708 EPROM through

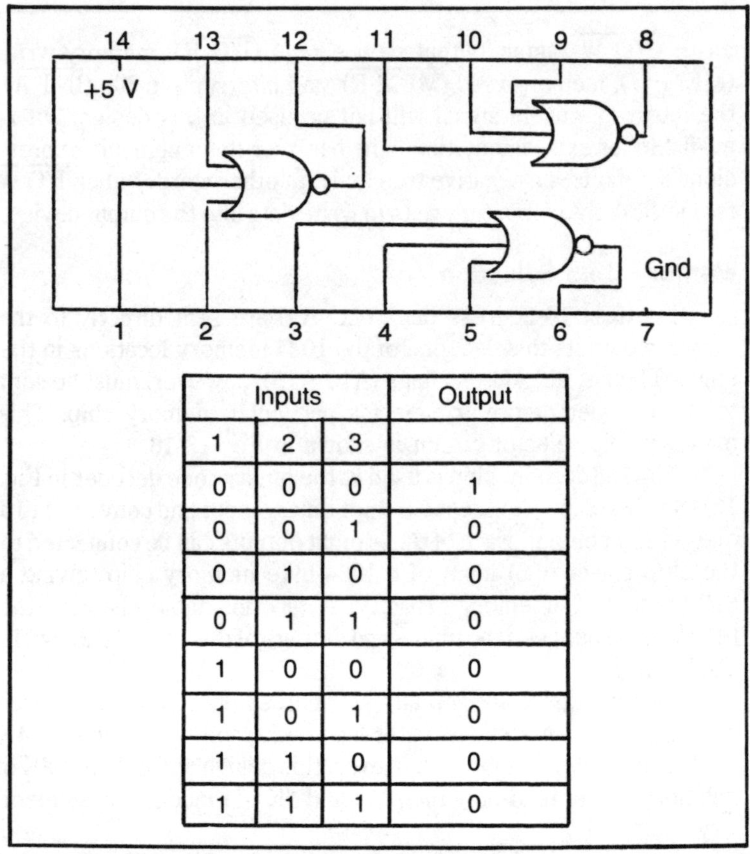

Inputs			Output
1	2	3	
0	0	0	1
0	0	1	0
0	1	0	0
0	1	1	0
1	0	0	0
1	0	1	0
1	1	0	0
1	1	1	0

Fig. 13-19. Pinout diagram and truth table for the 7427 triple 3-input NOR gate (Texas Instruments 1976, p. 5-12).

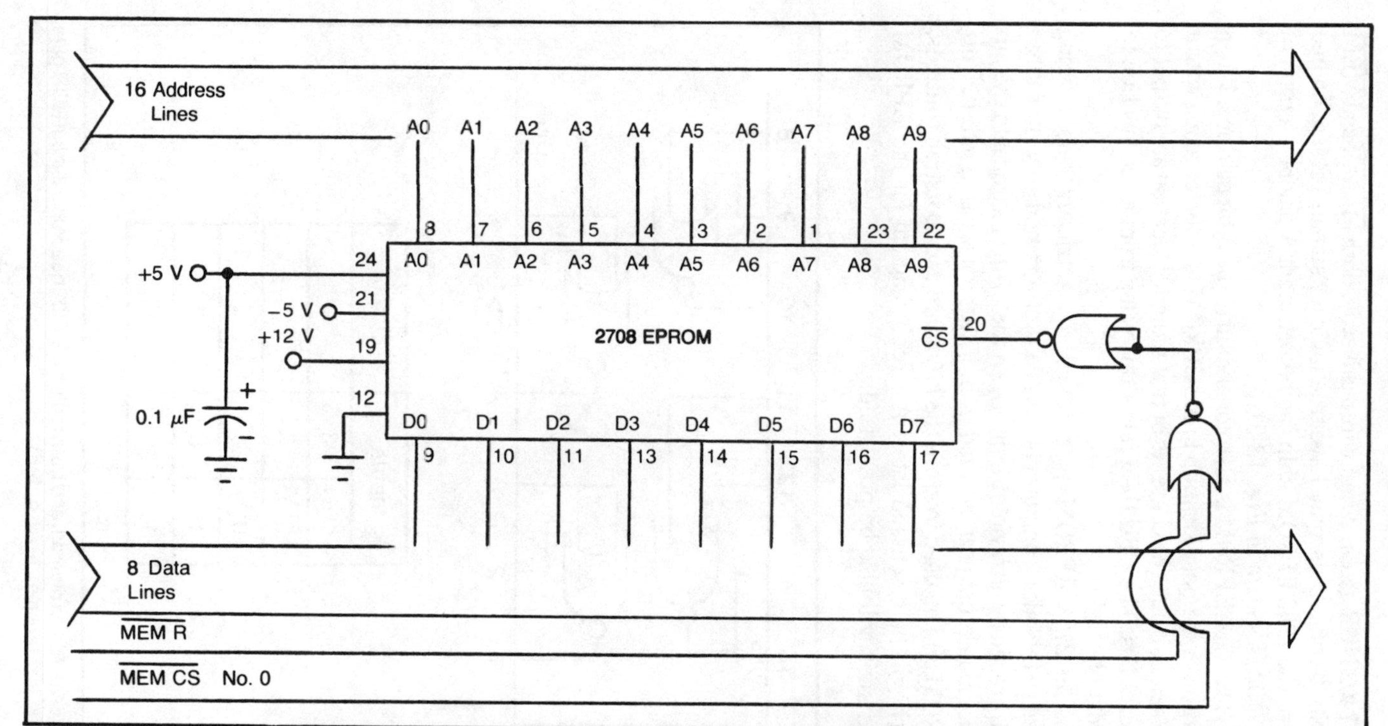

Fig. 13-20. Wiring diagram for the 2708 electrically programable read-only memory (EPROM).

two 7402 NOR gates. When wired this way, both $\overline{\text{MEM R}}$ and $\overline{\text{MEM CS}}$ No. 0 commands must have true logic (0) signals present on line before the 2708 EPROM chip can be selected. A pinout diagram of the 7402 is shown in Fig. 13-21.

The 2708 EPROM is also connected to the 8-bit data bus. When a particular memory location in the EPROM is selected along with a chip select input, the contents of that memory location are output to the data bus. The CPU then reads this value from its connection to the data bus.

The 2708 EPROM is programed electrically before being wired to the microcomputer. It can be reprogramed only after it has been erased by ultraviolet light. Once the 2708 is wired into the microcomputer, data can be read from it, but not written into it. This EPROM will be used to store the keyboard and bootstrap routines, which will be discussed later. A circuit diagram for the EPROM programer will also be presented.

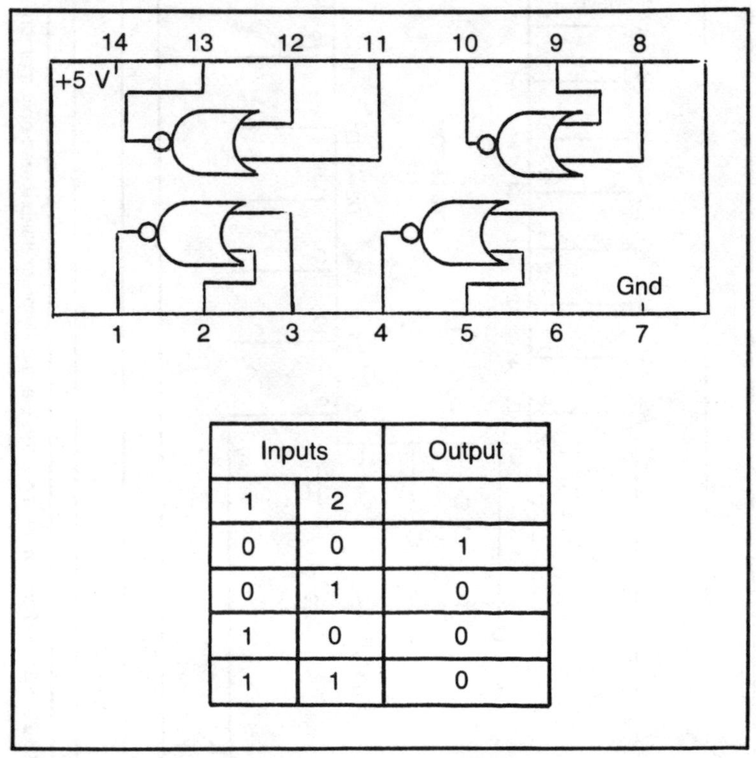

Inputs		Output
1	2	
0	0	1
0	1	0
1	0	0
1	1	0

Fig. 13-21. Pinout diagram and truth table for a 7402 quadruple 2-input NOR gate (Texas Instruments, 1976, p. 5-6).

Random-Access Memory (RAM)

The *random-access memory* contains the main decision-making program for the priority power distribution system. Its wiring diagram is shown in Fig. 13-22. Two 2114 L-3 RAM chips, each containing 1024 4-bit words of memory storage, are used in this circuit to produce 1024 8-bit words of memory. The first four bits from the data bus (D0-D3) are connected to one RAM chip, and the last four bits (D4-D7) are connected to the other RAM chip. When the same memory location is addressed on both chips, the two 4-bit words output from the two chips form one 8-bit word. The address lines A0-A9 are connected to both chips.

The RAM chips are both enabled by the memory chip select command No. 1. When the $\overline{WE}$ input is pulled low by the $\overline{MEM\ W}$ input, the chips will accept data on the data bus to be written into memory. When the $\overline{WE}$ input is high ($\overline{MEM\ W}$=logic 1), memory data will be output to the data bus. Before data can be written into or read from the RAM, however, the $\overline{CS}$ input must be pulled low by the chip select line.

Once the EPROM has been programed with the keyboard and bootstrap routines, the RAM can be programed through the keyboard. This program can be erased and reprogramed as many times as you wish.

Input/Output Port Chip Selection

The I/O port chip selector circuit (see Fig. 13-23) operates similarly to the memory chip selector circuit of Fig. 13-18. The only real difference is that the memory chip selector decodes an address from pins A10, A11, and A12 of the address bus, and the input/output port chip selector circuit decodes an address from pins A0, A1, and A2. If the first eight address lines were decoded by the circuit, 256 different input or output port chips could be selected. Because the first three are used, only eight input and eight output ports can be selected. Even this is more than will be used in this microcomputer design.

The five address lines not being decoded are applied to pin 6 of the 74138 decoder after passing through a 5-input NOR gate made from a 74LSO5 hex inverter with open-collector outputs. This circuit disables the decoder if any of the address lines A3-A7 have a high logic level (+5). A pinout diagram of the 74LSO5 hex inverter is shown in Fig. 13-24.

Each of the chip select output lines can be connected to one input port and one output port. This means there can be an input

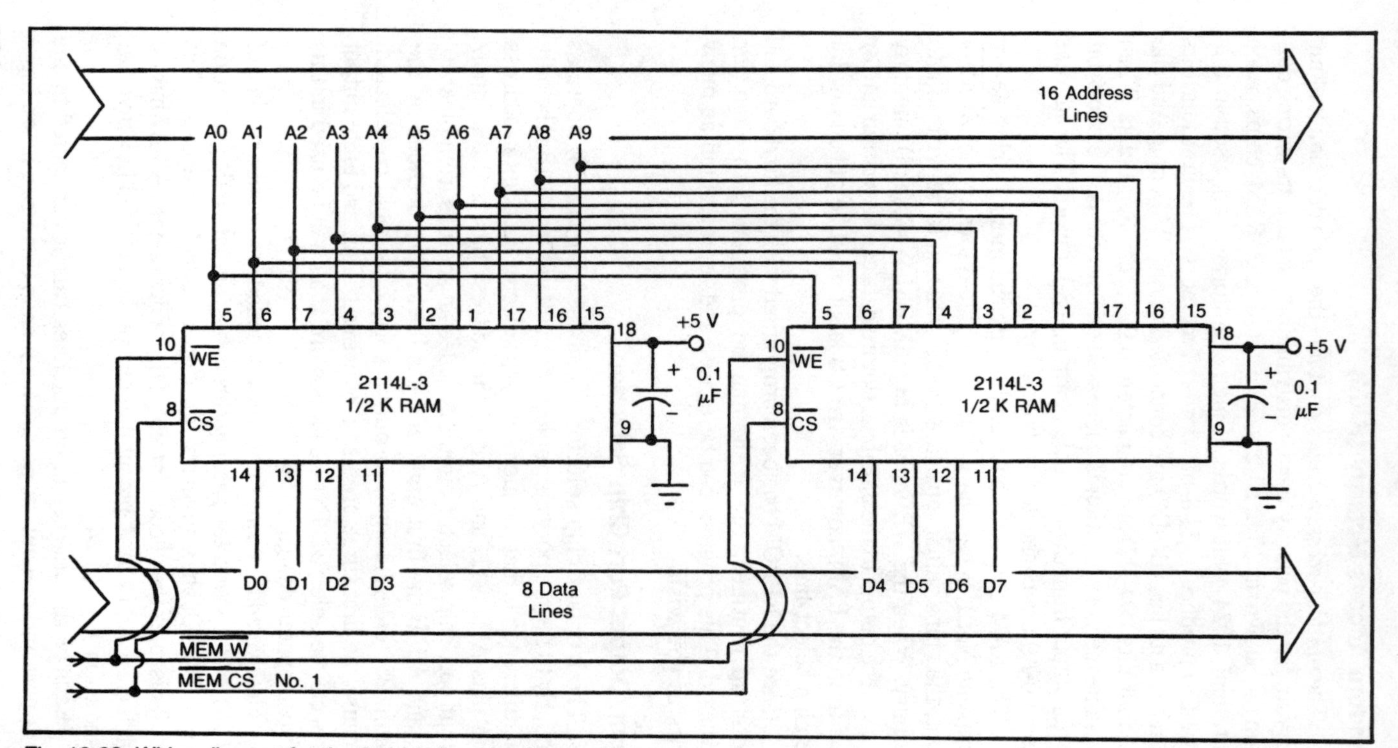

Fig. 13-22. Wiring diagram for the 2114 L-3 read-only memory.

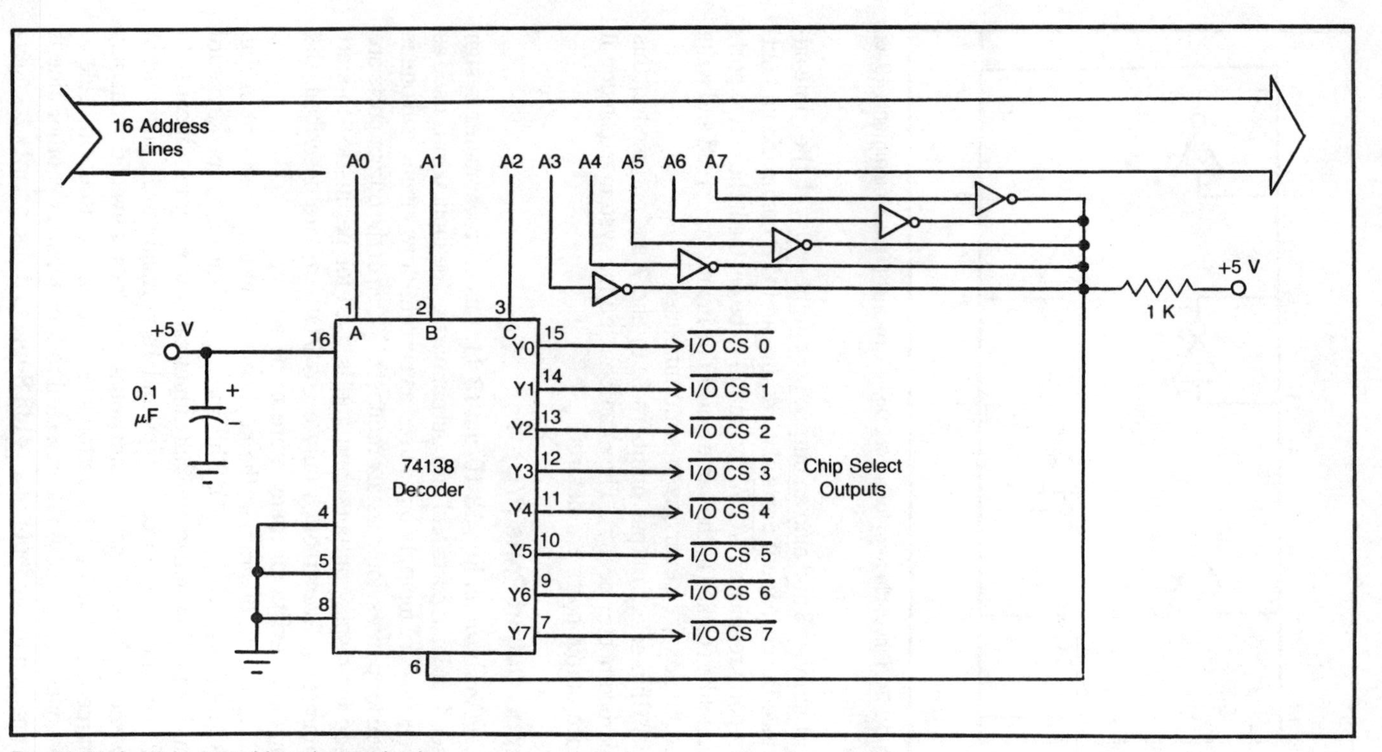

Fig. 13-23. Input/output chip selector circuit.

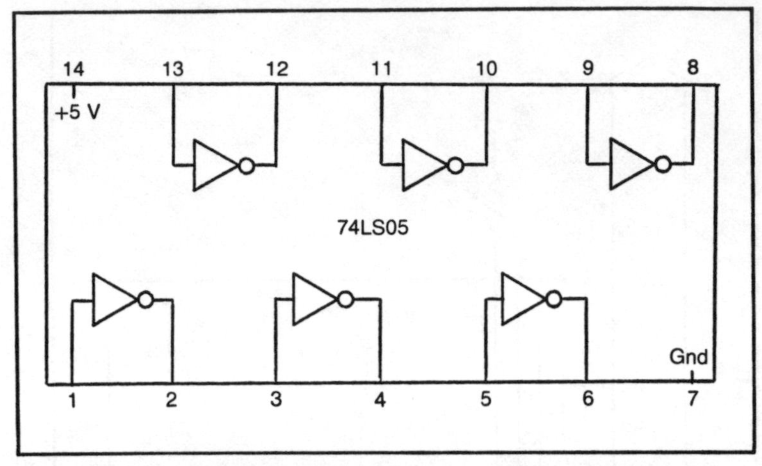

Fig. 13-24. Pinout diagram for a 74LS05 hex inverter (Texas Instruments 1976, p. 5-7).

port number 3 and also an output port number 3 with the same address. Because input ports also require an $\overline{\text{I/O R}}$ command and output ports require an $\overline{\text{I/O W}}$ command to be enabled, these two ports cannot be enabled at the same time. The input/output read and write commands are never given simultaneously.

Chip select output numbers 5, 6, and 7 are not used in this microcomputer design. They can be used for system expansion if more output ports are desired.

Input/Output Ports (I/O)

As shown in Fig. 13-10 and 13-11, this microcomputer design uses four input ports and four output ports. One input port is used as the keyboard input, two are used as current-draw inputs, and one is used for permission-to-start requests. Three of the output ports are used for display outputs, and one is used for permission-to-start commands. Even though only eight I/O ports are described, the capability exists for many more to be added.

A circuit diagram of the keyboard input port is shown in Fig. 13-25 (Titus 1976, p. 35). This input port is used by the keyboard routine to allow you to manually input data and programs. It consists of four integrated circuit chips and 15 normally open pushbutton switches. These can be individual switches salvaged from any source available, or a 16-button keyboard can be used. One side of every switch (or key) is connected to ground, and the other side is connected to one input of a 74148 8-line to 3-line priority encoder.

The priority encoders are used to convert the single bit of switch information into its own particular binary code, as shown in Table 13-4. For example, if switch 6 is pushed, a low level is applied to input 6, and the output becomes A2=A1=L, AO=H. Each switch closure has a different binary code associated with it.

Switches representing numbers 0 through 7 are connected to priority encoder number 1, and switches representing letters S, C, G, H, L, A, and B are connected to priority encoder number 2. Because both encoders are identical, there must be some way of

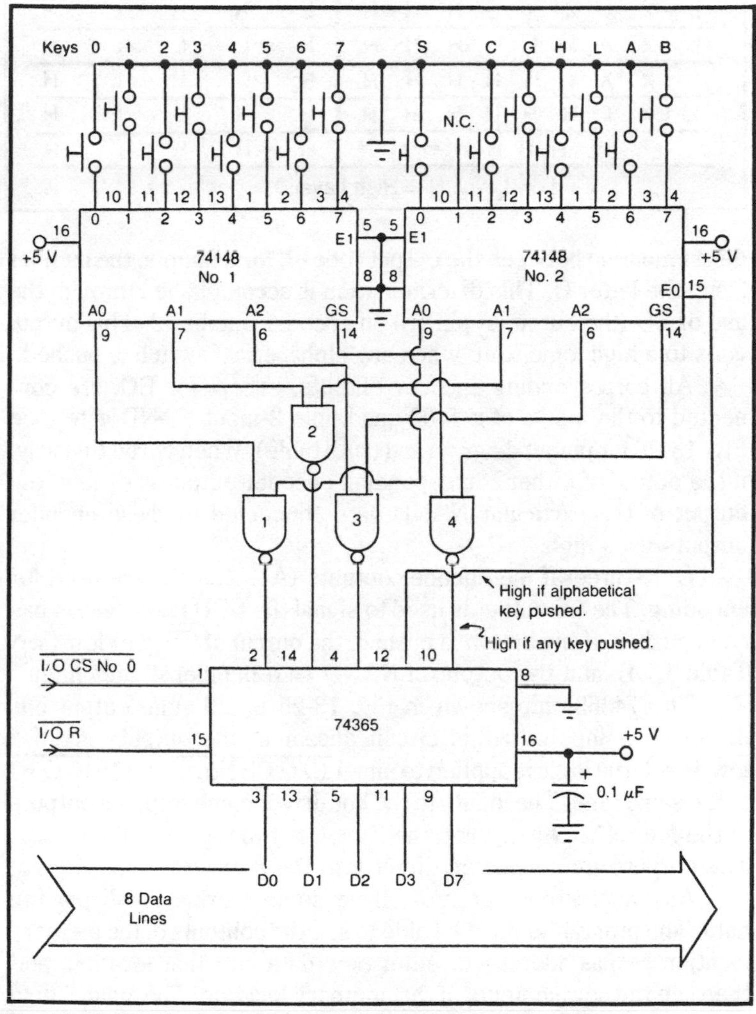

Fig. 13-25. Circuit diagram for the keyboard input port No. 0.

Table 13-4. Function Table for the 74148 Encoder (Texas Instruments 1976, p. 7-151).

Inputs									Outputs				
E1	0	1	2	3	4	5	6	7	A2	A1	A0	GS	E0
H	X	X	X	X	X	X	X	X	H	H	H	H	H
L	H	H	H	H	H	H	H	H	H	H	H	H	L
L	X	X	X	X	X	X	X	L	L	L	L	L	H
L	X	X	X	X	X	X	L	H	L	L	H	L	H
L	X	X	X	X	X	L	H	H	L	H	L	L	H
L	X	X	X	X	L	H	H	H	L	H	H	L	H
L	X	X	X	L	H	H	H	H	H	L	L	L	H
L	X	X	L	H	H	H	H	H	H	L	H	L	H
L	X	L	H	H	H	H	H	H	H	H	L	L	H
L	L	H	H	H	H	H	H	H	H	H	H	L	H

L = Low Level, H = High Level, X = Irrelevant

discriminating between the output code of, for example, the number 2 and the letter C. This discrimination is accomplished through the use of EO (the output of pin 15) on encoder number 2. This output goes to a high logic state when any alphabetical switch is pushed.

All corresponding encoder outputs, except for EO, are connected to the inputs of a 7400 quadruple 2-input NAND gate (see Fig. 13-26 for pinout diagram and truth table). When wired this way, if the output of either corresponding encoder output goes low, the output of the particular NAND gate connected to these encoder outputs goes high.

Only three of the encoder outputs (A1, A2, A3) are used for encoding. The GS output is used to signal the CPU that a button has been pushed. If any button is pushed, the output of GS goes low (see Table 13-4), and the output of NAND gate number 4 goes high.

The 74365 chip shown in Fig. 13-25 is a 3-state output bus driver. The inputs to this circuit appear at the output only if a low-level true logic is applied to pins 1 $\overline{\text{(I/O CS No. 0)}}$ and 15 $\overline{\text{(I/O R)}}$ at the same time. The inputs to the bus driver come from the outputs of the four NAND gates and the EO output of the second encoder. The outputs are connected directly to the 8-bit data bus.

Any microcomputer must have some method of displaying data. The programer must be able to see the contents of the memory location he has addressed, enter new data into that location, and then see the new contents of that memory location. The manual data input port was just discussed. The output port of Fig. 13-27 is used

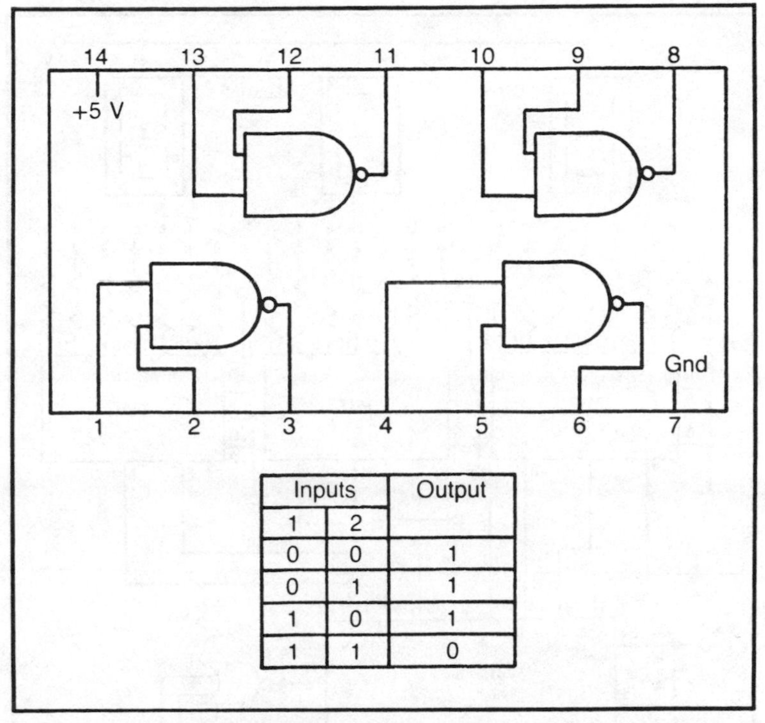

Inputs		Output
1	2	
0	0	1
0	1	1
1	0	1
1	1	0

Fig. 13-26. Pinout diagram and truth table for the 7400 NAND gate (Texas Instruments 1976, p. 5-6).

BCD-to-seven-segment decoder/drivers through the 74100 8-bit bistable latch. The $\overline{\text{I/O W}}$ and $\overline{\text{I/O CS}}$ control lines are input to a 7402 NOR gate, and the NOR gate output is then connected to the 1G and 2G inputs of the 8-bit latch. When both the $\overline{\text{I/O W}}$ and $\overline{\text{I/O CS}}$ control lines have a zero logic level, the output of the NOR gate goes high, and the 8-bit data on the data bus is transferred to the decoder/drivers. The decoder/drivers then apply the decoded data to the seven-segment displays. When the output of the NOR gate goes low, the output data of the 8-bit latch remains the same until another high level NOR gate output comes along.

Each 8-bit data word is decoded by the 7447 chips into a 3-digit octal number. Figure 13-28 shows how the data word is broken up, and it also shows the relationship between a 3-digit/bit binary number and a 1-digit octal number.

There are actually three of these 3-digit output ports (see Fig. 13-11). Port number 0 is used for the low address display, port number 1 for high address display, and port number 2 for data

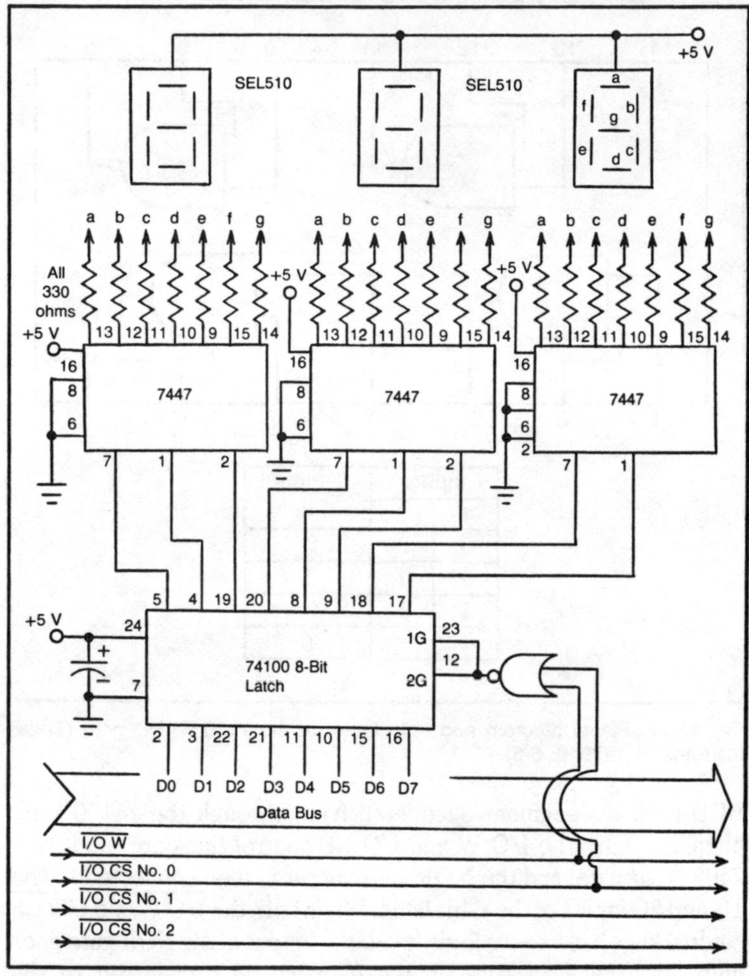

Fig. 13-27. Wiring diagram for the 3-digit octal data display output port.

display. All three ports are identical except for the $\overline{\text{I/O CS}}$ control line connections (see Fig. 13-27).

Output port number 3 is used to output permission-to-start commands to appliances and is shown in Fig. 13-29. This output port also uses a 74100 8-bit bistable latch and a 7402 NOR gate. The operation of these two circuits is the same, except that in this circuit the outputs of the latch go directly to the eight appliances.

The permission-to-start request input port is shown in Fig. 13-30. This circuit utilizes a 74244 3-state output line driver as the input device. Both the $\overline{\text{I/O R}}$ and $\overline{\text{I/O CS}}$ commands are input to the chip enable inputs (pins 1 and 19) via two NOR gates. When the two command lines go low at the same time, the $\overline{PTSR}$ signals from the appliances are placed on the data bus.

We must describe one more circuit before we can build a working microcomputer. This circuit, the power supply, is not usually considered a part of the computer circuitry, but certainly the computer cannot operate without it.

Power Supplies

This microcomputer requires three different voltage levels. All of the circuits require +5 volts, but some also require +12 and −5 volts. Because all circuits use +5 volts, most of the power consumption comes from this source.

The circuit in Fig. 13-31 shows how all three voltages can be obtained from one transformer. The transformer is usually the most expensive component in a power supply, and keeping the number and size of them to a minimum is important. The transformer used in this circuit has a 120-volt primary and a 24-volt, center-tapped secondary that will carry 4 amps. The center tap is connected to ground.

The output of the transformer is rectified by four 1N1612 (or equivalent) 5-amp diodes connected in a full-wave bridge configura-

8-Bit Binary Number --- XX XXX XXX
Equivalent, 3-Digit
Octal Number --- Y Y Y

3-Bit Binary Number	Equivalent, One-Digit Octal Number
000	0
001	1
010	2
011	3
100	4
101	5
110	6
111	7

Fig. 13-28. 8-bit binary to 3-digit octal number conversion.

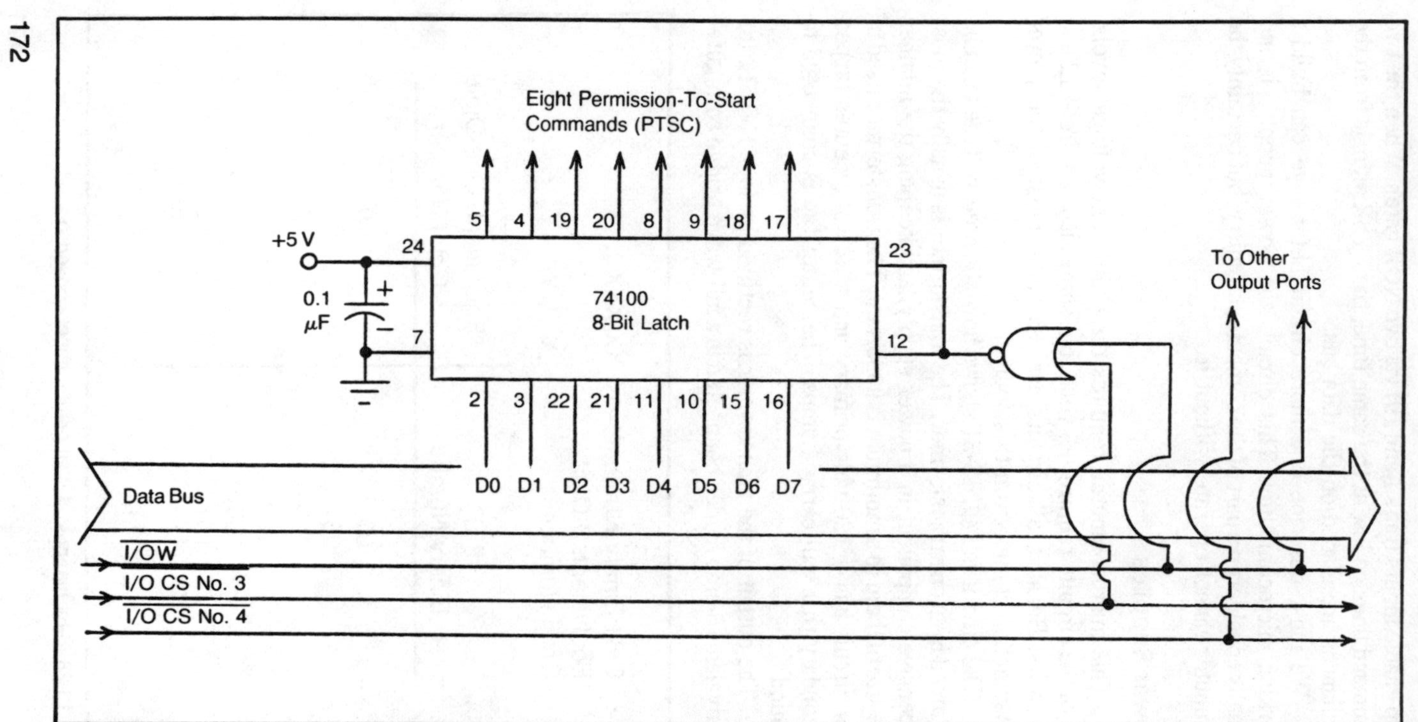

Fig. 13-29. Wiring diagram for the permission-to-start command output port.

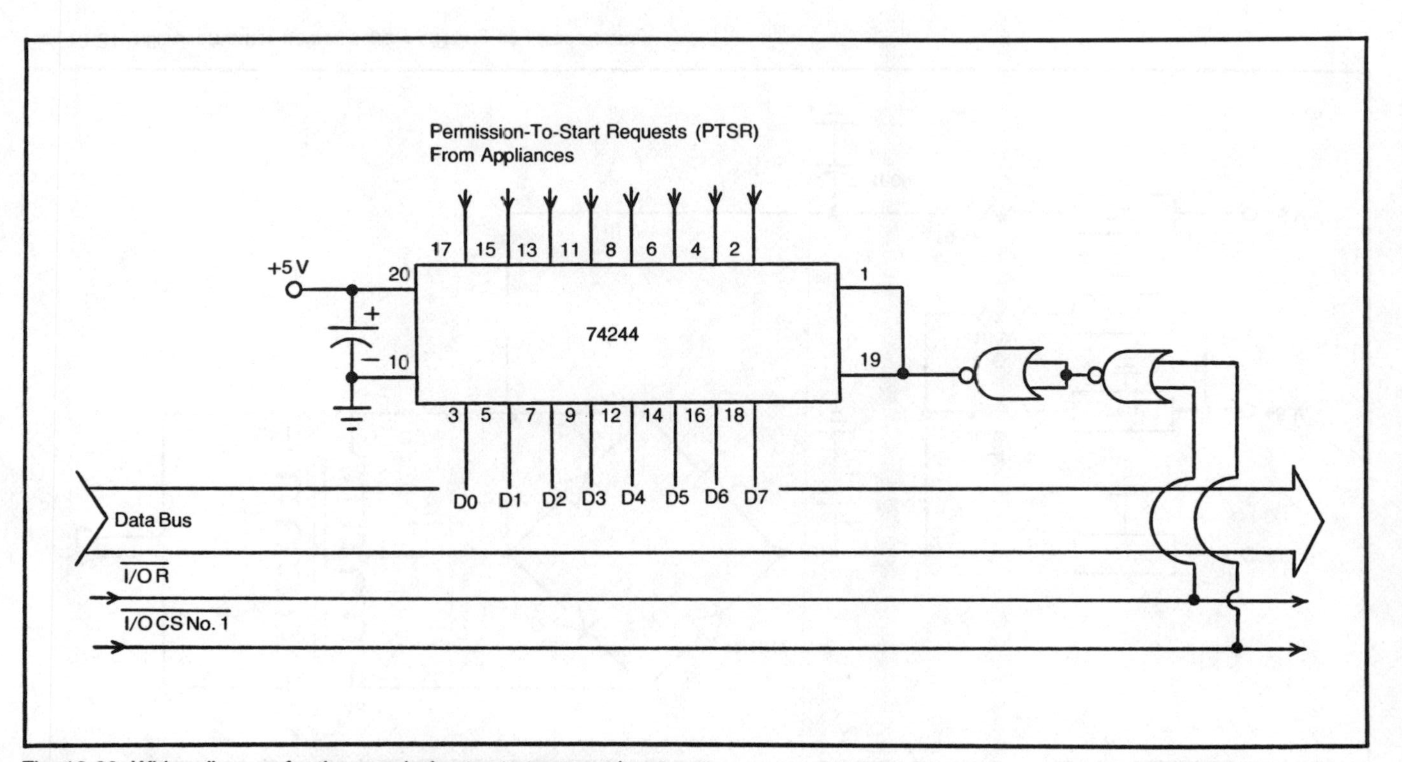

Fig. 13-30. Wiring diagram for the permission-to-start request input port.

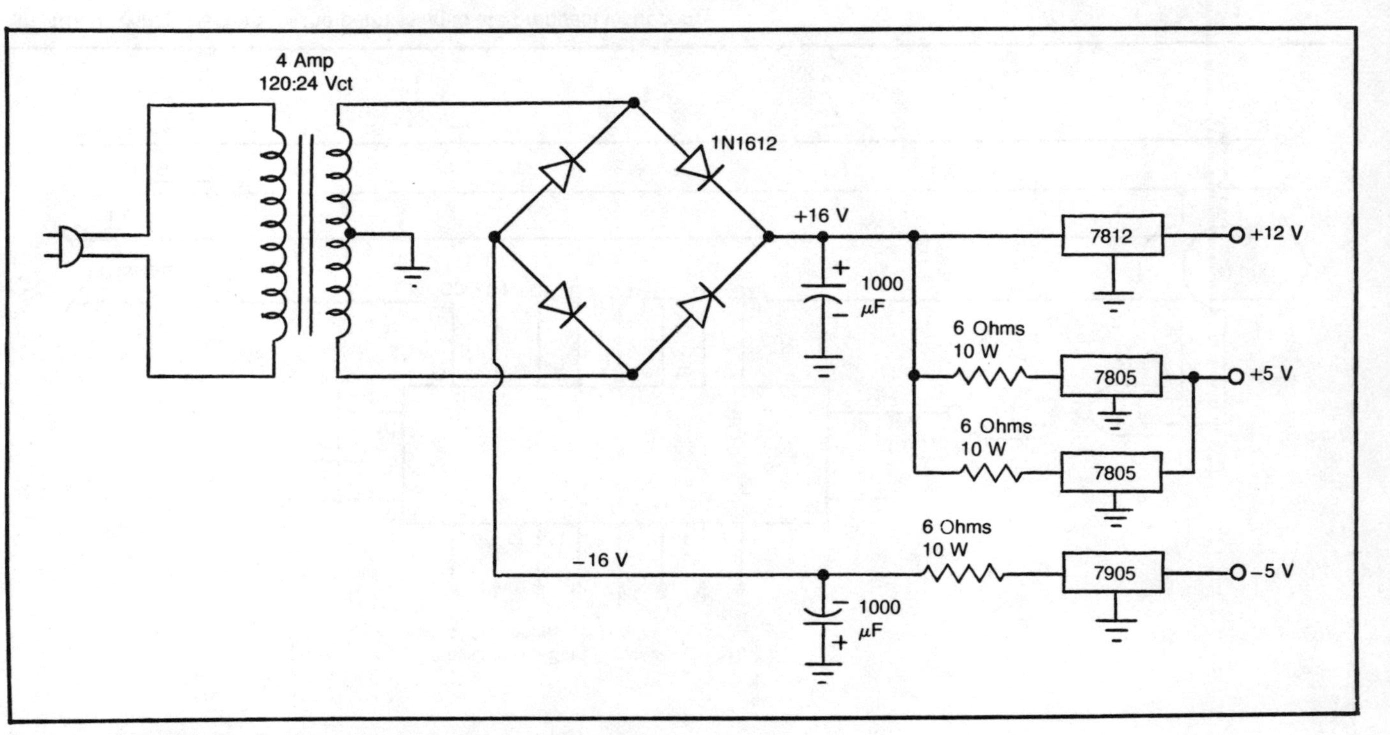

Fig. 13-31. Microcomputer power supply utilizing one transformer.

tion. The bridge produces two outputs: +16-Vdc and −16-Vdc. After the bridge rectifies the plus and minus 12 volts rms ac input voltages, the two resulting dc output voltages are equal to 1.414 times the input rms voltages, or approximately +16 volts and −16 volts dc. These values are, however, the no-load voltages. The voltages will be lower when a load is connected to the power supply output.

To keep the input voltages to the computer from varying too much as the current load varies, each voltage must be filtered by a capacitor and passed through a voltage regulator before it is connected to the computer. The +16 volts is input to three regulators: one +12-volt (7812) and two +5-volt (7805) in parallel. The −16 volts is input to one −5-volt regulator (7905).

Each regulator is rated at 1 amp. If two regulators are paralleled, twice as much current can be supplied with the same regulator output voltage. The three 5-volt regulators require large (in terms of lower) series resistors. These are needed because of the relatively large input voltage as compared to the output voltage. If these resistors were not used, each 5-volt regulator would have to dissipate approximately 11 watts (1 amp × 11 volts) at rated current output. With the resistors in, this reduces to 5 watts, because the resistors will dissipate 6 watts.

The circuit of Fig. 13-31 uses only one transformer, but a lot of energy is wasted in the resistors required at the regulator inputs. If lower voltages could be used to drive these three regulators, the resistors would not be needed. One way to get these lower voltages is to use two transformers, as shown in Fig. 13-32.

The two transformers in Fig. 13-32 have their primaries wired in parallel. Both primaries are rated at 120 Vac rms. Transformer No. 1 has a 24-volt center-tapped, 1-amp secondary. Its output is rectified, filtered, and then regulated by a 7812 12-volt regulator.

Transformer No. 2 has a 12-volt, center-tapped, 2-amp secondary. Its output is rectified and filtered to produce one +8-Vdc output and one −8-Vdc output. These voltages are then regulated by their respective regulators into +5 Vdc and −5 Vdc. Because the input voltages to these regulators are low, no voltage-dropping resistors are needed, and no unnecessary power loss exists.

This concludes the description of the actual microcomputer hardware. The integrated-circuit parts list is shown in Table 13-5. The circuitry was broken down into manageable subsections for ease of understanding and has hopefully been an aid to the reader. The next section describes the keyboard and bootstrap routines that are used to input data to the microcomputer and initiate program executions.

KEYBOARD AND BOOTSTRAP ROUTINES

Before a computer can be useful, there must be some method, for entering programs and data and for executing the programs after they have been entered. The keyboard and bootstrap routines are used for this purpose. This program is burned into an electrically programable read-only memory (EPROM) before it is installed into the computer. The EPROM is a static memory device, which means the program stored in memory cannot be lost by turning the power off.

The flowchart for the keyboard and bootstrap routines is shown in Fig. 13-33 (Titus 1976, p. 86). The program starts by setting the stack pointer to the highest address in RAM plus one. The stack is a block of memory in RAM used to temporarily store contents of registers. Placing it at the end of the RAM keeps it separated from the working memory space. When writing programs you must avoid this area of memory space.

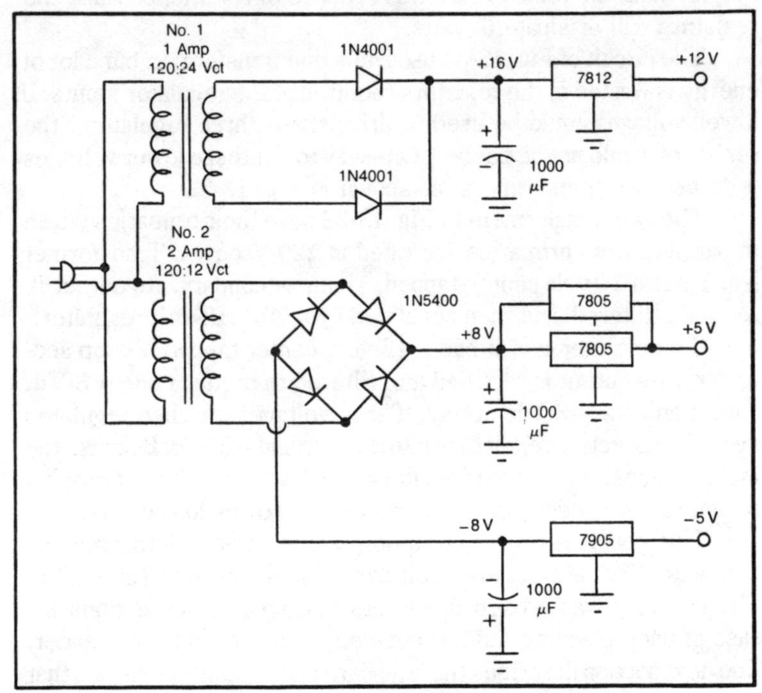

Fig. 13-32. Microcomputer power supply utilizing two transformers.

Table 13-5. Integrated Circuits Parts List for the PPDS Microcomputer.

Part No.	Noun	Brand	Qt.
8080A	Microprocessor CPU	Intel	1
8224	Clock generator and driver	Intel	1
74244	Octal buffers/line drivers	TI	4
8228	System controller and bus driver	Intel	1
74138	3-to-8 line decoder	TI	2
7427	Triple 3-input positive NOR gate	TI	1
2708	1024 × 8-bit EPROM	Intel	1
2114	1024 × 4-bit static RAM	Intel	4
74LS05	Hex inverter with open-collector outputs	Intel	1
74100	8-bit bistable latch	TI	5
7402	Quadruple 2-input positive NOR gate	TI	3
7447	BCD-to-seven-segment Decoder/driver	TI	9
SEL510	Common anode seven-segment display	Radio Shack	9
74365	Hex bus driver	TI	1
7400	Quadruple 2-input positive NAND gate	TI	1
74148	8-line to 3-line octal priority encoder	TI	2
7812	+12-volt regulator	Radio Shack	1
7805	+5-volt regulator	Radio Shack	2
7905	−5-volt regulator	Radio Shack	1

The program next addresses the first memory location in RAM, displays the address at output ports 0 and 1, and then displays the content of that memory location at output port 2. After the address and data have been displayed, the program calls the keyboard input subroutine (see Fig. 13-34) and waits for a key input from the operator.

When you push a key, its octal code is stored in register A and the program returns from the subroutine. The program must then decide whether a numeric key or an alphabetical key has been pushed. Table 13-6 contains the key codes. If an alphabetical key has been pushed, its octal code will be greater than or equal to 010. If a numeric key has been pushed, the code will be less than 010 and equivalent to the key number (i.e., key 6 will have an octal code of 006). If the key pushed was numeric, copies of the memory content and the key octal code are combined to make a new octal number

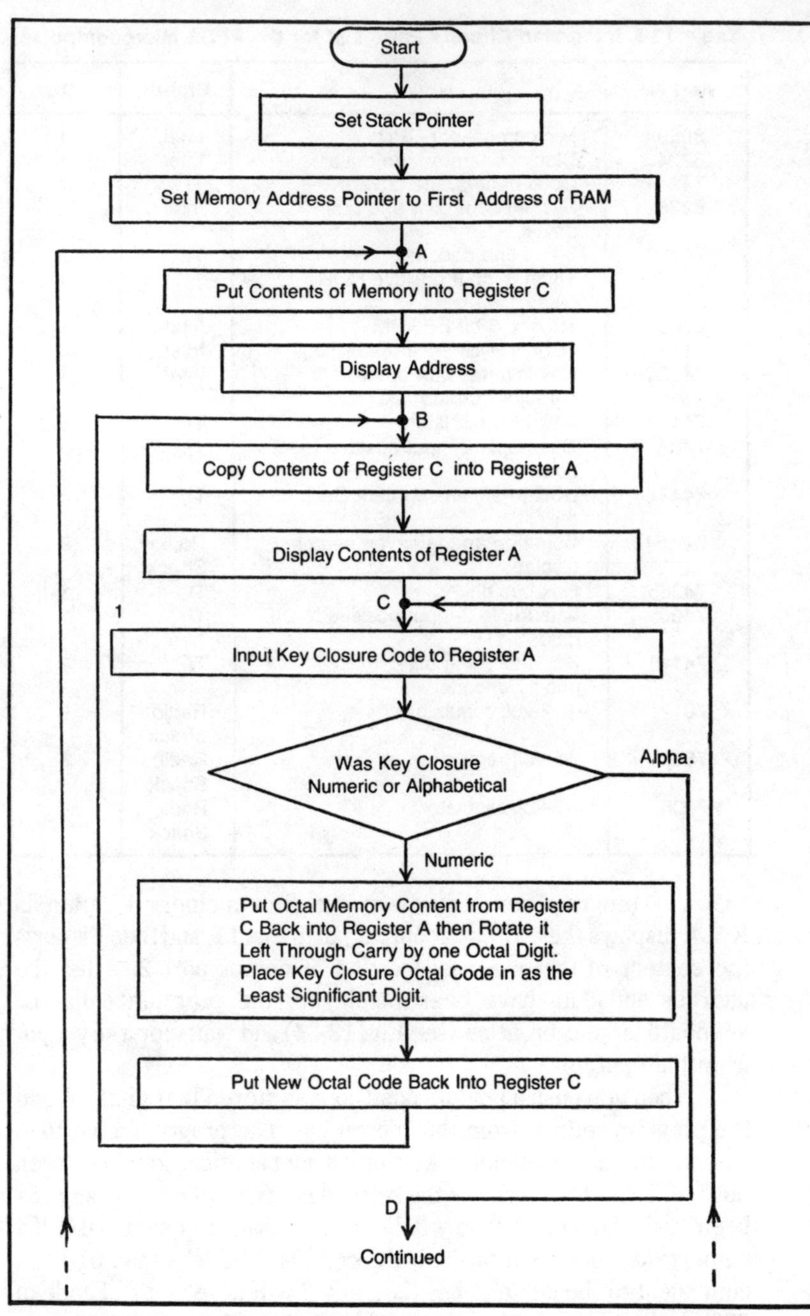

Fig. 13-33. Keyboard and bootstrap routine flowchart. (Continues through page 180.)

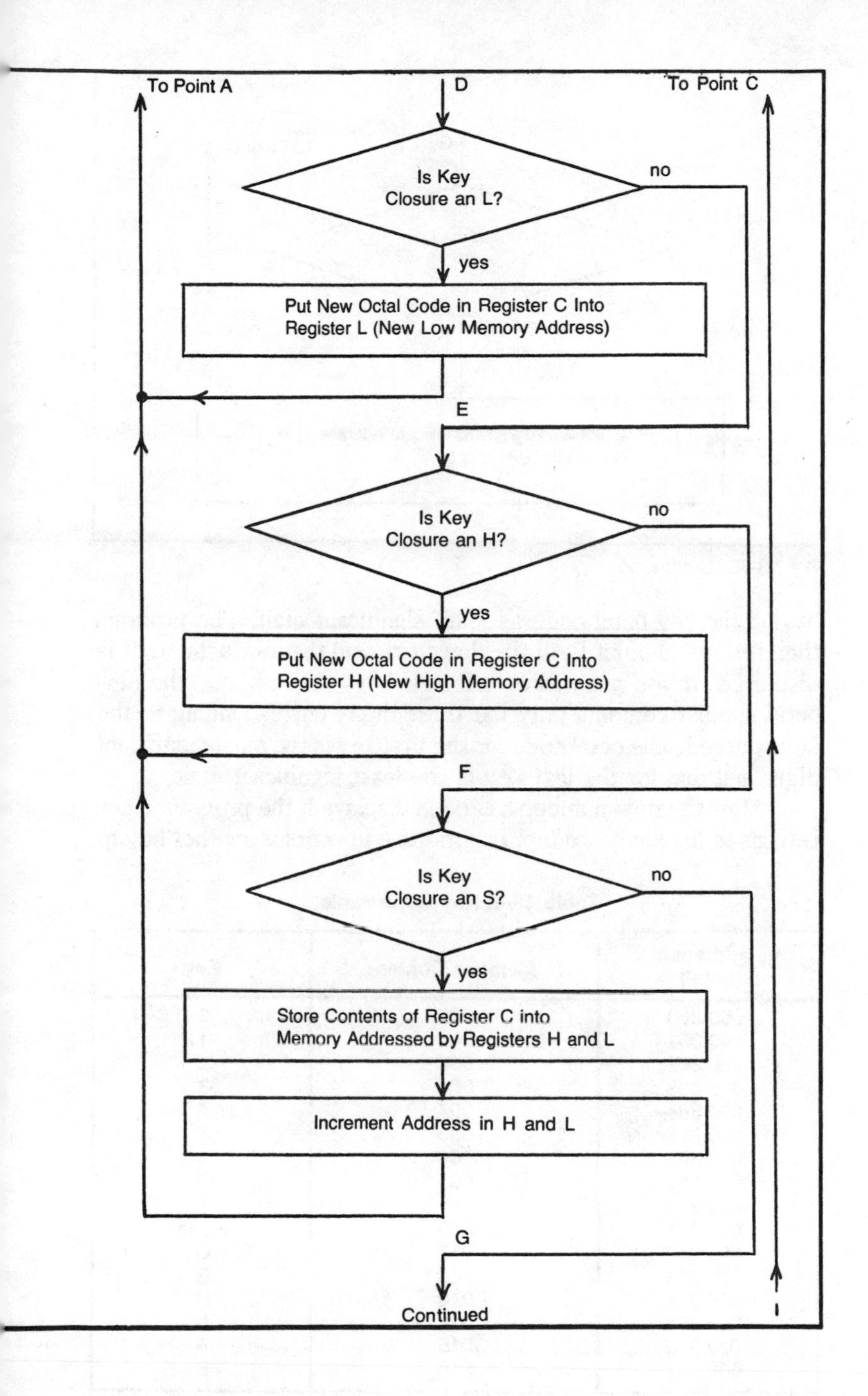
To Point A
D
To Point C
Is Key
Closure an L?
no
yes
Put New Octal Code in Register C Into
Register L (New Low Memory Address)
E
Is Key
Closure an H?
no
yes
Put New Octal Code in Register C Into
Register H (New High Memory Address)
F
Is Key
Closure an S?
no
yes
Store Contents of Register C into
Memory Addressed by Registers H and L
Increment Address in H and L
G
Continued
I

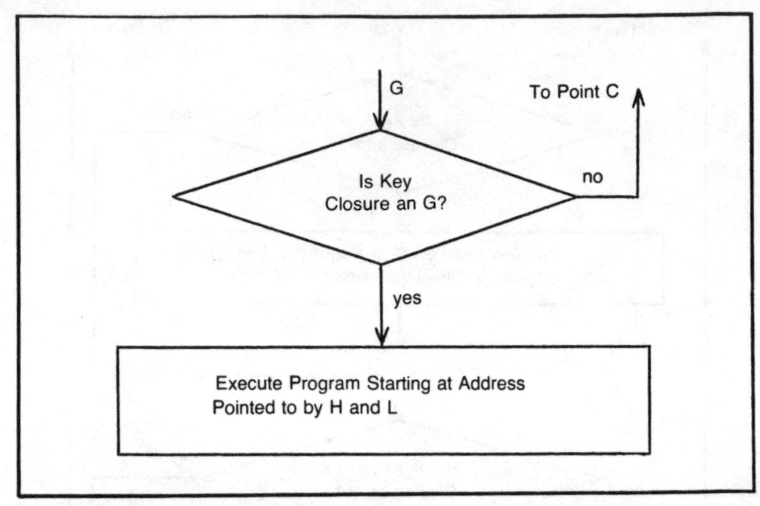

Fig. 13-33. Continued

having the key octal code as least significant digit. The program then returns to point B on the flowchart, and the new octal digit is displayed. If you push three numeric keys sequentially, the new octal number contains only the three digits corresponding to the keys pushed. The octal code for the first key is the most significant digit, and that for the last key is the least significant digit.

After the new number has been displayed, the program again returns to the keyboard input subroutine to wait for another key to

Table 13-6. Key Code Table.

Address (octal)	Memory Content	Key
000 360	000	0
000 361	001	1
000 362	002	2
000 363	003	3
000 364	004	4
000 365	005	5
000 366	006	6
000 367	007	7
000 370	013	S
000 371	000	None
000 372	017	C
000 373	012	G
000 374	010	H
000 375	011	L
000 376	015	A
000 377	016	B

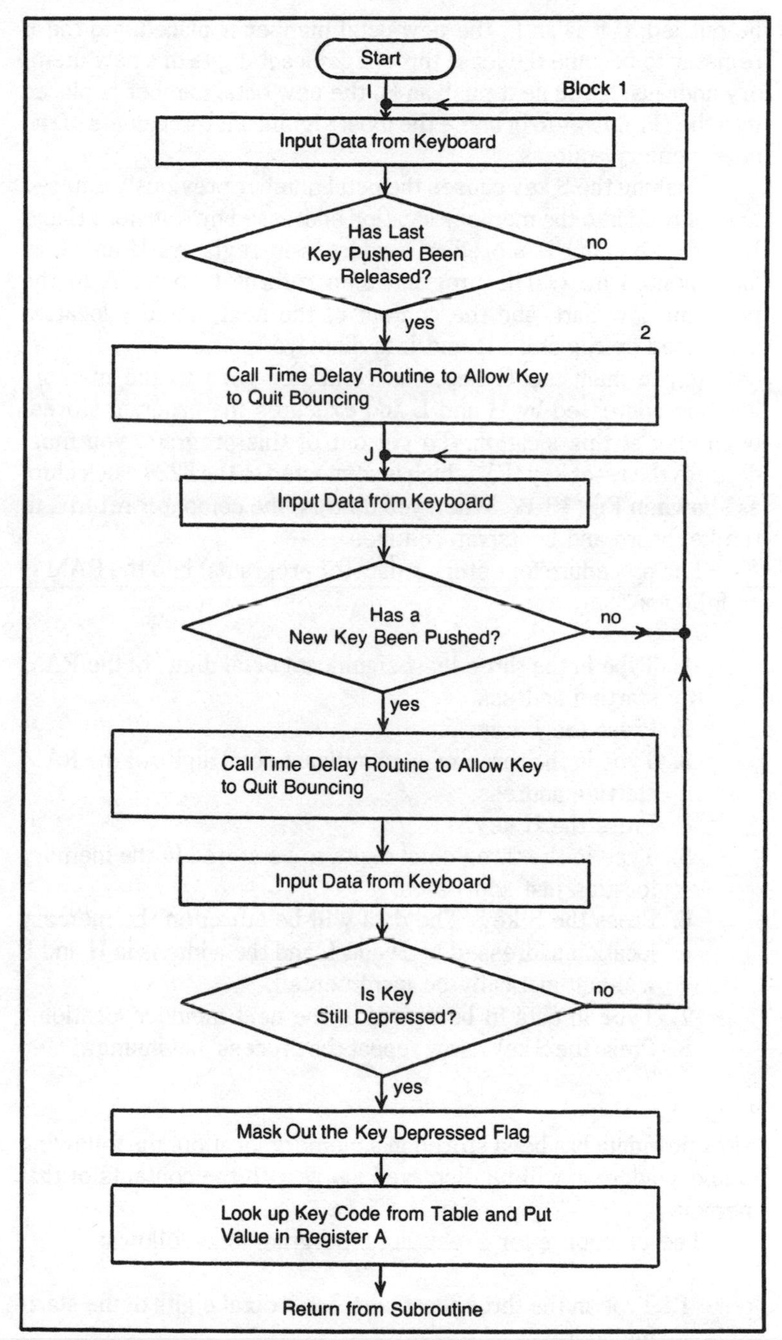

Fig. 13-34. Keyboard input subroutine flowchart.

be pushed. If it is an L, the new octal number is placed into the L register to become the least three significant digits of a new memory address. If you next push an H, the new octal number is placed into the H register to become the most significant three digits of the new memory address.

Pushing the S key causes the octal number previously entered to be stored into the memory location addressed by registers H and L. After the data is stored, the address in registers H and L is incremented by 1. The program then returns to point A in the program flowchart, and the content of the next memory location addressed by registers H and L is displayed.

If you push the G key, the computer goes to the memory location addressed by H and L and executes the program stored, beginning at this location. To get out of this program, you must depress the reset key (R), which is connected to the 8224 clock chip, as shown in Fig. 13-17. When you do this, the computer returns to the keyboard and bootstrap routines.

The procedure for entering data (or programs) into the RAM is as follows:

1. Type in the three least significant octal digits of the RAM starting address.
2. Press the L key.
3. Type in the three most significant octal digits of the RAM starting address.
4. Press the H key.
5. Type in the three octal digits to be stored in the memory location just addressed.
6. Press the S key. The data will be stored in the memory location addressed by H and L and the address in H and L will automatically be incremented.
7. Type in data to be stored in the next memory location.
8. Press the S key. Now repeat the process, beginning at step 7.

After new data has been stored in a memory location, the following memory address will be displayed along with the contents of that memory.

The procedure for executing a program is as follows:

1. Type in the three least significant octal digits of the starting address.

2. Press the L key.
3. Type in the three most significant octal digits of the starting address.
4. Préss the H key.
5. Press the G key.

The actual keyboard and bootstrap machine language program is listed in Appendix C. It is a modified version of a keyboard routine that appeared in *Radio Electronics* (Titus 1976, p. 86). This program can be stored in a 2708 EPROM using the EPROM programer described in the next section. When this is done you can build a working microcomputer.

2708 EPROM PROGRAMER

The programer described in this section was designed by Lhomond Jones especially for programing 2708 EPROMs. It is fully manually operated and programs one memory location at a time. The programing steps are time consuming and monotonous, but the price is right. The parts for this programer total only about $40 at 1982 prices.

There are three steps to follow when programing a 2708 EPROM. First, the address of the memory location to be programed must be placed on the EPROM address inputs. Next, the actual data to be stored in that memory location must be placed on the data input/output lines. A 27-Vdc programing pulse must then be input to pin 18. This pulse causes the data input to be stored in the memory location addressed.

The programing pulse width (tpw) should be 0.1-1 milliseconds wide and occur N times, where N × tpw = 100 milliseconds. The correct way to program the 2708 EPROM is to apply one programing pulse to one memory location, and then step to the next location until all locations have received one programing pulse. Repeat this process until N loops have been completed (Engineer Staff Intel 1977, p. 6-86), however, if enough time is given between each programing pulse, it is possible to apply all N pulses to one location before going to the next. This 2708 EPROM programer uses the latter technique.

Addressing Circuit

The addressing circuit for the programer is shown in Fig. 13-35. The binary address for the EPROM is supplied by two 74LS193 up-down counters. The address can be incremented by

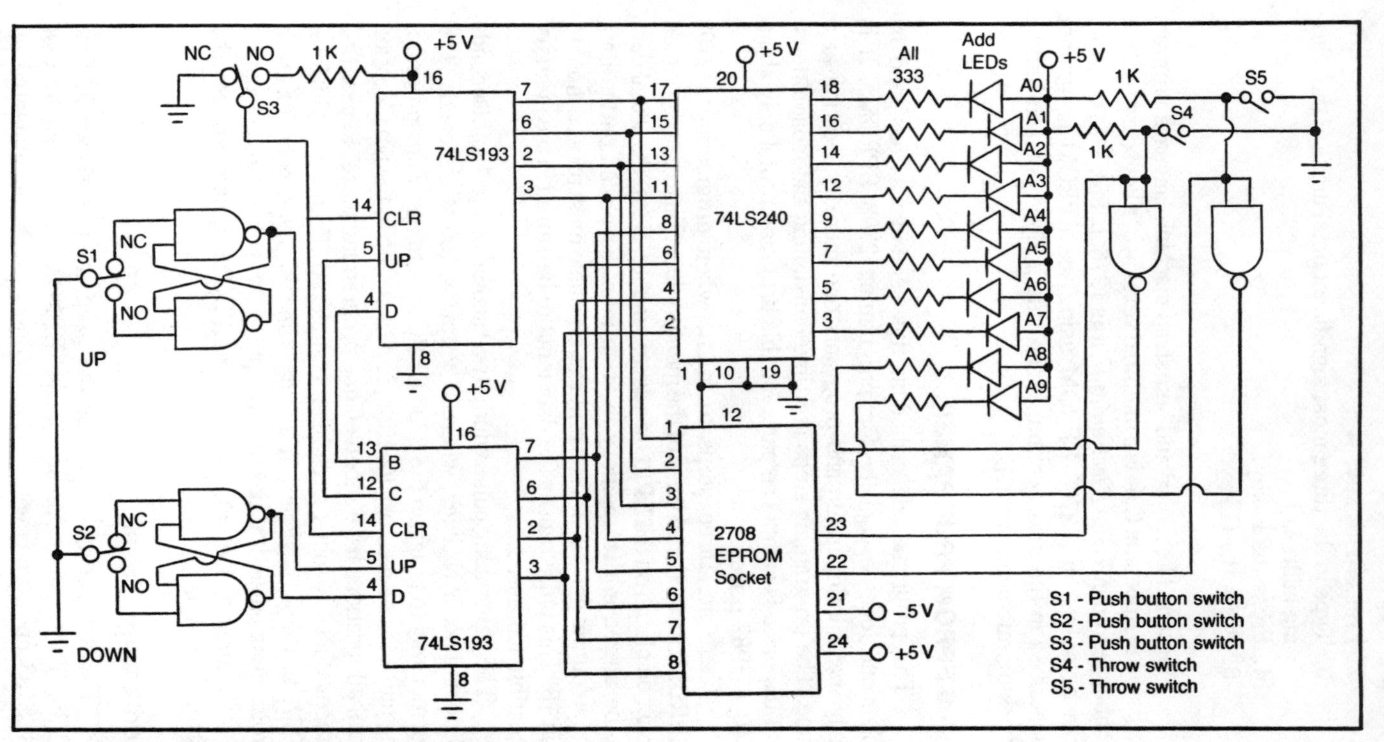

Fig. 13-35. 2708 EPROM programer addressing circuit.

depressing the pushbutton switch S1, or decremented by depressing switch S2. Pressing switch S3 clears the address and sets it to zero. The two most significant address bits are input manually by setting switches S4 and S5.

The 74LS240 inverted-output line driver is also used in the addressing circuit, along with eight resistors and eight LEDs, to give a display of the actual address being input to the EPROM by the 74LS193 counters. The two manual address switches also have LEDs to display their output.

The address, power, programing pulse, and data connections are input to a 24-pin socket. The 2708 EPROM can then be plugged into the socket for programing.

Data Input Circuit

The data input portion of the programer is shown in Fig. 13-36. When the double-pole-double-throw switch (S15) is in the up position, the data input by switches S7-S14 can be programed into the EPROM. When S15 is in the down position, the actual data stored in the EPROM is displayed by the 74LS240 and LEDs. The input switch data is applied to the EPROM through a 74LS244 noninverted-output line driver.

Programing Pulse Generator

The programing pulse is applied to the 2708 EPROM on pin 18, as shown in Fig. 13-36. The programing pulse generator is shown in Fig. 13-37. This circuit uses two NE555 timers to generate a series of 100 1-millisecond wide, 27-volt pulses.

When S6 is depressed, the voltage at point B goes low for 10 seconds (see Figs. 13-37 and 13-38). This low voltage allows the 1-millisecond wide negative pulse train present at point A to pass through the NOR gate where they are also inverted and changed to positive pulses. The two transistors, connected to the output of the NOR gate, amplify the 5-volt pulses into 27-volt pulses, which are then applied to pin 18 of the EPROM.

Power Supply

The +27 volts, along with the +12, +5, and −5 volts, are produced by the power supply shown in Fig. 13-39. The +12-, +5-, and −5-volts section is similar to the circuit in Fig. 13-31. The +27 volts is produced by a transistorized regulator that uses two transistors, three resistors, and one zener diode. The voltage at point D

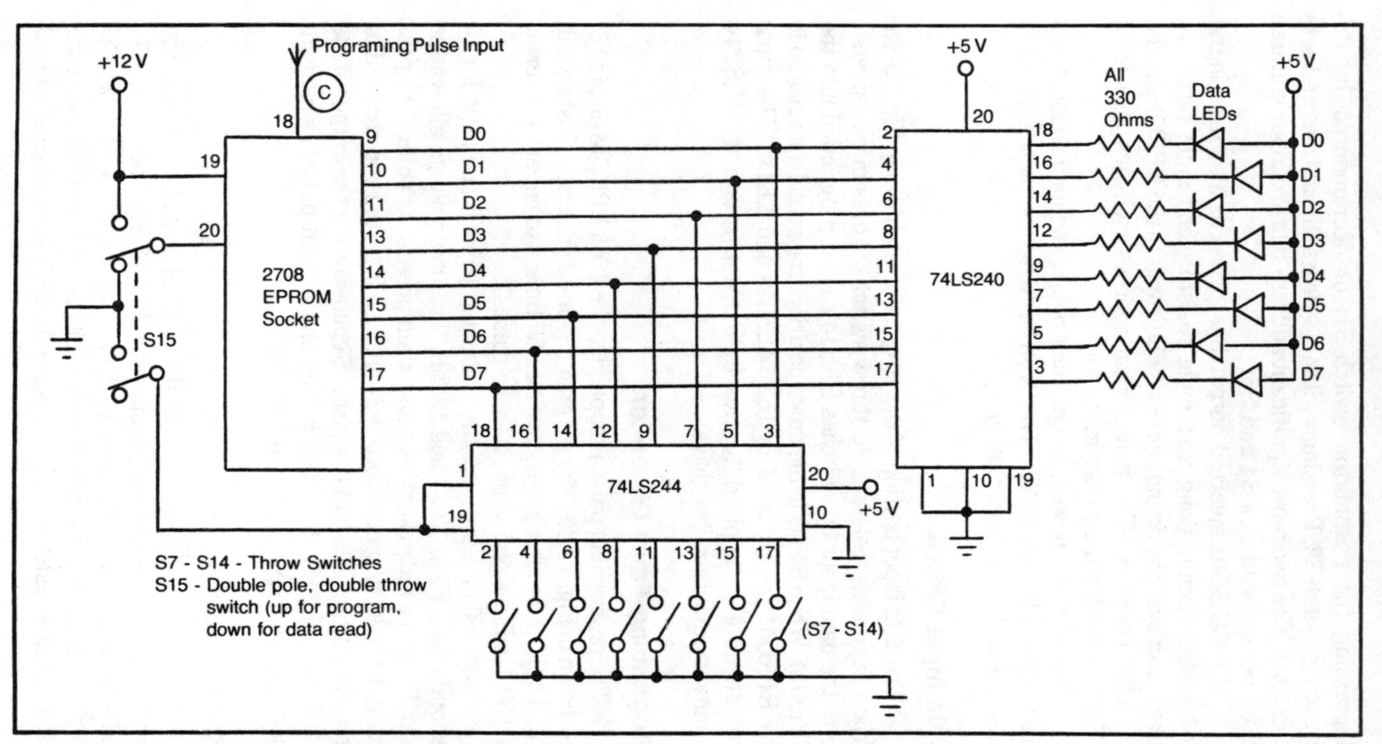

Fig. 13-36. 2708 EPROM programer data input circuit.

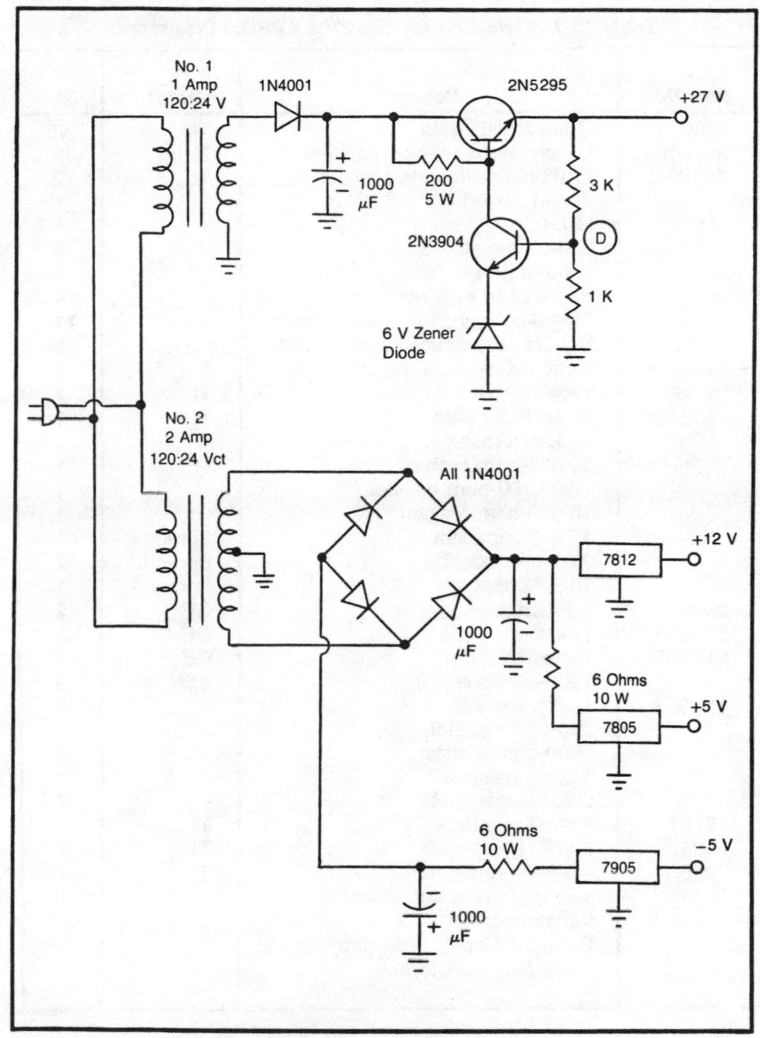

Fig. 13-39. Power supply for the 2708 EPROM programer.

in the 27-volt regulator is always equal to one-fourth the output voltage. When the output voltage equals 27 volts, the voltage at point D is 27/4 or 6.75 volts. At this time the 2N3904 transistor begins to conduct current, turns the 2N5295 transistor off, and prevents the output voltage from going higher than 27 volts.

The parts list for the 2708 EPROM programer is shown in Table 13-7 The cost of these parts should be between $30 and $50 at 1982 prices.

Table 13-7. Parts List for the 2708 EPROM Programer.

Part No.	Noun	Brand	Qt.
7400	Quad NAND gate	TI	2
74LS193	Synchronous up-down counter	TI	2
74LS240	Octal buffer (inverted output)	TI	2
	24-pin socket (for EPROM)		1
	LEDs		18
	330-ohm resistors		18
	1000-ohm resistors		7
	Pushbutton switches		4
	Single-pole-double-throw switches		11
	Double-pole-double-throw switch		1
74LS244	Octal buffer (noninverted output)	TI	1
NE555	Timers	TI	2
7402	Quad NOR gate	TI	1
	300-ohm resistor		1
	30,000-ohm resistor		1
	1,000,000-ohm resistor		1
	10,000-ohm resistor		1
	4.7-μF capacitor	Sprague	1
	0.01-μF capacitor	Sprague	2
	10-μF capacitor	Sprague	1
2N3904	Transistor	GE	2
2N3906	Transistor	GE	1
2N5295	Transistor	GE	1
	1000-μF capacitor	Sprague	3
1N4001	1-amp diodes		5
	200-ohm resistor		1
	3000-ohm resistor		1
	6-ohm resistors		2
	6-volt zener diode		1
7812	12-volt regulator	TI	1
7805	5-volt regulator	TI	1
7905	−5-volt regulator	TI	1
	1-Amp, 24-volt secondary transformer		1
	2-Amp, 24-volt, center-tapped secondary transformer		1

Operating Procedure

The following are step-by-step procedures for programing a 2708 EPROM with this programer:

1. Turn the power on and press the address clear pushbutton S3.
2. Plug the EPROM chip into the 24-pin socket, being careful not to put it in backwards.
3. Increment the address to first location to be programed if it is not zero.

4. Set S15 to program position. Data LEDs will now read data to be stored in EPROM.
5. Set input in binary form using S7-S14. Verify correct data by reading data LEDs.
6. Press the program sequence start switch (S6). One hundred, 1-millisecond program pulses will be applied to the EPROM when this switch is depressed and released. The program sequence will last for approximately 10 seconds and will be indicated by the program sequence LED shown in Fig. 13-37. When the light of the LED goes out, the sequence is finished.
7. When the programing sequence is finished, place S15 in the data-read position. The data LEDs will read the actual data content of the memory location addressed. It should correspond with the value just stored.
8. Increment the address by pressing and releasing S1.
9. Repeat the process, beginning at step 4, until last location has been programed.

Using this EPROM programer is slightly time consuming, but its low cost should more than make up for the extra time used. After its usefulness has run out for the builder, it can then be rented out to other computer enthusiasts who need their 2708 EPROMs programed.

You should now have enough information to build and operate a microprocessor-based, priority power distribution system. This will be time consuming but not overly expensive. The money you save by purchasing a smaller alternate energy power system will more than pay for the parts to build the PPDS.

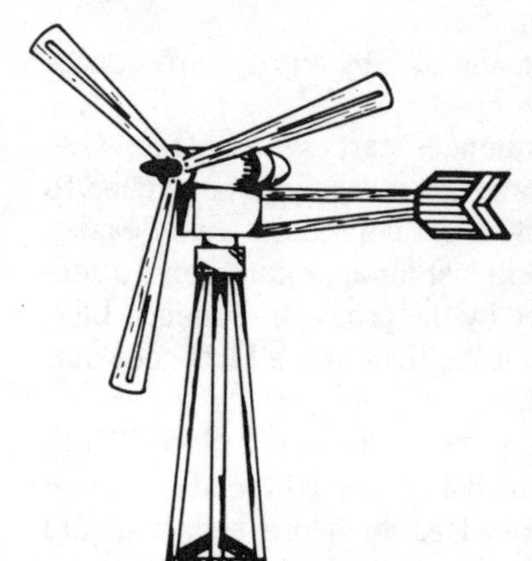

14

Circuits for the Control, Monitoring, and Support of Electric Vehicles

Most people do not realize that the electric motor has been in existence longer than the gasoline engine. There are basically two reasons why the gasoline engine has been more popular for propelling the automobile. The first—and major—reason was the low cost of gasoline. The second reason was the small energy-to-weight ratio of the early storage batteries that would have caused an automobile to have a very short range. The first argument is no longer valid, and the second is rapidly becoming invalid.

With the cost of gasoline gradually increasing, it is no longer cheaper to drive a gasoline-powered vehicle than an electric-powered vehicle. Even though the cost of utility power is also increasing, it is not increasing at the same rate as gasoline. This leaves only one remaining argument—the short range caused by the inability of present day batteries to store sufficient energy for sustained driving. If current research continues on storage batteries, new batteries may soon be developed that will give electric cars a range somewhat compatible with gasoline automobiles.

Recently, many experimenters have been building hybrid electric cars with advantages of both gasoline and electric automobiles. This type of car is powered by an electric motor that receives its power from a small battery bank and an on-board gasoline-powered generator, which provides the increased range while the battery bank provides backup power for acceleration. A major advantage of this type of system is that the gasoline engine can run at a constant rpm chosen for maximum efficiency. Some people have reported achieving more than 100 miles per gallon with these hybrid electric cars (for more information on hybrid electric cars write to *The Mother Earth News*, P.O. Box 70, Hendersonville, North Carolina 28739).

BASIC ELECTRIC-VEHICLE THEORY

An *electric vehicle* has three basic components: the battery bank, the dc electric motor, and the speed control system. Figure 14-1 shows the functional block diagram of a typical system, along with voltmeter, ammeters, and a couple of optional features.

The battery bank can contain any number of batteries connected in series, parallel, or series-parallel combinations, however, the total voltage and maximum current capacity must match, or be close to, the ratings of the motor. For example, if a 36-volt, 400-amp motor is to be used and there are six 12-volt, 200-amp batteries available, the battery configuration of Fig. 14-2 could be used. Because voltage adds when batteries are in series and current adds when batteries are in parallel, this configuration will produce 36 volts and have a maximum current of 400 amps. Also notice in Fig. 14-2 that the original 12-volt battery for the car electrical system is connected to the battery bank through a diode. Thus, it can be charged along with the battery bank, but it will not discharge into the motor. A voltmeter across the bank will give an indication of the charge level of the batteries. This is analogous to the fuel-level gauge in a gasoline-powered vehicle.

As an optional feature, Fig. 14-1 also shows an on-board generator. This can be powered by a small gasoline engine. Because some dc generators will also operate as motors, you might need to put a diode in series with the output of the generator to prevent current from reversing through the generator and motorizing it. You can also place an ammeter in series with the output of the generator to give an indication of how much current is supplied by the generator.

The heart of an electric vehicle is the dc motor. Obviously, an ac motor should not be used because there is no way to store alternating current power, and converting dc to ac will waste valuable energy. The three types of dc motors available are *series-wound, shunt-wound,* and *compound-wound.* These are discussed later.

The power rating of the dc motor chosen should be adequate to propel the automobile at the desired maximum speed. A good

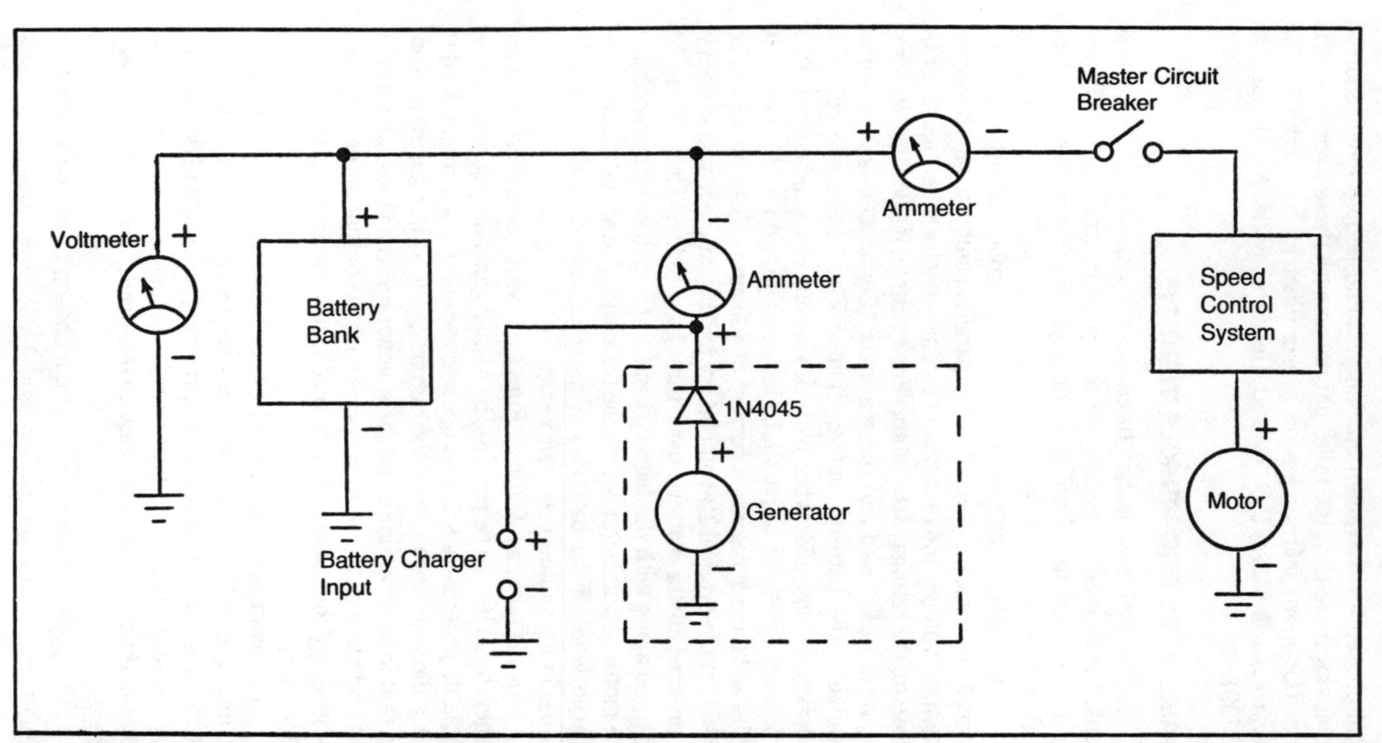

Fig. 14-1. Basic electrical diagram of a typical electric vehicle.

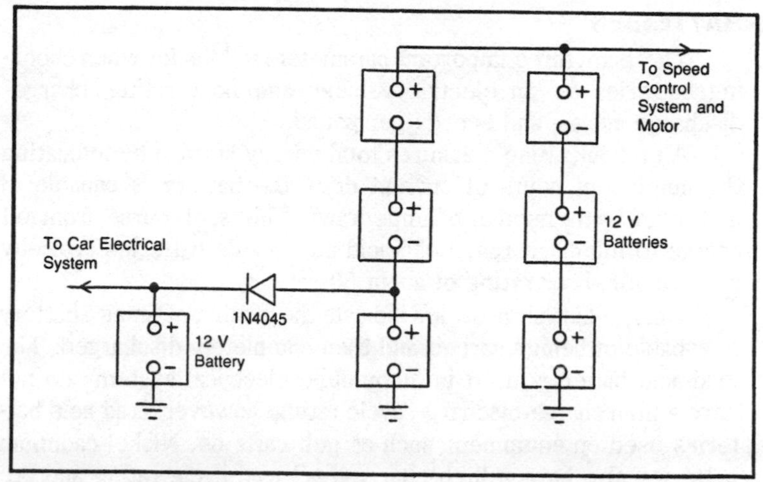

Fig. 14-2. Typical battery-bank configuration for an electric vehicle.

general rule is to have at least 1 horsepower for every 200 pounds of car. If the aerodynamics of the car is good, this should give the car a maximum speed of 55 miles per hour.

Also, the rpm range of the motor should be matched closely to the rpm of the transmission shaft when driving at normal speeds. The following formula gives the relationship between motor rpm and vehicle speed in miles per hour:

$$\text{Vehicle mph} = \text{Motor rpm}\left[\frac{\text{Wheel radius in feet}}{\text{Gear ratio of transmission}}\right] \times 0.0714 \qquad \textbf{Eq. 14-1}$$

If the wheel radius is 1 foot, the gear ratio of the transmission is 3, and the motor is rated at 3600 rpm, the vehicle's top speed should be 85.68 miles per hour if the motor has sufficient power rating. When driving at 55 miles per hour, the motor in this car would be turning at 2311 rpm. This is well within the safe operating rpm of the motor, but may or may not be the most efficient rpm for operation of the motor. This depends on the particular motor used.

Naturally, all electric vehicles must have some sort of speed control system. Figure 14-1 shows the speed control system placed between the motor and battery bank. It could, however, just as easily be placed between the motor and ground. There are four types of speed control techniques: *battery switching, resistor switching, shunt field control*, and *solid-state pulsing.* All of these techniques are discussed later. The following sections discuss in more detail the three basic components of an electric vehicle.

BATTERIES

There are three important parameters to look for when choosing batteries for an electric vehicle: amp-hour rating, charge-discharge cycles, and energy per pound.

Amp-hour rating measures total energy storage by indicating the number of hours of current draw the battery is capable of multiplied by the number of amps drawn. This is, of course, from full charge to full discharge. Lead-acid automobile batteries typically have an amp-hour rating of about 50.

Charge-discharge cycles indicate the number of times a battery is capable of being charged and then completely discharged. The lead-acid batteries used in automobile electrical systems do not have a high charge-discharge cycle rating; however, lead-acid batteries used on equipment such as golf carts do. Nickel-cadmium batteries also have a high charge-discharge cycle rating and are easier to maintain because, unlike the lead-acid type, they are dry cell batteries, but they are very expensive.

For increased range, an electric vehicle requires a lot of energy storage and as little weight as possible. Therefore, choose batteries with a large energy-per-pound rating. Energy per pound can be found by multiplying amp-hour rating by battery voltage and dividing by battery weight. Batteries typically range from 10 to 20 watt-hours per pound, where watts is defined as volts × amps.

You must carefully choose the batteries to be used on an electric vehicle. Batteries are a major cost in the total budget and can be a major factor in the failure of the electric vehicle project if chosen incorrectly.

DC MOTORS

As stated earlier the three types of dc motors are series-wound, shunt-wound, and compound-wound (Fitzgerald, Kinsley, and Kusko 1971, p. 266). Figure 14-3 shows the wiring relationship of the field coil and armature for these motors. The field coil and armature are in series for the series-wound motor and in parallel for the shunt-wound motor. The compound motor, however, has both a series and a parallel field coil. All three types have their own set of characteristics, some of which are advantageous for electric vehicle use, and some of which are not.

The series-wound motor has a drooping speed versus load characteristic and a torque proportional to the square of the current. These characteristics give the electric vehicle much the same feel as a gasoline engine. It also has excellent starting torque, while

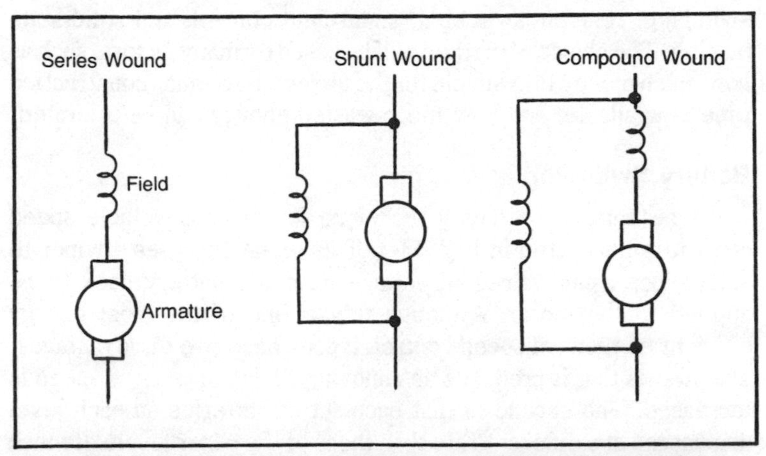

Fig. 14-3. Electrical diagram for series-wound, shunt-wound, and compound-wound dc motors.

most small gasoline engines do not. The only disadvantage of the series-wound motor is that the no-load speed is dangerously high. This is not a problem unless the original clutch is left in the vehicle. If this is the case, the motor must never be clutched with the accelerator pad depressed.

The shunt-wound motor speed tends to remain nearly constant as load is added, and torque is almost proportional to armature current. This motor is not normally recommended for electric vehicle use because of its low starting torque. The one advantage of the shunt-wound motor is that speed can be controlled over a wide range by varying the relatively small current in the field coil. Speed is maximum when field current is minimum and minimum when field current is maximum. This field current can be controlled by simply placing a large potentiometer in series with the field coil.

The compound-wound motor has the advantages of both the series- and shunt-wound motors, but neither of the disadvantages. The no-load speed is limited by the shunt field, yet it still retains the good starting torque. Its speed can also be controlled to a certain extent by using a potentiometer in series with the shunt field similar to that used on the shunt-wound motor. The compound-wound motor is definitely recommended for electric vehicle use, but may require an additional speed control system, four of which are discussed in the following section.

SPEED CONTROL TECHNIQUES

As stated before, the four types of speed control are battery switching, resistor switching, shunt-field control, and solid-state pulsing. The choice of method will be based on many factors, such as how much money is available for the project, how much construction time is available, and how much wasted energy can be tolerated.

Battery Switching

One battery switching technique for electric vehicle speed control is illustrated in Fig. 14-4. This technique uses a wiper to select increasing values of voltage from the battery bank to be applied to the motor. Although this is one of the most energy efficient methods of speed control, it does have two disadvantages. The first is that it produces an annoying series of jerks as speed is increased. The second is that each set of batteries on each level discharges at a different rate than those of the other levels. If much slow city driving is done, the battery connections will have to be changed very often to allow all the batteries to discharge at an even rate.

The wiper in Fig. 14-4 can be connected to the original accelerator pedal. The contact plates can be made from copper and recessed in wood so the surfaces of the wood and copper are flush. The copper plates must be spaced far enough apart so the wiper cannot make contact with two plates at the same time. If two plates are contacted at the same time, one set of batteries will be shorted and fireworks will follow. This wiper arm assembly is simple to build, and if only highway driving is done, its simplicity makes this speed control system an obvious choice.

There is another battery switching technique that does not have the disadvantage of uneven battery discharge. This technique is illustrated in Fig. 14-5. Depending on which set of switches is closed, this battery bank will apply 12 volts, 24 volts, or 48 volts to the motor. Table 14-1 lists the switch closures for each of the output voltages.

The only problem now is how to get only these switches closed at any particular time. Figure 14-6 shows how this can be done with a wiper arm similar to the previous one and a series of relays. The diodes are placed in the circuit to prevent current from flowing into relays not selected.

There are three disadvantages to this battery switching technique. First, it requires 11 high-current relays, which are very expensive; however, if no more than 200 amps will be drawn, you can use automobile starter solenoids, which you buy can at a local salvage yard for a small fraction of the retail cost. Second, this

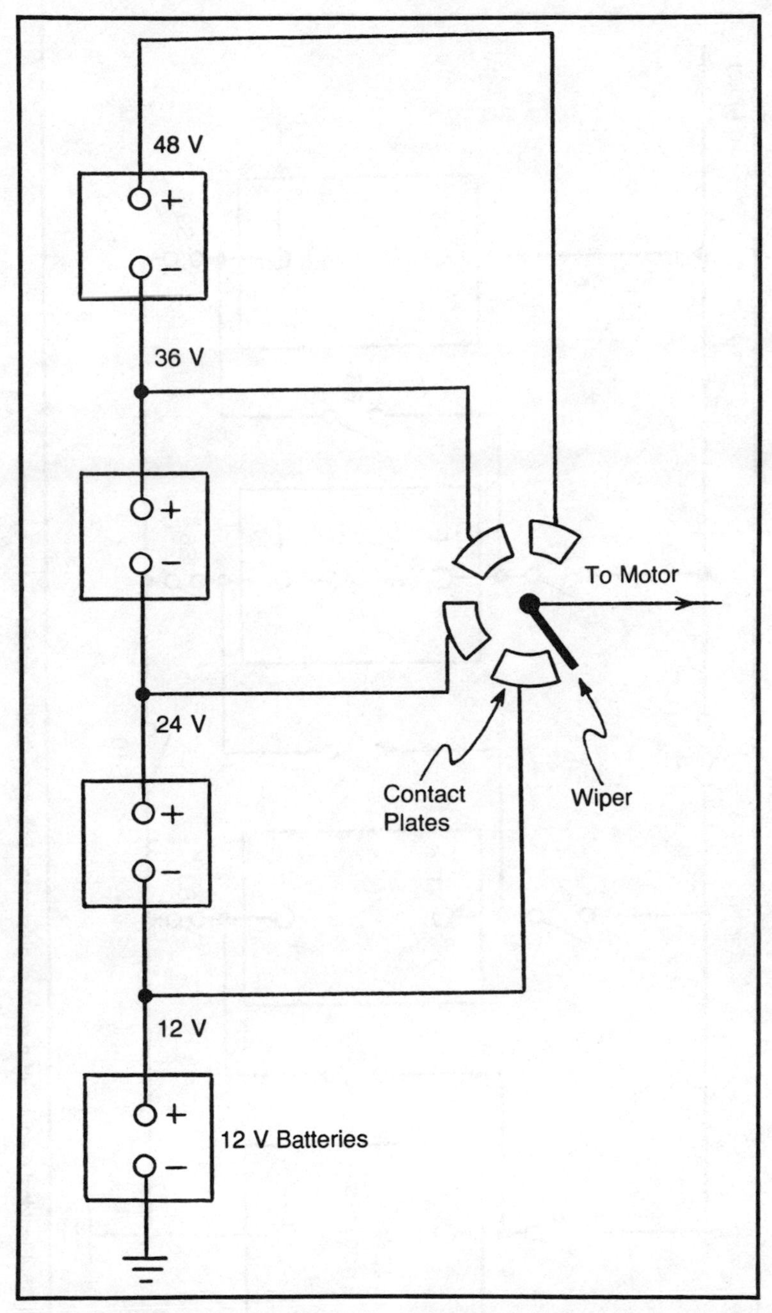

Fig. 14-4. Electromechanical diagram of a simple battery switching technique of speed control.

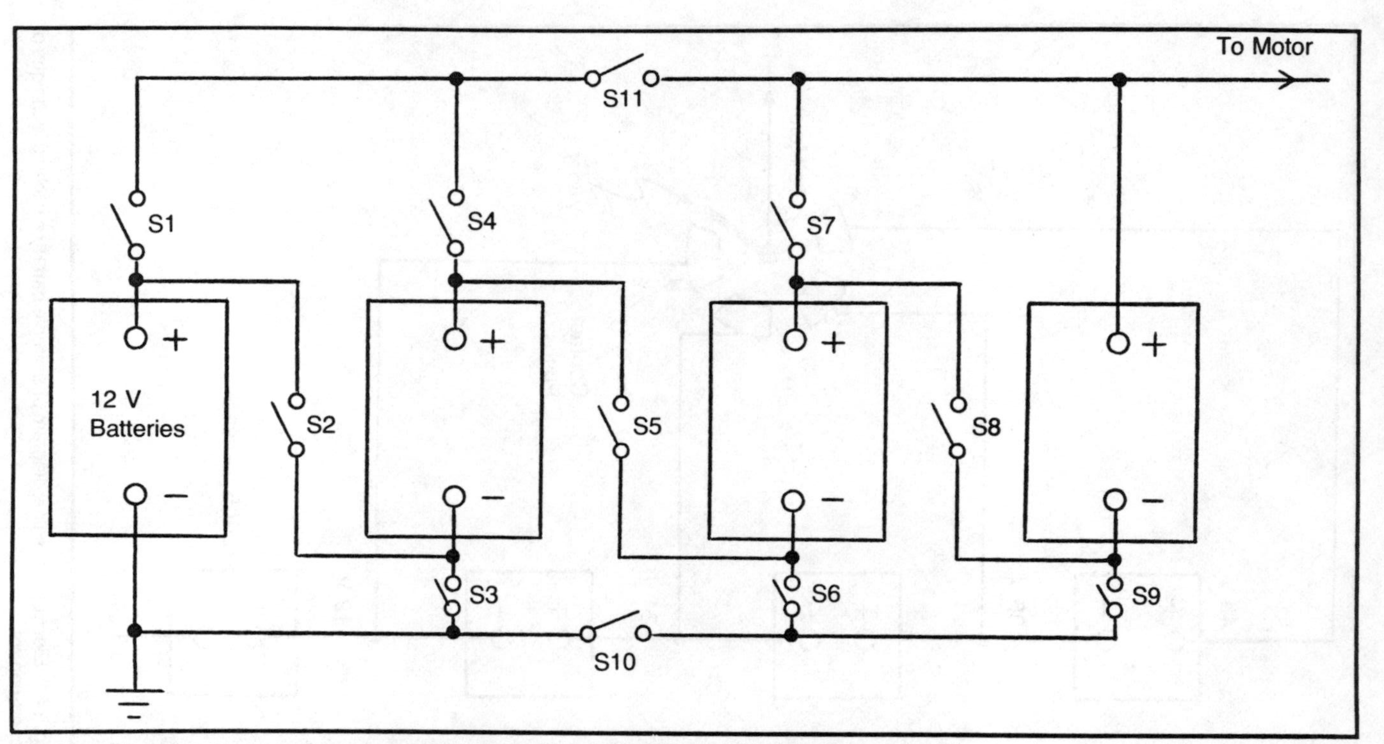

Fig. 14-5. Electrical diagram of a more sophisticated battery switching, speed control system.

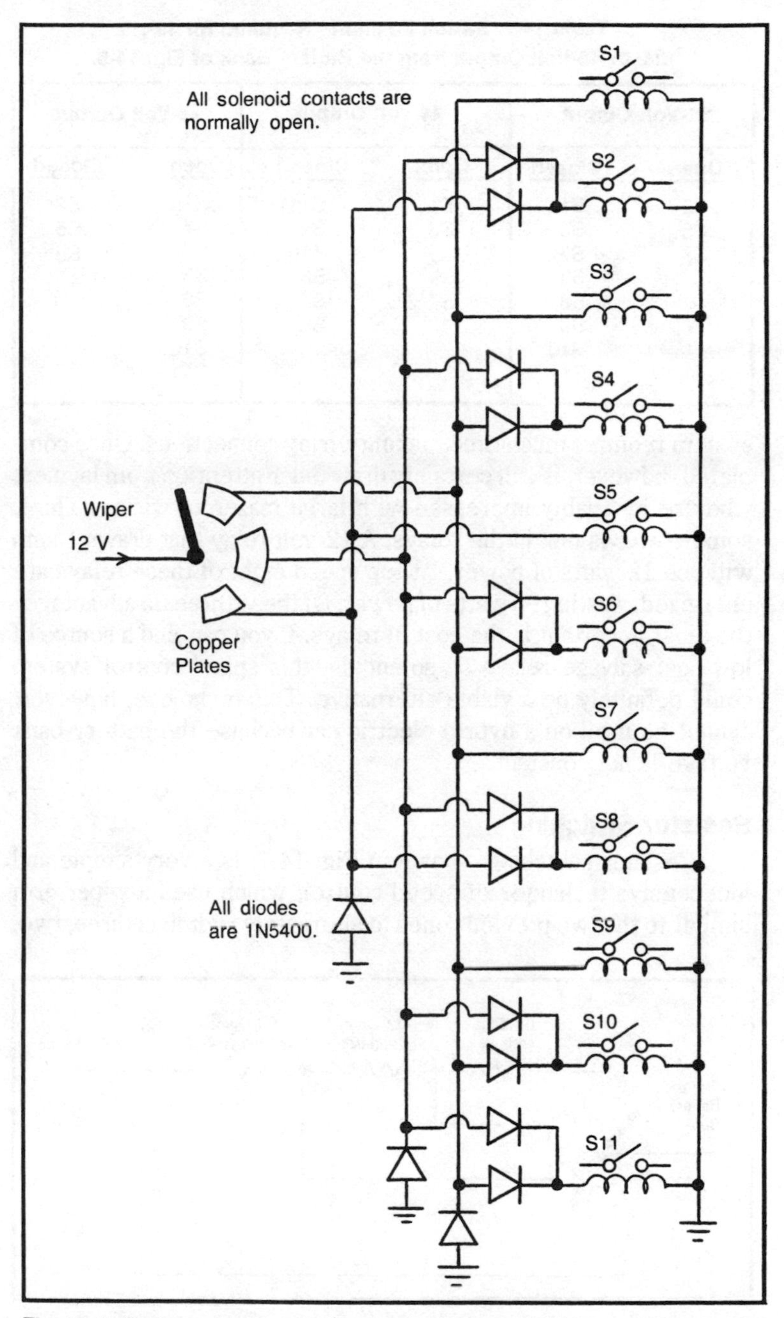

Fig. 14-6. Switch selection circuitry for the battery switching speed controller shown in Fig. 14-5.

Table 14-1. Switch Positions Required for 12-, 24- or 48-Volt Output from the Battery Bank of Fig. 14-5.

12-Volt Output		24-Volt Output		48-Volt Output	
Open	Closed	Open	Closed	Open	Closed
S2	S1	S1	S11	S1	S2
S5	S4	S3	S2	S4	S5
S8	S7	S5	S10	S7	S8
	S3	S9	S8	S3	
	S6	S7	S4	S6	
	S9		S6	S9	
	S10			S10	
	S11			S11	

system requires much time making wiring connections. Once completed, however, it will certainly draw much attention from laymen, who are invariably impressed with large mazes of wiring. Third, some power is lost in the relays. A 12-volt relay that draws 1 amp will use 12 watts of power. At top speed eight of these relays are energized, wasting 96 watts of power. Of these three disadvantages the most important is the cost of relays. If you can find a source of low-cost salvage relays or solenoids, this speed control system could definitely be a viable alternative. This technique, however, cannot be used on a hybrid electric car because the battery-bank voltage is not constant.

Resistor Switching

Resistor switching, shown in Fig. 14-7, is a very simple and inexpensive technique of speed control, which uses a wiper arm similar to the two previous ones to alternately switch in three, two,

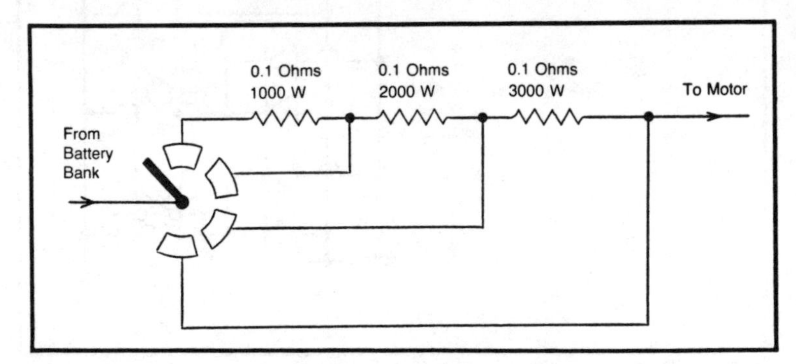

Fig. 14-7. Electromechanical diagram showing the resistor switching technique of speed control.

one, or zero resistors between the battery bank and the motor. The size and power rating of these resistors depends on the current required to operate the motor. The resistors in Fig. 14-7 were chosen for a maximum current of 400 amps and a battery-bank voltage of 36 volts.

Finding resistors of this size may be difficult, however, they can be made from any type of wire. The wire should be rated at about one-half the current to be conducted. After cutting a length of wire with the required resistance, coil it and mount it in a well-ventilated location. For example, if you use 0.01-ohms-per-foot wire, then you need a 10-foot section. A larger diameter wire should be used for the resistor requiring higher power rating. Because wire resistance is inversely related to diameter, a longer section must be used with larger wire.

The disadvantage of using the resistor switching technique is that a lot of energy is wasted in the resistors. as much as 3000 watts (4 horsepower) can be dissipated here. Of course, if most of the driving is done at top speed there will be little energy wasted. Also, during winter months, air can be blown across these resistors and circulated into the car for heating. As before, the adequacy of this speed control technique is dependent on the particular use of the vehicle and the simplicity desired.

Shunt Field Control

If the motor being used is shunt-wound or compound-wound, a potentiometer placed in series with the shunt field can be used for speed control over a certain range. Figure 14-8 illustrates this method.

For this system both the switch and potentiometer in Fig. 14-8 are connected to the accelerator pedal. When the accelerator pedal is depressed, the switch closes and the wiper arm begins to move

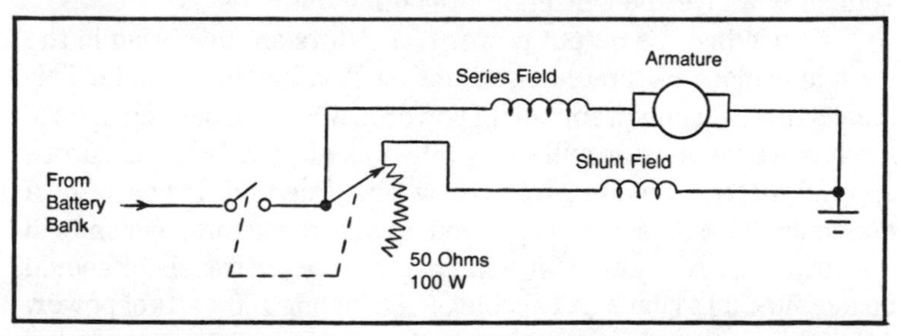

Fig. 14-8. Electromechanical diagram showing the shunt field control method of speed control.

from minimum resistance toward maximum resistance. There will be a jerk when the accelerator pedal is first depressed because the shunt field resistance cannot regulate speeds at the low end. This is also a simple and inexpensive speed control technique, but it cannot be used on a series-wound motor.

Solid-State Pulsing

The solid-state pulsing technique can be used on any type of dc motor and provides a smooth regulation over almost the full range of speeds. The circuit diagram for the solid-state pulser is shown in Fig. 14-9, with speed being controlled by varying the resistance of potentiometer R4.

The output voltages (V1 and V2) of both 555 timer circuits in Fig. 14-9 are shown in Fig. 14-10. Output voltage V1 is a series of 0-volt pulses occurring at a rate of 144 pulses per second. Output voltage V2 is a series of positive pulses also occurring at a rate of 144 pulses per second, but having a variable width (t3). The width of this pulse is controlled by varying potentiometer R4, with a minimum value of 0.00002 seconds and a maximum valve of 0.0069 seconds or tl+t2. When voltage (V2) is high, it turns on the output power transistors and applies current to the motor. When the width of the positive pulse of V2 is minimum or 0.00002 seconds, current is applied to the motor 0.00002 seconds out of every 0.0069 seconds or 0.29 percent of the time. This represents the minimum-speed position of R4 and is analogous to the idle on a gasoline-powered vehicle. When the width of the positive pulse of V2 is maximum or 0.0069 seconds, current is applied to the motor 0.0069 seconds out of every 0.0069 seconds or 100 percent of the time. This represents the full-speed position of potentiometer R4. Be careful not to exceed this position on potentiometer R4. If this happens, the width of the positive pulse of V2 will suddenly drop to one-half, and the vehicle will give the sensation of stalling out.

Notice that the output power transistors are operating in the switching mode (saturated or off) rather than the linear mode. This means that a minimum amount of power is wasted in the transistors. Each power transistor will carry 50 amps of current and dissipate approximately 50 watts of power when turned on. If you want a maximum of 400 amps, you need eight transistors, which will dissipate approximately 400 watts of power. Each transistor should be mounted on a heat sink capable of dissipating 100 watts of power. Also, the wire connecting the output power transistors in parallel should be the same length. At these high currents, wire resistance

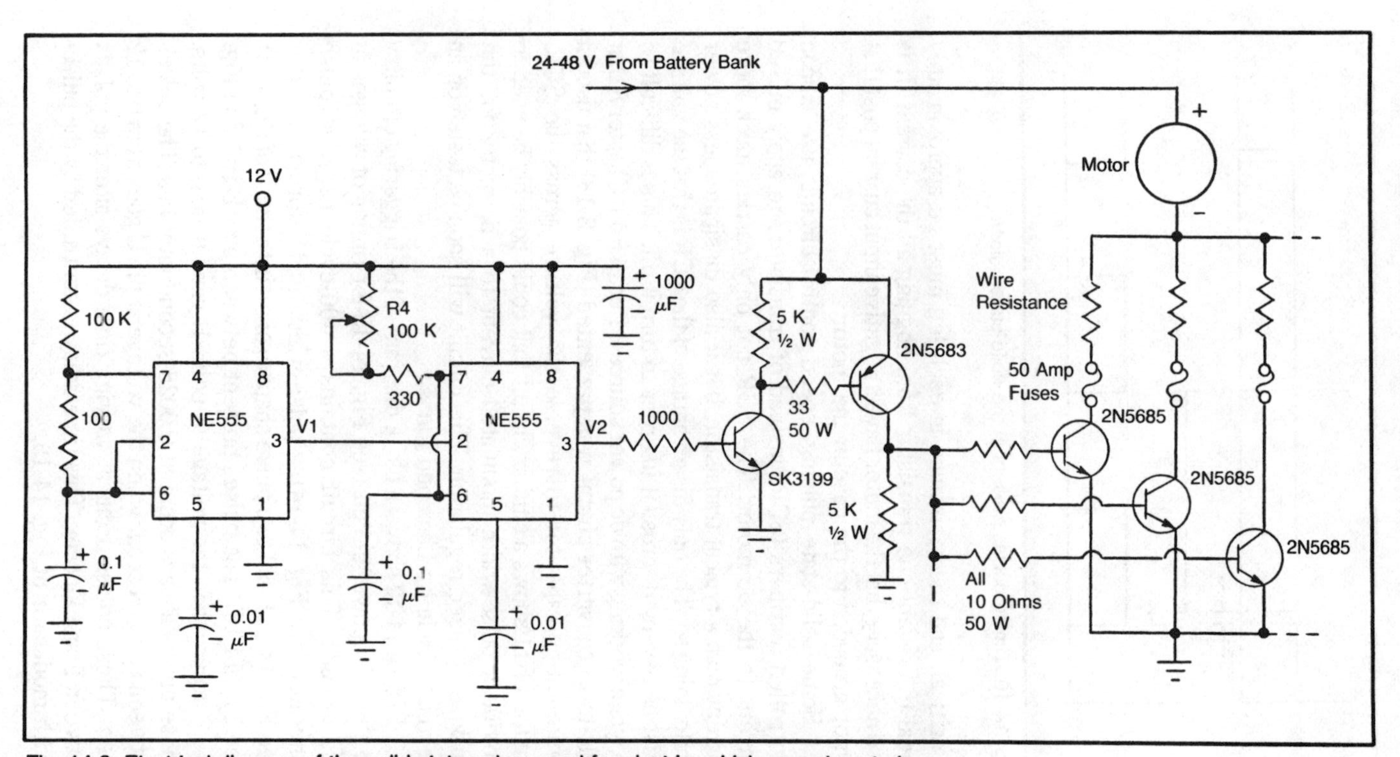

Fig. 14-9. Electrical diagram of the solid-state pulser used for electric vehicle speed control.

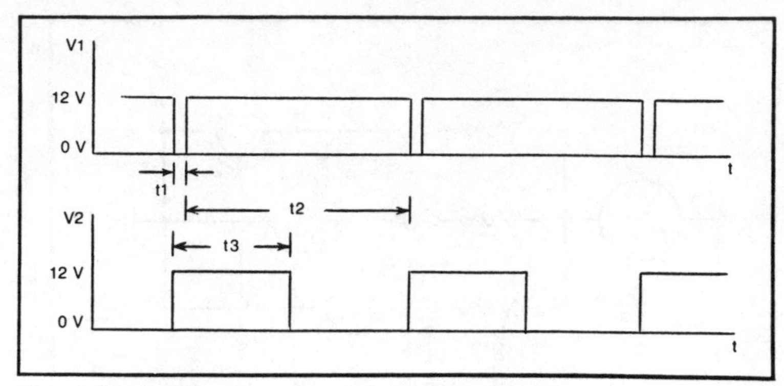

Fig. 14-10. Timing pulses present in the solid-state pulser.

is critical, and the resistance in each leg must be approximately equal to prevent one transistor from hogging all the current. It is also necessary that the motor have a negative input current post that is not grounded to the case of the motor.

Some solid-state pulsing, speed control circuits use silicon controlled rectifiers (SCR) instead of transistors to apply pulsed current to the dc motor. The SCR not only carries much more current than a typical transistor, but it also dissipates less power while doing so. The only disadvantage of the SCR is that the voltage must be reversed across it in order to turn it off. This is difficult to do when working with dc power sources. Recall from Chapter 8 that a dc-to-ac converter circuit was presented (Fig. 8-18) that used a commutating capacitor to reverse the voltage across the SCR. Figure 14-11 shows a similar circuit that could possibly be used to provide a pulse-width modulated current to a dc motor for the purpose of speed regulation. This circuit will soon be tested on an electric vehicle by Lhomond Jones.

Figure 14-12 shows a block diagram of the triggering circuitry for the SCR power circuit (see Fig. 14-13 for timing of voltages in this circuit). This circuit contains a 400-cycle negative pulse generator (see Fig. 14-14), a pulse-width modulator and inverter (see Fig. 14-15), and two pulse shapers similar to those of Fig. 8-21 (see Fig. 14-16). The pulses present between pins E and F in Fig. 14-12 occur when the voltage at point B goes from 0 to 12 volts. These pulses always occur at 0.0025-second intervals. The pulses between G and H occur when the voltage at pin C goes from 0 to 12 volts. These pulses occur at varying time delays after the pulses between E and F occur. This time delay is controlled by the pulse-width modulator of Fig. 14-15.

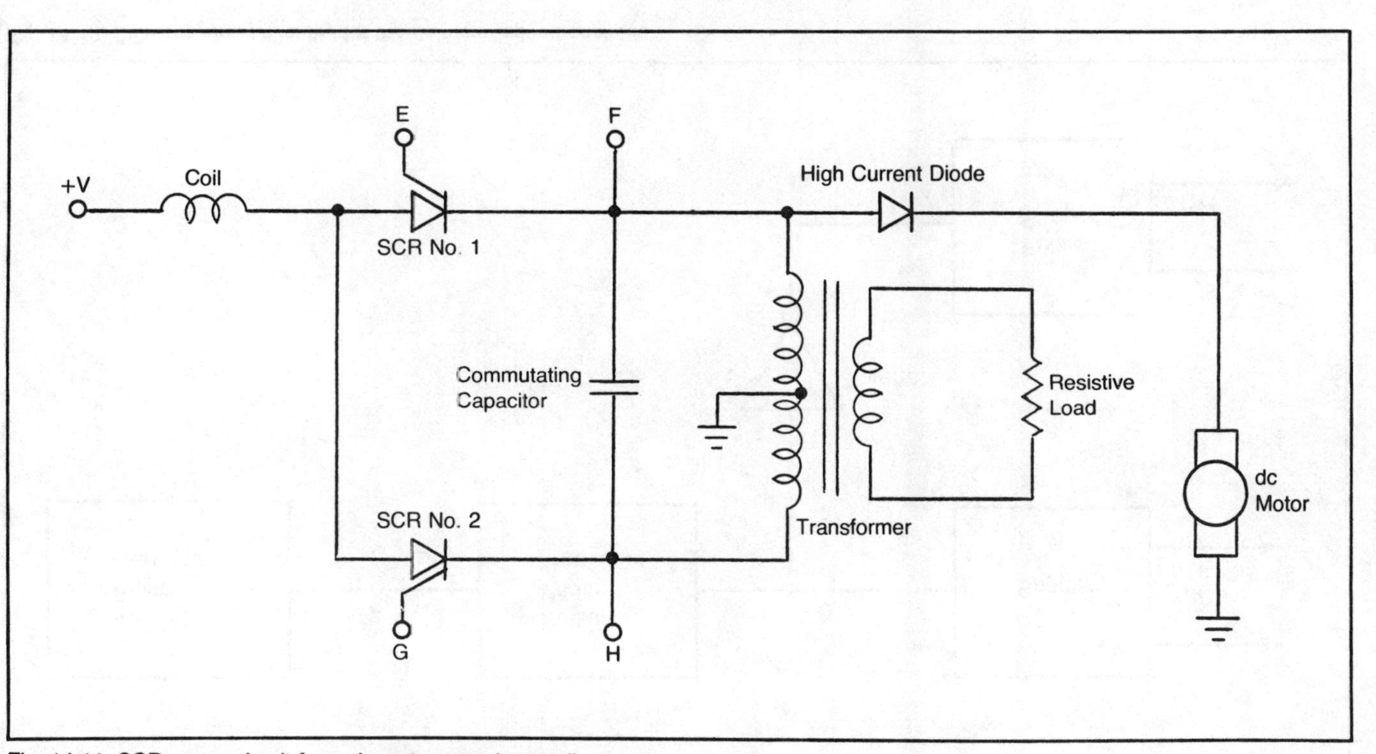

Fig. 14-11. SCR power circuit for a dc motor speed controller.

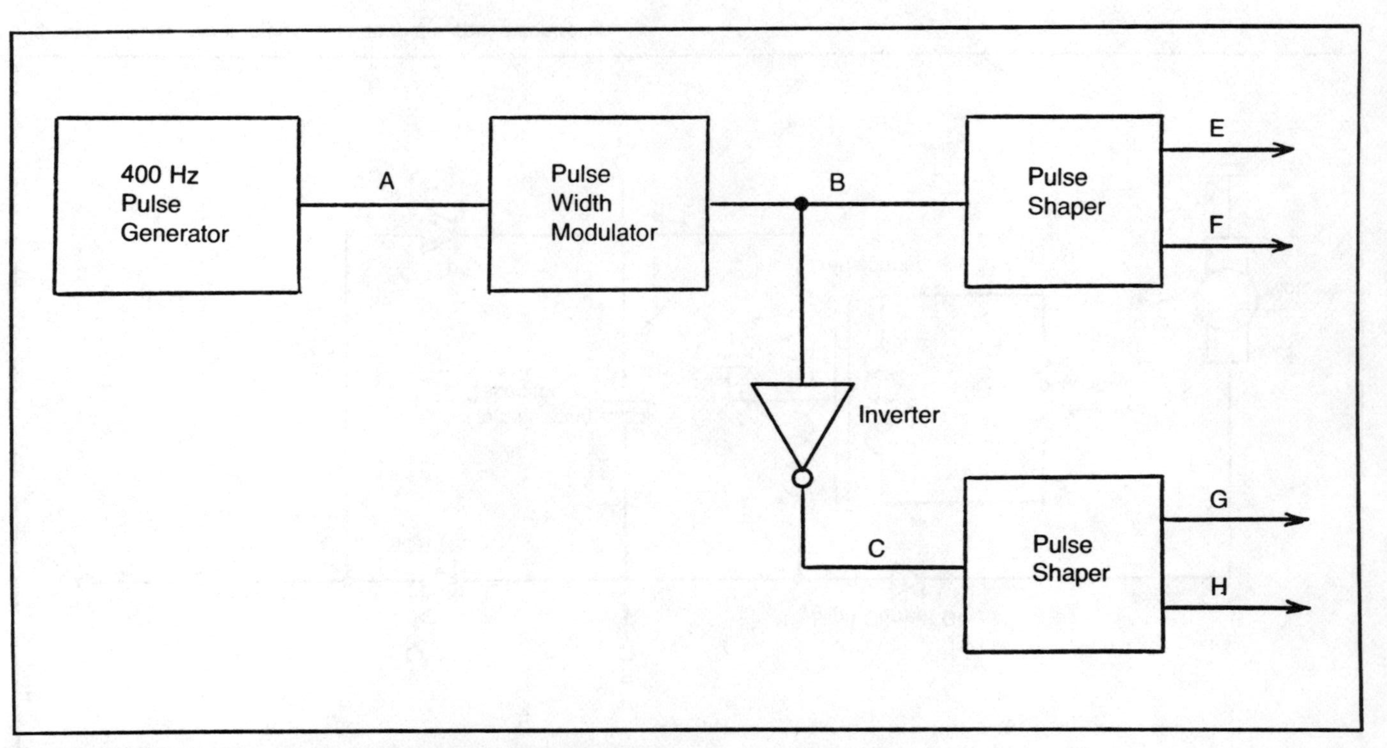

Fig. 14-12. SCR triggering circuitry for the dc motor speed controller.

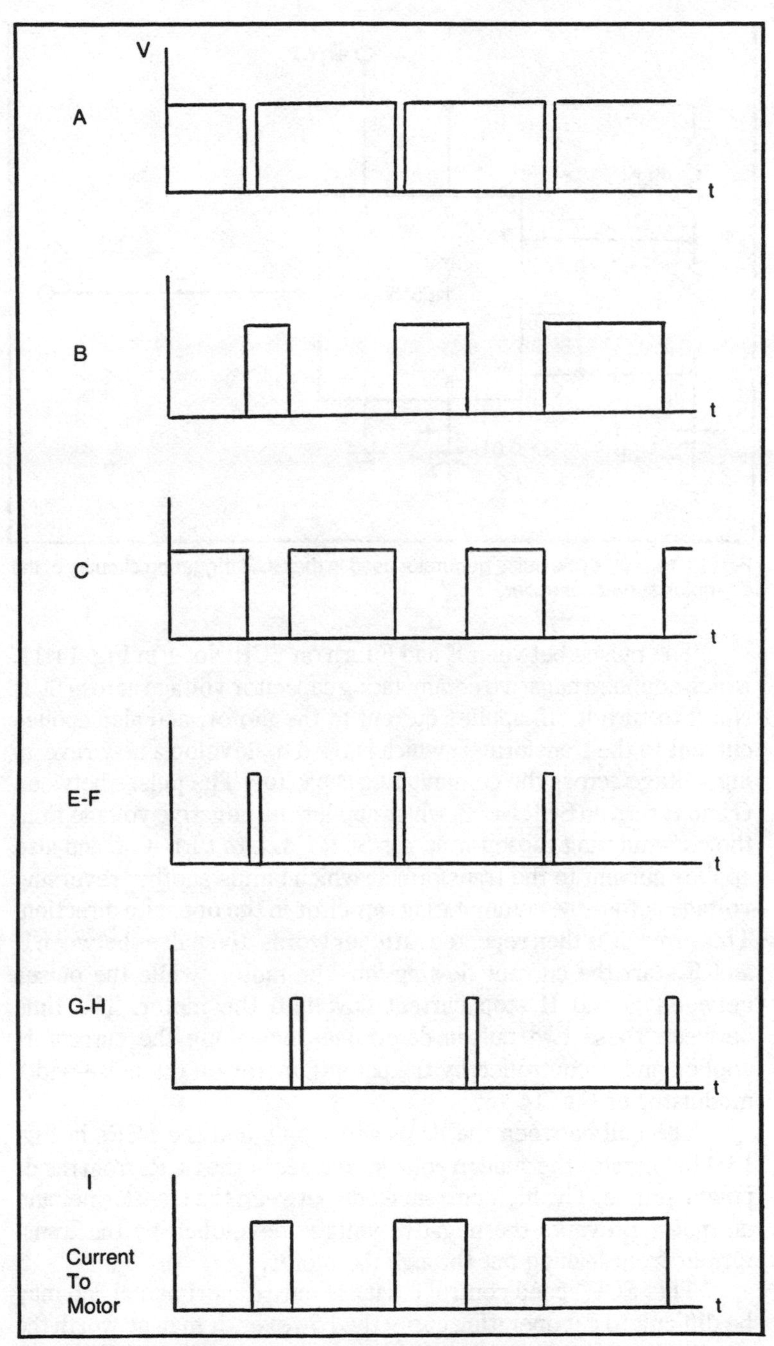

Fig. 14-13. Timing pulses present in the SCR triggering circuitry.

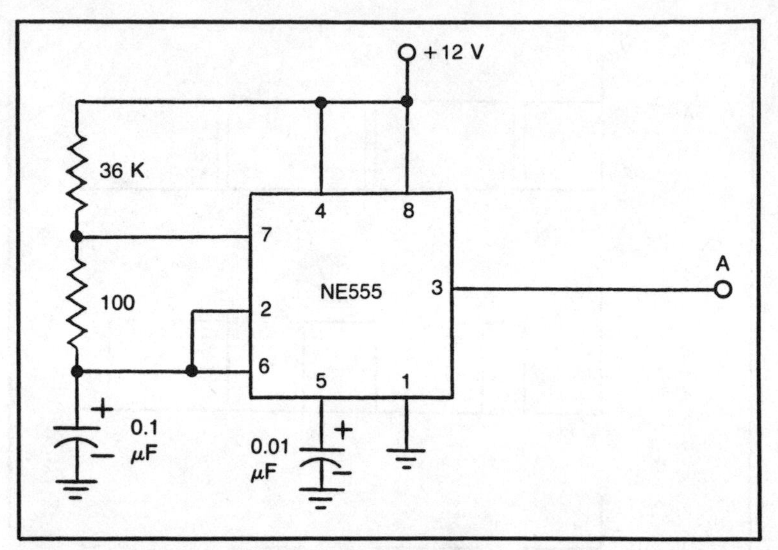

Fig. 14-14. 400-cycle pulse generator used in the SCR triggering circuitry of the dc motor speed controller.

The pulses between E and F turn on SCR No. 1 in Fig. 14-11, which applies a negative commutating capacitor voltage across SCR No. 2 to turn it off, applies current to the motor, and also applies current to the transformer, which is used to develop a new reversing voltage across the commutating capacitor. The pulses between G and H turn on SCR No. 2, which applies the negative voltage from the commutating capacitor across SCR No. 1 to turn it off and also applies current to the transformer, which builds another reversing voltage across the commutating capacitor in the opposite direction. This process is then repeated. In other words, the pulses between E and F start the current flowing into the motor, while the pulses between G and H stop current flow into the motor. The time between these two pulses determines how long the current is applied and is controlled by the potentiometer on the pulse-width modulator of Fig. 14-15.

The coil between the dc power supply and the SCRs in Fig. 14-11 separates the sudden voltage changes at the SCRs from the dc power source. The high-current diode between the transformer and dc motor prevents the negative voltage developed by the transformer from leaking out through the motor.

This SCR speed control circuit is only experimental and may be difficult to get operating correctly; however, it may be worth the trouble because the SCRs are inexpensive and efficient.

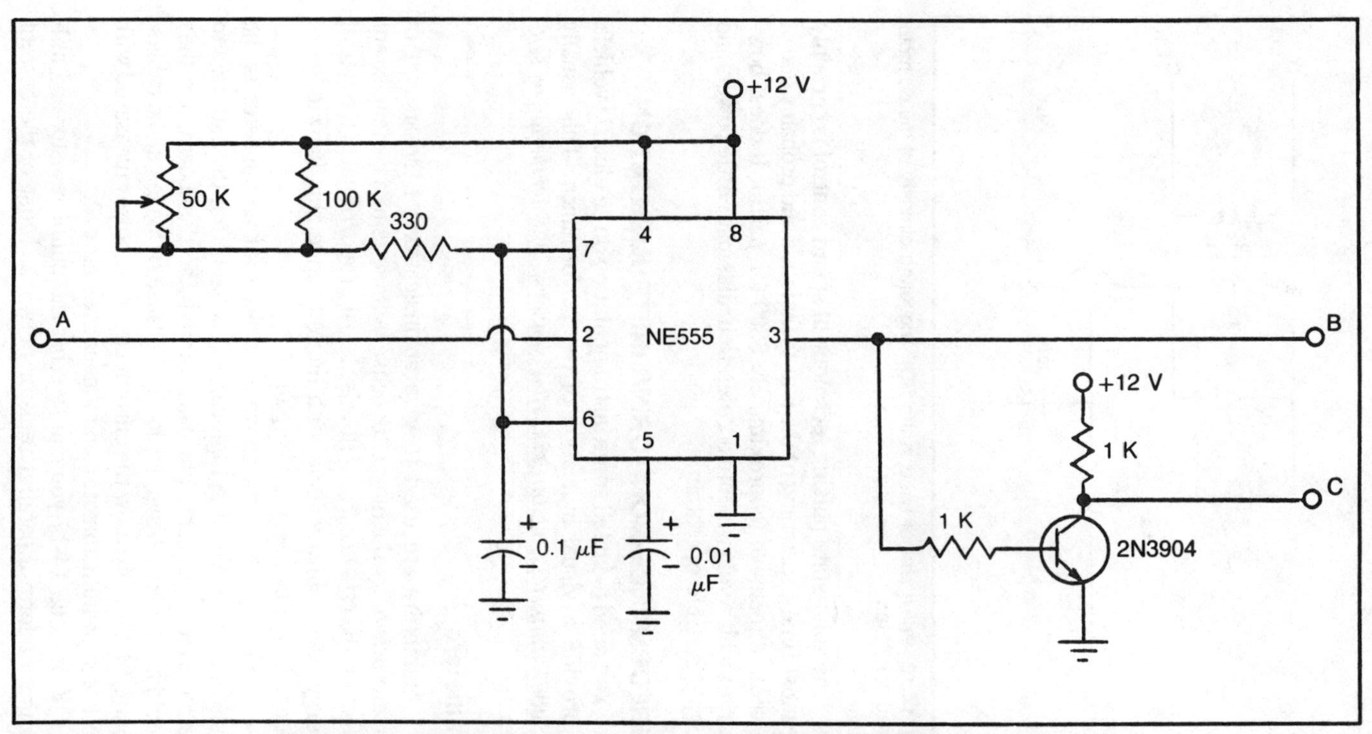

Fig. 14-15. Pulse-width modulator and inverter used in the SCR triggering circuitry of the dc motor speed controller.

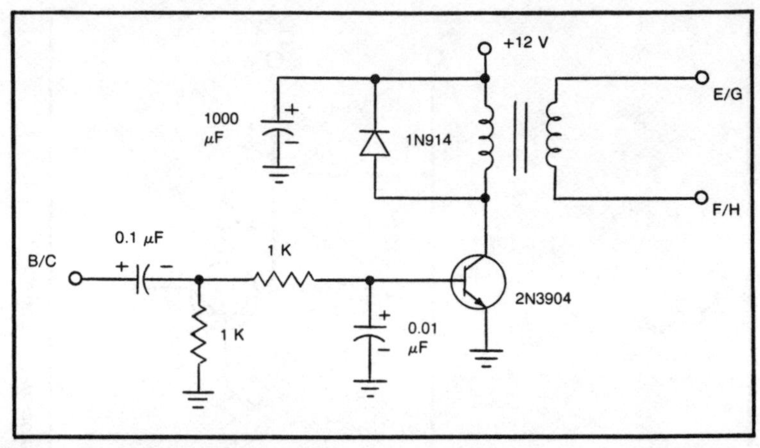

Fig. 14-16. Pulse shaper used in the SCR triggering circuitry of the dc motor speed controller.

The solid-state pulsing technique of speed control is certainly the most time-consuming of the four techniques and probably one of the most expensive (approximately $200 for parts). It does, however, exhibit a very smooth acceleration, dissipate little power, and provide reliable operation.

PERFORMANCE-MONITORING INSTRUMENTATION

You need to install some instrumentation on an electric vehicle to provide a visual indication of the performance. This usually includes *ammeters, voltmeters, rpm sensors,* and *temperature sensors.*

Ammeters

Ammeters are used to give an indication of the amount of current drawn by the motor. Because voltage is to a certain extent constant, the rate of current flow is almost proportional to the rate of energy usage. In a vehicle with limited range this knowledge of energy usage can be very helpful.

Obviously, the most convenient place for an ammeter is the dashboard; however, the high-current cables from the batteries to the motor will probably not pass through the dashboard, and they should not be lengthened to do this because of losses in the cables. Figure 14-17 shows how the current can be remotely measured with the use of a milliammeter and a shunt resistor.

From Fig. 14-17 you can see that the shunt resistor and milliammeter form a parallel resistor network. Most of the current passes through the low resistance of the shunt resistor, and a very small amount is diverted to the relatively high-resistance meter. Equation 14-2 shows how the two currents relate to the two resistances:

$$i = (I+i) \frac{R1}{R1+R2} \qquad \textbf{Eq. 14-2}$$

where I is the current through the shunt resistor, i is the current through the meter, R1 is the resistance of the shunt resistor, and R2 is the resistance of the meter.

Suppose a 10-milliamp, 1000-ohm meter is available for use on the electric vehicle. Because the meter is numbered from 1 to 10 on the front panel, it would be logical to have the meter read full scale for 1000 amps (half-scale for 500 amps, etc.). Letting i equal 0.01 amps, I+i equal 1000 amps, and R2 equal 1000 ohms, we can use Eq. 14-2 to calculate the resistance of the shunt resistor needed. This is 0.01 ohms. These shunt resistors can be purchased or fabricated from exact lengths of wire in much the same way as described earlier in the section on resistor switching speed control. You can calculate the power rating required for the shunt resistor from the formula

$$\text{Power rating} = I^2 R1 \qquad \textbf{Eq. 14-3}$$

Voltmeters

As a battery losses its charge, its voltage decreases. A

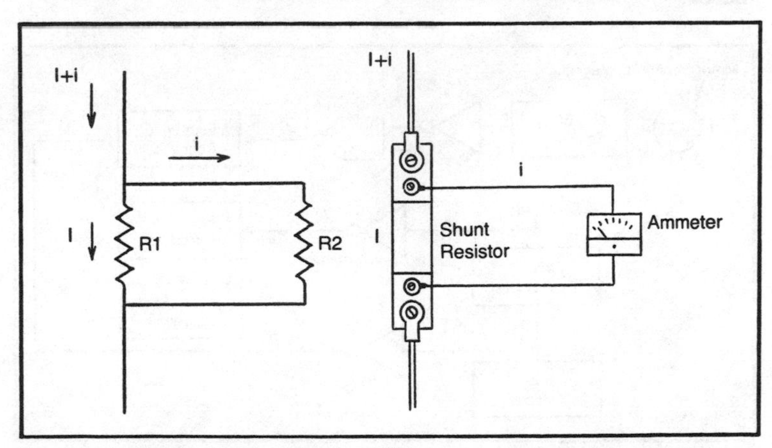

Fig. 14-17. Electrical and mechanical diagram of a high-current ammeter.

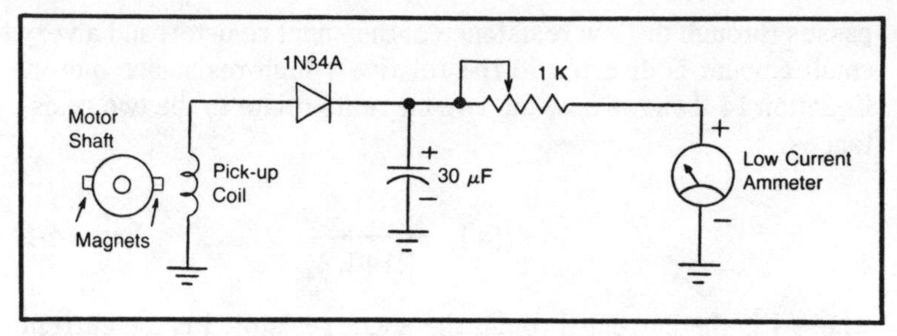

Fig. 14-18. Electromechanical diagram of the analog motor rpm sensor.

dashboard voltmeter with its electrical connections across the battery bank will give a good indication of battery charge. Battery voltage also decreases with increasing current due to internal battery resistance. This voltage drop must not be confused with battery discharge. The electrical location of the voltmeter is illustrated in Fig. 14-1.

Rpm Sensors

All dc motors have an operating rpm in which optimum performance occurs. You could achieve an increase in electric vehicle range and efficiency if you had a motor rpm indication to reference.

Figure 14-18 shows a simple, but not so accurate, motor rpm indicator. This circuit uses magnets on the motor shaft and a pickup coil similar to that of Fig. 3-3. The ac output of the pickup coil is rectified, filtered, and applied to a low-current dc ammeter through a potentiometer. The potentiometer can be adjusted to calibrate the meter.

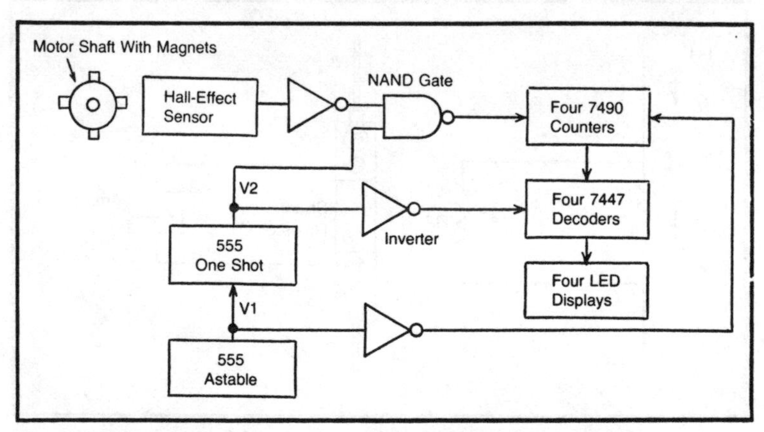

Fig. 14-19. Functional block diagram of a digital motor rpm sensor.

If you want an exact value of rotation rate, use the circuit of Fig. 14-19. This is exactly the same as the wind-speed indicator of Fig. 2-4, except for the changes shown in Fig. 14-20. This circuit should be calibrated by checking the width of the positive pulse on V2 (see Fig. 14-21) with an oscilloscope and adjusting it to 0.25 seconds with potentiometer R35. When the positive pulse width is 0.25 seconds, the output is calibrated in cycles per second (rpm/60) rather than revolutions per minute and requires two digits rather than four. This circuit also requires four magnets on the motor shaft for correct calibration.

Temperature Sensors

If an electric motor is loaded beyond its capacity, it can become too hot and burn up. Therefore, you must have some type of indication of hazardous temperature levels.

If a simple, red light warning of high motor temperature is sufficient, you can use the circuit of Fig. 14-22, which uses a thermistor mounted on the motor, a red LED mounted in the dash for high-temperature warning, and a 50K-ohm potentiometer for calibration. The potentiometer can be adjusted so the warning light will come on when the motor reaches a predetermined temperature. Some method of heating the thermistor and monitoring actual temperature will have to be established for the calibration procedure. Once the temperature of the thermistor is brought up to the desired warning light temperature, you can adjust the potentiometer until the LED burns brightly. The circuit is then calibrated.

If you want a constant temperature readout, use the circuit in Fig. 14-23. This also requires a heat source and an accurate method of temperature measurement for calibration. As temperature increases on the thermistor, actual temperature values can be written on the meter face. The 50K-ohm potentiometer can be adjusted to change the range of the thermometer if it is inadequate.

If a 10-milliamp meter is used, the total resistance of the meter and resistor R must be 1200 ohms. If it is a 1-milliamp meter, the total resistance must be 12,000 ohms. These resistance values will produce a full-scale reading on the meter when the transistor is completely turned on.

The four preceding instruments (ammeters, voltmeters, rpm sensors, and temperature sensors) should be all that is necessary for monitoring the mechanics of the electric vehicle. If the original transmission is used in the vehicle, speed can be measured normally with the standard speedometer.

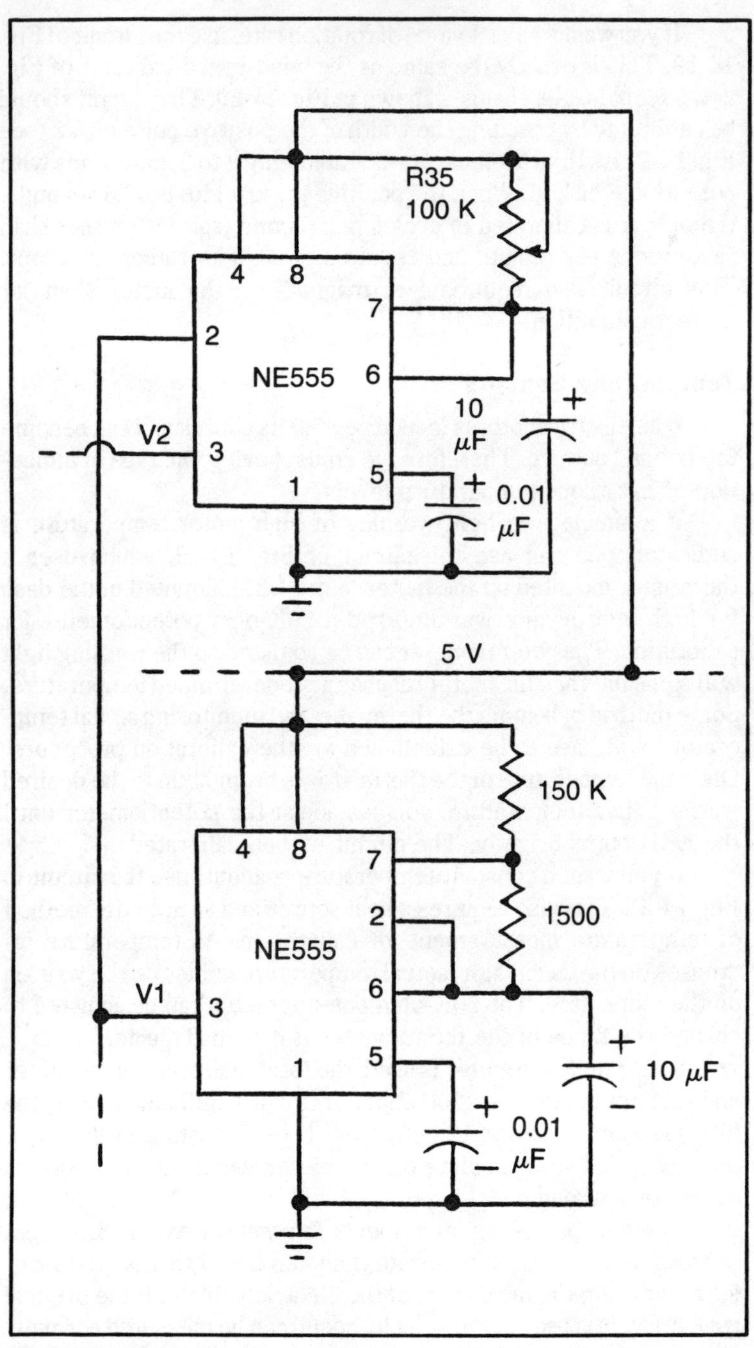

Fig. 14-20. Timing pulse generator for the digital motor rpm sensor.

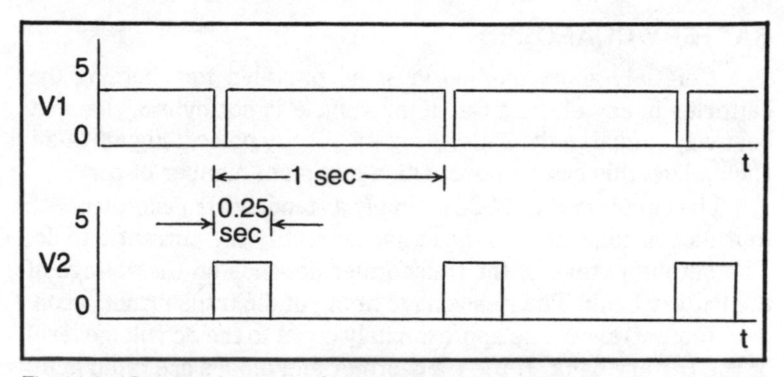

Fig. 14-21. Output voltage waveforms produced by the timing pulse generator in Fig. 14-20.

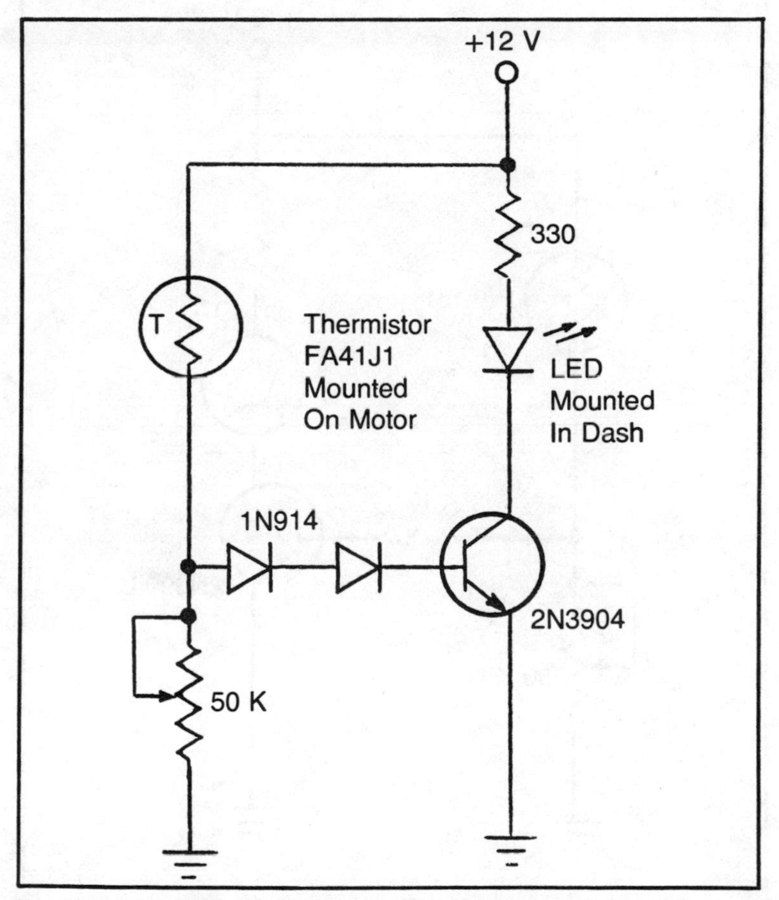

Fig. 14-22. Circuit diagram for the motor high-temperature warning light.

BATTERY CHARGERS

Certainly some method must be provided for charging the batteries in any electric car. If the vehicle is not hybrid, the next best way to charge the batteries is with utility power. Figure 14-24 shows how this can be done with a minimum number of parts.

The circuit in Fig. 14-24 is simply a stepdown transformer with four diodes connected to the output to rectify the current into dc. The output voltage of the transformer depends on the voltage of the battery bank. The rms voltage rating of the transformer secondary (output) should be approximately equal to the dc voltage level of the battery bank. If the transformer and diodes are rated at 40

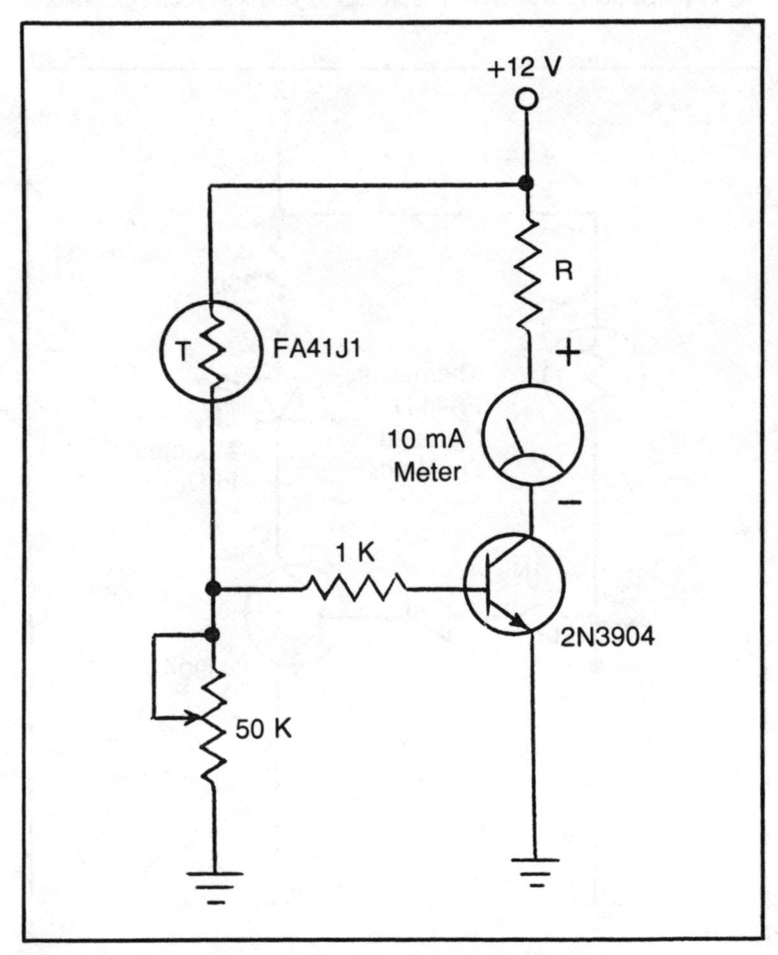

Fig. 14-23. Circuit diagram for the electric motor thermometer.

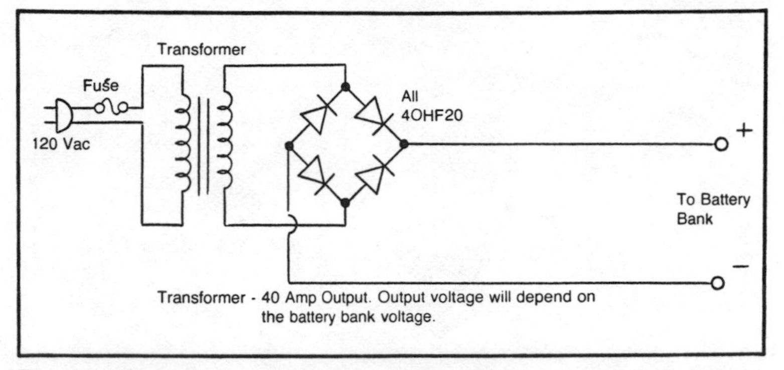

Fig. 14-24. Circuit diagram for the general purpose, high-current battery charger.

amps, this charger will provide approximately 320 amp-hours of charge in an 8-hour period, depending on the losses in the circuitry and batteries. If a faster charge is desired, the transformer and diodes must have higher current ratings.

This chapter, although brief, should provide enough information for the average person to build an energy-saving electric vehicle as long as he has a basic knowledge of electricity. All the circuits presented in this book have not been overly technical and have purposely been kept simple (in engineering terms), so that a minimum amount of electronics knowledge is necessary to understand them. For those who need more help in understanding the electronics involved, an index has been added.

Appendix A
Basic Electronics

For those of you who need some electronics background, this Appendix should help. The basic concepts of resistance, capacitance, inductance, and semiconductors are covered. The treatment is not exhaustive but it will supply information that will be helpful in building the projects.

RESISTANCE AND RESISTORS

Resistance is the ability of a material to oppose current flow through itself when a voltage is placed across it. Resistance can be compared to the friction between a box and the floor when someone tries to push the box across the floor. The force applied to the box by the person is analogous to the voltage across the resistive material, and the actual movement of the box is analogous to the current flow through the material.

There is a mathematical formula, called Ohm's law, that describes the relationship between voltage, current, and resistance. This formula is shown in Eq. A-1.

$$I = \frac{V}{R} \qquad \textbf{Eq. A-1}$$

Where V is the voltage across the material, I is the current flow through the material, and R is the resistance of the material measured in units called ohms. If the resistance is 10 ohms and the voltage is 10 volts, the current flow through the material is 1 amp.

Resistance in a material also causes the material to dissipate energy when current is flowing. Power is the rate of energy use and is usually measured in watts (volts × amps). The rate of energy dissipation in the resistance can then be determined by multiplying the voltage across the resistance by the current flow through the resistance. This is shown in Eq. A-2.

$$P = VI \qquad \textbf{Eq. A-2}$$

Equations A-1 and A-2 can now be combined to give the following formula for power dissipation in a resistance:

$$\begin{aligned} P &= VI \\ &= V \left[\frac{V}{R}\right] \\ &= \frac{V^2}{R} \end{aligned} \qquad \textbf{Eq. A-3}$$

We can also show that Eq. A-4 is true for power dissipation in a resistance.

$$P = I^2R \qquad \textbf{Eq. A-4}$$

In both of these equations P represents power or the rate of energy use.

If ac voltages are used, the rms (root-mean-square) value of the voltage and current must be used. Otherwise, the value for power dissipated will be incorrect.

All materials have resistance, but, when building electrical circuits, components called *resistors* are normally used. These components can be purchased with preset resistance values and power dissipation ratings. Standard resistor values range from 0.3 ohms to 500,000,000 ohms. Power dissipation ranges start at ¼ watt and go up. Table A-1 gives a list of the most commonly used resistor values and resistor power ratings.

Resistors normally use a color code painted around the body of the resistor to display the resistance value. Figure A-1 shows a sketch of a resistor with its color code. The first two colored rings represent the first two digits of the resistance value. The third ring represents the number of zeros to be added to the first two digits to complete the number. Each color represents a different number, as shown in Table A-2. For example, the resistor in Fig. A-1 has a red ring, a black ring, and an orange ring. From Table A-2, red repre-

Table A-1. Some Commonly Used Resistor Values and Resistor Power Ratings.

Standard Resistances	Standard Power Ratings in Watts
0.3	0.25
0.5	0.50
1.0	1
1.5	2
3.3	5
4.7	10
5	25
10	50
15	100
20	225
27	
33	
47	
82	
100	
150	
200	
220	
250	
330	
500	
750	
1000	
2000	
2500	
3000	
3300	
4700	
5000	
10000	
33000	
100000	
500000	
1000000	
500000000	

sents "2," black represents "0," and orange represents "3." The resistance then is 20,000 ohms.

The fourth ring in the color code represents the tolerance of the resistor. In other words, it represents the accuracy of the stated resistance value. If the ring is gold, the tolerance is 10 percent; if silver, 5 percent. The sample resistor in Fig. A-1 has a tolerance of 5 percent; therefore, the actual resistance of the resistor is guaranteed to be within 5 percent of the stated value. If there is no fourth ring, the tolerance is 20 percent.

Some resistors have variable resistance. They are called

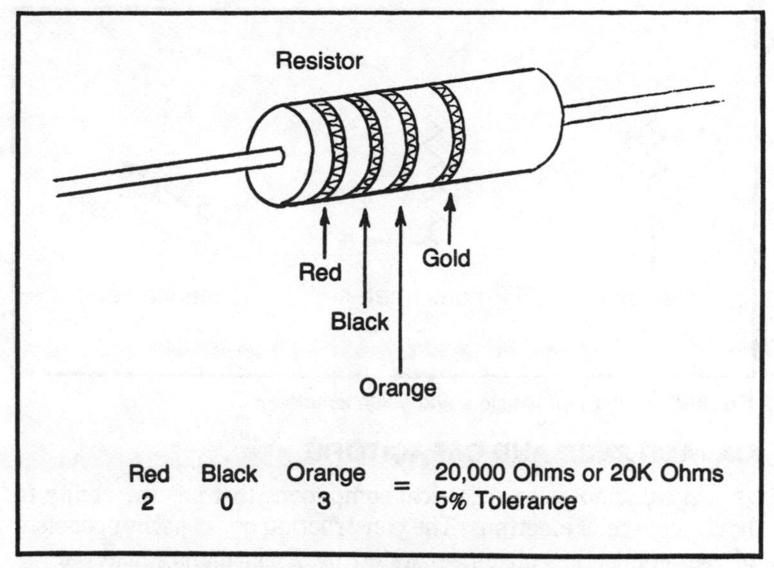

Fig. A-1. An example of resistor color codes.

potentiometers. The potentiometer has a wiper arm that can be moved from one end of the resistor to the other, selecting any value of resistance from zero to its maximum value. Figure A-2 shows the symbol for a resistor, and two symbols that can be used to represent the potentiometer in a circuit diagram.

When large-resistance potentiometers or resistors are referenced in circuit diagrams or in communications, their resistance value is usually abbreviated. For example, the 20,000-ohm resistor in Fig. A-1 is abbreviated as 20K-ohms. 20,000,000 ohms is abbreviated as 20M-ohms. The K represents a multiplication of 1000, and the M represents a multiplication of 1,000,000.

Colors	Number
Black	0
Brown	1
Red	2
Orange	3
Yellow	4
Green	5
Blue	6
Violet	7
Gray	8
White	9

Table A-2. Colors and the Numbers They Represent in Resistor Color Codes.

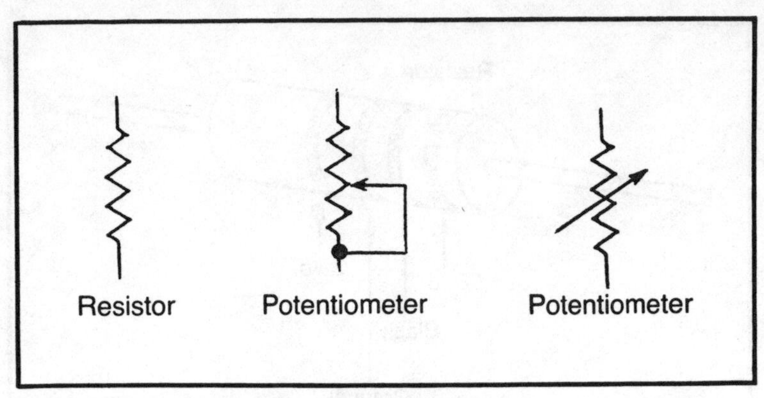

Fig. A-2. Symbols for resistors and potentiometers.

CAPACITANCE AND CAPACITORS

A *capacitor* is an electrical component that has the ability to hold a charge of electrons. The construction of a capacitor consists of two conducting plates separated by an insulating material, as shown in Fig. A-3. When a voltage is placed across this component, positive charges (lack of electrons) build on the plate facing the positive voltage, while negative charges build on the plate facing ground or negative voltage. These charges continue to build until the voltage across the capacitor equals the applied voltage. The flow of electrons to the negative plate and the flow of electrons away from the positive plate contribute to the current flow in the capacitor.

The amount of *charge* a capacitor can hold is given by the following formula:

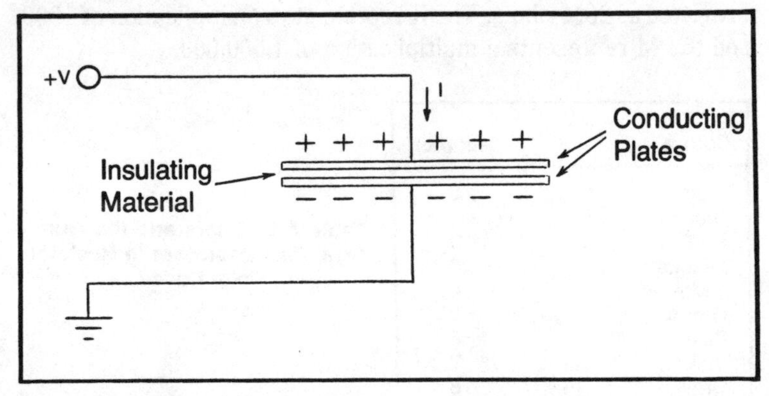

Fig. A-3. Simplified representation of a capacitor with an external voltage applied.

$$Q = CV \quad \textbf{Eq. A-5}$$

Where V is the voltage placed on the capacitor, C is the *capacitance* in *farads* of the capacitor, and Q is the charge of the capacitor in *coulombs* (1 coulomb = 6.28×10^{18} electrons).

The capacitance of a capacitor is determined by three things: the area of the plates, the distance between the plates, and the dielectric constant of the insulating material between the plates. Equation A-6 gives the relationship of these three parameters to capacitance.

$$C = \frac{8.858 \times 10^{-14} KA}{D} \quad \textbf{Eq. A-6}$$

where A is the surface area in square centimeters of each plate, D is the distance in centimeters between the plates, and K is the dielectric constant of the insulating material. For example, $K = 1.0006$ for air and $K = 80$ for water. Obviously, water is a better choice than air because the same size capacitor has 80 times the capacitance.

The energy stored in a capacitor is given by Eq. A-7.

$$E = \frac{CV^2}{2} \quad \textbf{Eq. A-7}$$

Where V is the voltage across the capacitor, C is the capacitance of the capacitor, and E is the energy stored in the capacitor measured in watt-seconds. A 1-microfarad (0.000001 farad) capacitor with 10 volts across it will store 0.00005 watt-seconds of energy. In other words, it can provide 0.00005 watts for 1 second or 5 watts for 0.00001 seconds when discharged into a load.

When a voltage is placed across a capacitor, current flows into the capacitor to charge the plates. This current is equal to the capacitance of the capacitor multiplied by the rate of change of voltage across the capacitor. This is illustrated in Eq. A-4.

$$I = C\frac{dV}{dt} \quad \textbf{Eq. A-8}$$

Where dV/dt represents the instantaneous change in voltage per change in time. This equation insinuates that if a voltage is quickly applied to a capacitor, the capacitor will draw a large current for that short time. This property of capacitors makes them well suited for use as filters. When placed between a voltage reference and ground, they absorb the quick voltage variations by quickly drawing current. This same property makes them useful in passing ac signals while, at the same time, blocking dc.

Figure A-4 shows the symbols used to represent capacitors in circuit diagrams. Notice the different symbols for polarized and nonpolarized capacitors. Polarized capacitors must have their positive terminal connected to positive voltage and their negative terminal connected to negative voltage or ground. These capacitors usually contain some type of liquid or gel insulating material, and if connected backwards they allow a small leakage current to pass through them. The nonpolarized capacitors have ceramic insulation and can be connected in either direction.

The unit of capacitance is the farad; however, capacitors of 1 farad or more are very uncommon. They are typically measured in microfarads (μF), where 1 μF = 0.000001 F. Some are measured in nanofarads (0.000000001 farads with symbol nF) or picofarads (0.000000000001 farads with the symbol pF). The capacitance value is usually written on the capacitor along with its maximum dc working voltage.

INDUCTANCE, INDUCTORS, AND TRANSFORMERS

Inductance is the property of a coil of wire that causes it to

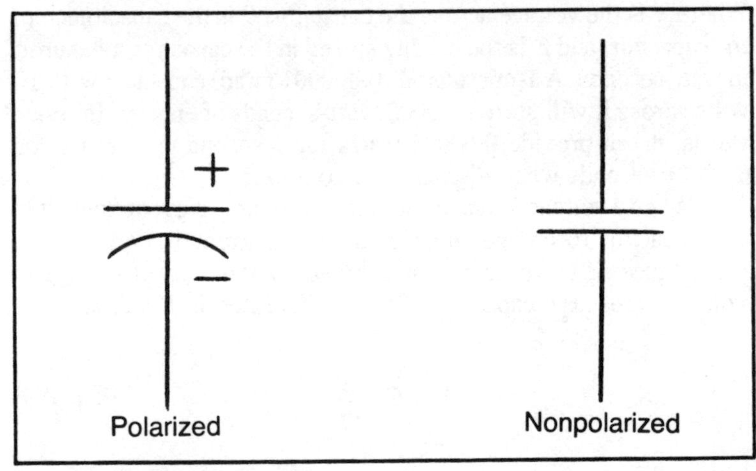

Fig. A-4. Capacitor symbols.

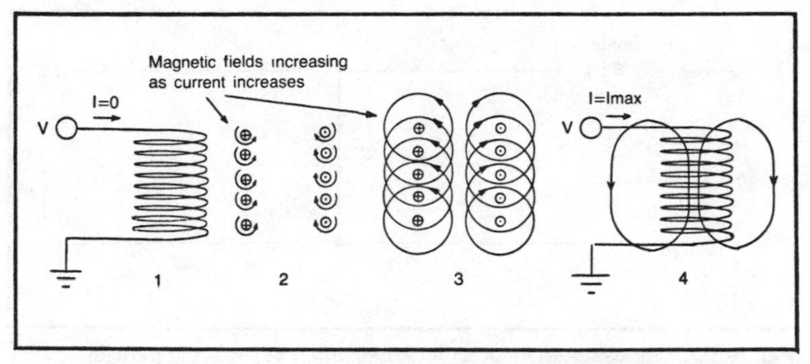

Fig. A-5. An inductor or coil with its magnetic field increasing as its current increases.

oppose a change in current flow even though there may be no resistance in the wire. When a voltage is placed across a coil (or inductor), current begins to flow through the wire. This current causes a magnetic field to build up around each conductor, as shown in Fig. A-5. These magnetic fields cut across the adjacent conductors, inducing a voltage in the opposite direction to the applied voltage. This opposing voltage is called *back emf.*

This opposition to change in current flow is illustrated mathematically by Eq. A-9.

$$V = L \frac{dI}{dt} \qquad \textbf{Eq. A-9}$$

Where L is the inductance of the coil in henrys (H), dI/dt is the rate of current change in amps per second, and V is the back emf of the coil in volts. When the current in the coil reaches its maximum value, it is no longer changing, and there is no back emf developed. The only voltage developed across the coil for this constant current is due to the resistance of the wire.

From this discussion it is apparent that inductors oppose alternating current, but will conduct direct current well once the magnetic field builds up. This property also makes the coil suitable for filter circuits in much the same way as the capacitor. Since the inductor passes dc but not ac, it can be placed in series with the applied voltage to filter out the ac. The capacitor passes ac and blocks dc; therefore, it should be placed in parallel with the applied voltage to conduct the ac to ground while allowing the dc to continue to flow. If both the inductor and the capacitor are used in a filter circuit as shown in Fig. A-6, a very smooth dc signal can be obtained.

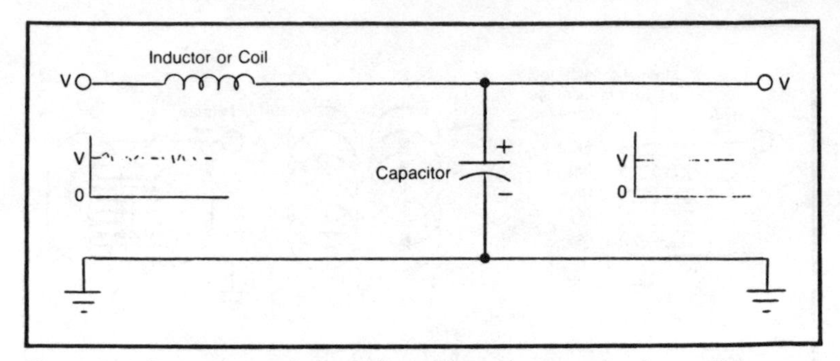

Fig. A-6. A filter using an inductor in series and a capacitor in parallel.

Also notice in Fig. A-6 that the symbol used for the inductor is simply a series of loops.

Like the capacitor, the inductor also has the ability to store energy. The inductor stores its energy in the magnetic field it builds around the coil. The amount of energy stored in this field is given by Eq. A-10.

$$E = \frac{LI^2}{2} \qquad \textbf{Eq. A-10}$$

where L is the inductance of the coil, I is the current flowing in the coil measured in amps, and E is the total energy stored in the magnetic field. When the current is shut off, the magnetic field collapses and releases its energy by trying to keep the current flowing. If the current is turned off quickly, the coil, trying to keep this current flowing, will develop a large negative voltage. The voltage developed by the collapsing field can be calculated from Eq. A-9. An automobile ignition system is a good example of high-voltage generation by a collapsing field in a coil. Before electronic ignition, mechanical points were used to apply and remove current from the ignition coil. This is now done electronically.

Inductors can also be used as pickup coils in rotation sensing circuits. In these circuits the moving magnetic field comes from a magnet as it passes by the coil. The magnetic lines of force crossing the conductors in the coil induce a voltage across the coil in the same way as the collapsing or growing field in the coil induces its voltage. These induced voltages can be counted by a counter to give a measure of rotation, or they can be rectified and filtered to produce a dc voltage whose magnitude is proportional to rotation.

If two coils are coiled together without any electrical connection, the changing magnetic field in one coil can be used to induce a voltage in the other coil. A component built this way is called a *transformer.* Depending on the ratio of the number of turns in one coil to the number of turns in the other coil, you can make the output voltage of the second coil larger or smaller than the input voltage of the first coil. This ratio is illustrated by Eq. A-11.

$$\frac{V_{out}}{V_{in}} = TR = \frac{\text{Turns of secondary coil}}{\text{Turns of primary coil}} \quad \textbf{Eq. A-11}$$

The input voltage is applied to the *primary* coil, and the output voltage is taken from the *secondary* coil. In Eq. A-11, TR is the turns ratio of the transformer.

The symbols for standard and center-tapped transformers are shown in Fig. A-7. This figure also illustrates the turns-ratio symbology used. The standard transformer in Fig. A-7 has a turns ratio of 0.1, which is written 10:1 or 120:12. This means that if a 120-Vac voltage is placed on the input (primary coil), the output voltage on the secondary coil will be 12 Vac. The two parallel lines between the coils of the transformer symbol represent the metal core about which the coils are wrapped. All transformers use a laminated metal core because metal allows for a stronger magnetic field in the transformer. The symbol for the center-tapped transformer shows a third wire output from the secondary coil. This wire simply connects to the middle of the secondary coil to allow two output voltages rather than one.

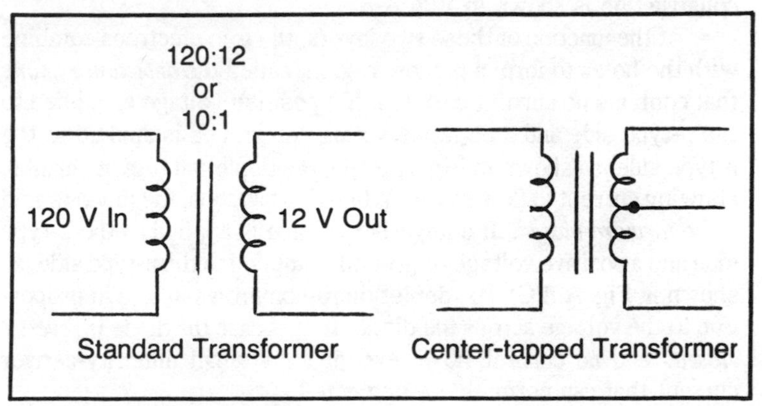

Fig. A-7. Transformer symbols.

Remember that power is equal to volts multiplied by amps. Because a transformer cannot create energy, the output power must equal the input power, except for a few losses inside the transformer. This means that if the output voltage is increased by the turns ratio, the output current must be decreased by the same ratio to keep the input and output powers equal. This is illustrated by Eq. A-12.

$$\frac{Iout}{Iin} = \frac{1}{TR} \qquad \textbf{Eq. A-12}$$

As an example, suppose the standard transformer in Fig. A-7 is in a circuit where the input voltage is 12 Vac and the input current is 1 amp ac. The output voltage of the transformer should be 1.2 Vac, and the output current should be 10 amps ac. The input power is 12 watts and the output power is 12 watts.

Inductors are rarely used in modern electronics because of their size and weight. Transformers are also large and heavy, but are used extensively because there is usually no way to design around them. They are the only components that can alter the magnitude of an ac voltage without wasting large amounts of energy.

DIODES

Diodes are semiconductor devices that conduct current in one direction only. These devices are made from a single crystal of silicon or germanium. One side of the crystal is doped with n-type impurities containing free electrons, and the other side is doped with p-type impurities containing holes (missing electrons). This construction is shown in Fig. A-8.

At the junction of these two layers, the free electrons combine with the holes to form a narrow region, called the *depletion region,* that contains no current carriers. If a positive voltage is applied to the p-type side and a negative voltage or ground is applied to the n-type side, as shown in Fig. A-8 (B), the depletion region shrinks, allowing current to flow freely. When this happens, the diode is said to be *forward biased.* If a negative voltage is applied to the p-type side and a positive voltage or ground is applied to the n-type side, as shown in Fig. A-8(C), the depletion region grows in size in proportion to the voltage across the diode. In this case the diode is *reverse biased,* and no current flows except for a small minority-carrier current that can normally be neglected.

Before a diode will conduct current, the forward-biasing

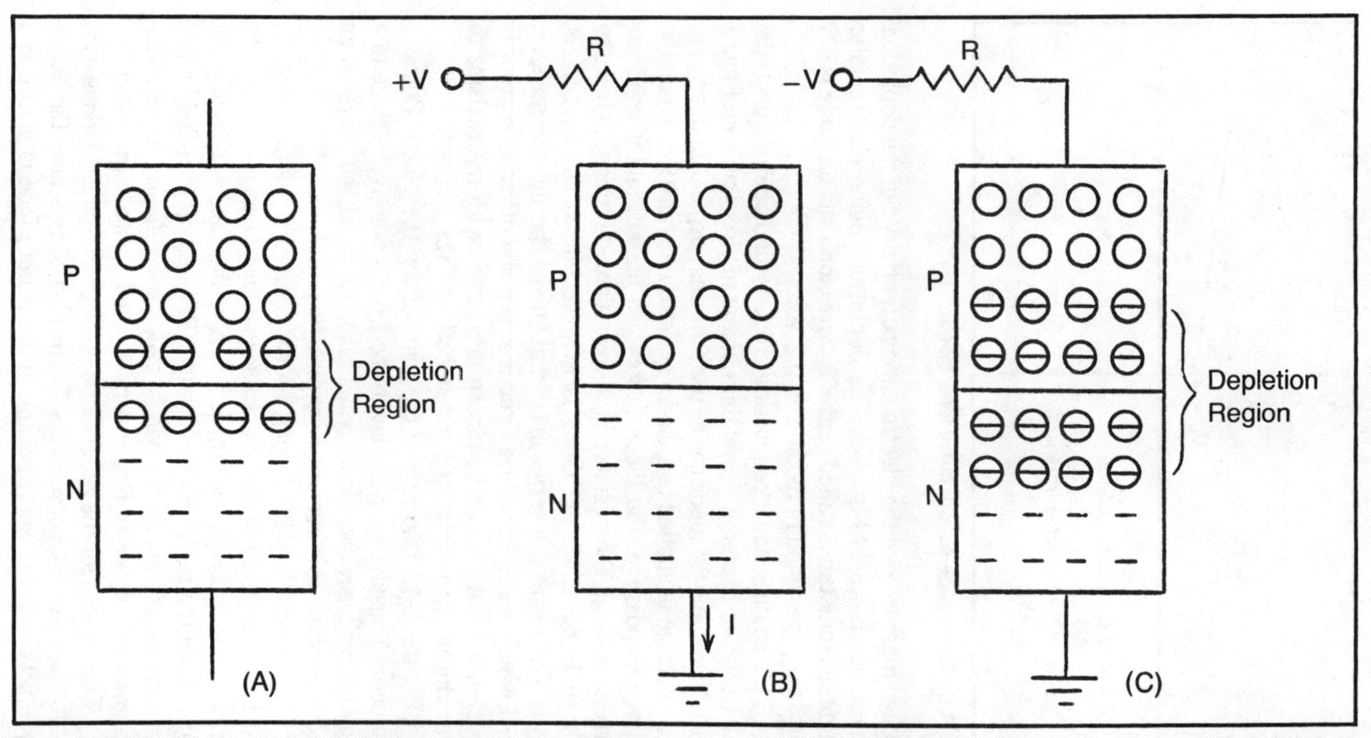

Fig. A-8. Simple sketches showing the depletion region of a diode for (A) no voltage (B) forward-biasing voltage (C) reverse-biasing voltage.

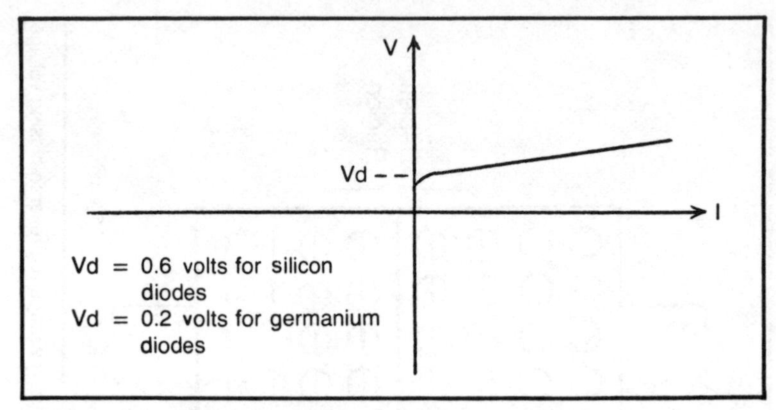

Fig. A-9. The voltage-current curve for a diode.

voltage must exceed the electrostatic potential Vd of the depletion region, as shown in Fig. A-9. The depletion region electrostatic potential (sometimes called *juction voltage*) for a silicon diode is 0.6 volts, for a germanium diode it is 0.2 volts.

This unidirectional conducting property of diodes makes them well suited for changing alternating current to direct current. Figure A-10 shows how the diode can be used for this purpose. In this figure an ac voltage is applied to a diode/resistor combination. When the voltage is positive, the diode is forward biased, and it conducts current through the resistor in a downward direction. When the voltage is negative, the diode is reverse biased and no current flows. The result is a series of current pulses through the resistor, all of which are in the same direction. If a smoother dc current is desired, a capacitor can be placed in parallel with the resistor to filter the voltage variations and smooth out the current.

Figure A-10 also shows the symbol for the diode. Because engineers consider current flow to be from a positive voltage to a negative voltage, the diode symbol was made in the shape of an arrow showing this current flow direction.

All diodes, besides having a forward-biased junction voltage, also have a reverse-biased *breakdown voltage.* These voltages are normally much higher than the junction voltage and will be slightly higher than the peak inverse voltage (PIV) rating supplied by the diode manufacturer. Some diodes, called *zener diodes,* are actually designed to operate in this reverse-biased breakdown region.

The symbol for the zener diode is shown in Fig. A-11 along with its voltage-current curve. This figure also shows how the zener diode acts as a voltage regulator. If the zener diode is placed in a

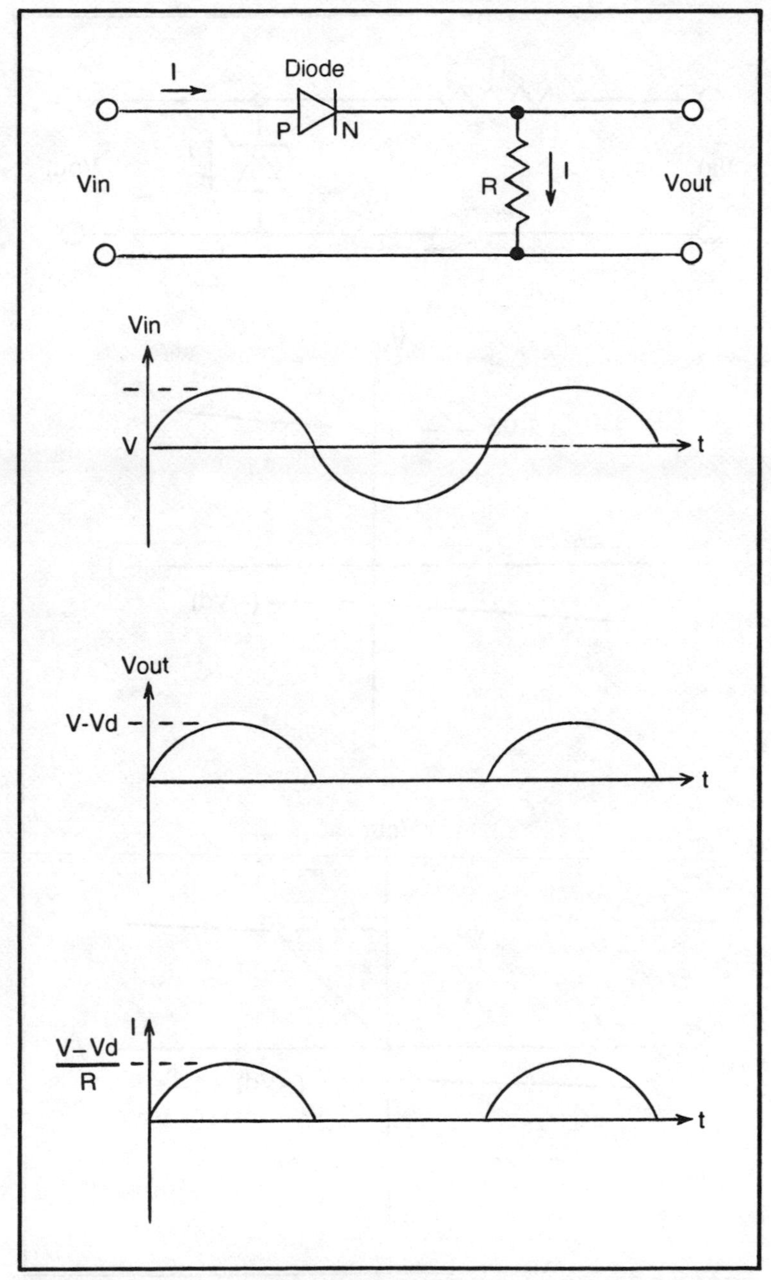

Fig. A-10. Circuit for using a diode to change an ac signal to a pulsating dc signal. Also shown are the input and output voltage waveforms and the output current waveform.

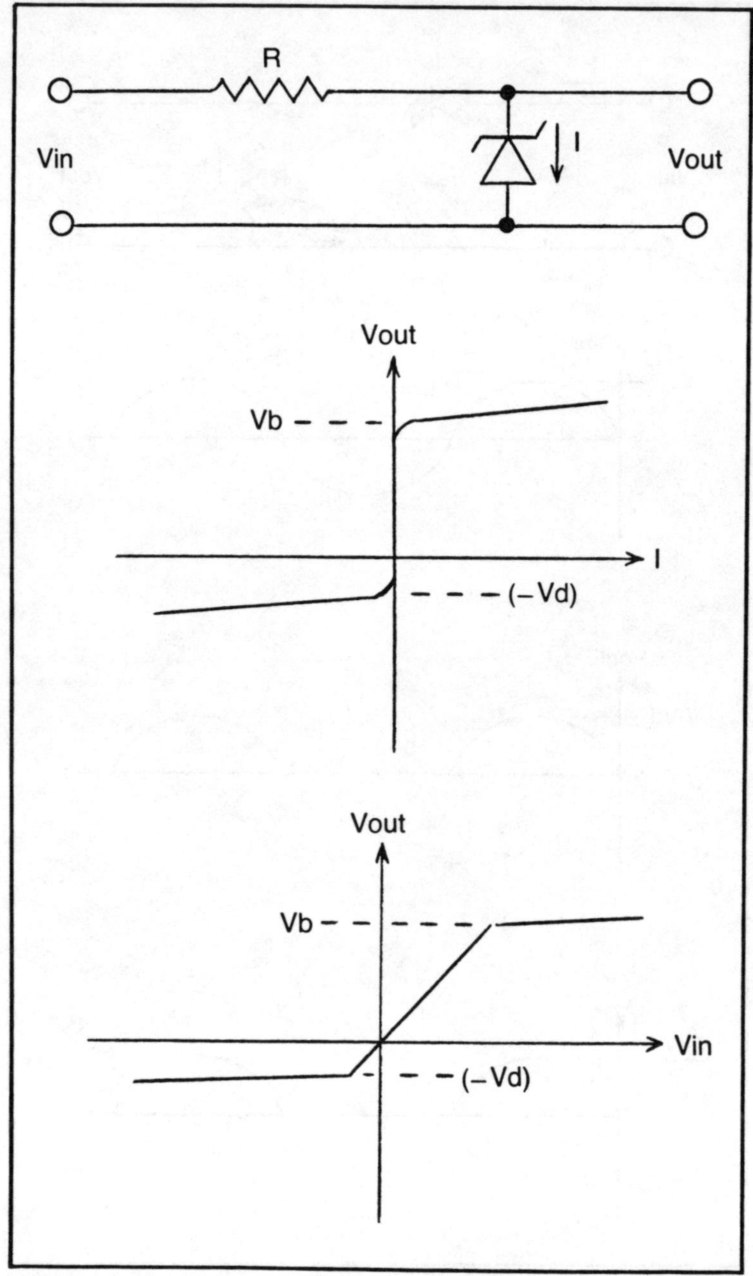

Fig. A-11. Circuit for using a zener diode as a voltage regulator. Also shown are graphs relating output voltage to diode current and output voltage to input voltage.

Circuit as shown in Fig. A-11, it acts as a normal diode when the input voltage is reversed. When the input voltage increases in the positive direction, the output voltage follows the input voltage until the zener-diode breakdown voltage Vb is reached. When this happens, the zener diode begins to conduct most of the current, preventing the output voltage from rising above Vb. The output voltage is then effectively regulated at a voltage equal to the zener-diode breakdown voltage. These characteristics of zener diodes make them very useful in voltage-regulator circuits.

Both zener diodes and regular diodes come in a variety of sizes, where size is measured in current capacity. You can buy regular diodes with different PIV ratings and zener diodes with different breakdown voltages.

TRANSISTORS

Transistors, developed in the 1950s, have revolutionized the electronics industry. These incredibly small current-amplifier devices have made it possible to build electronic circuits the size of a penny that once would have been, using vacuum tubes, the size of a large building.

There are several types of transistors, but this book only references the *bipolar junction* transistors. These are made from either silicon or germanium crystals and can be purchased in npn or pnp form. A silicon npn bipolar junction transistor consists of a thin p-type silicon crystal sandwiched between two n-type silicon crystals. A silicon pnp bipolar junction transistor consists of a thin n-type silicon crystal sandwiched between two p-type silicon crystals. Germanium bipolar junction transistors are constructed in the same way except that germanium crystals are used. The construction of both of these transistors is shown in Fig. A-12.

When a positive voltage is applied to the upper n-type crystal (*collector*) of the npn transistor and a negative voltage or ground is applied to the lower n-type crystal (*emitter*), the upper pn junction is reverse biased and the lower pn junction is forward biased. Because one junction is reverse biased, no current can flow from the collector to the emitter. If a positive voltage is now applied to the center p-type crystal (*base*), the lower junction will conduct current from the base to the emitter. Some electrons which enter the base from the emitter (electron flow is from negative voltage to positive voltage and is opposite hole flow, which is from positive voltage to negative voltage) fall into the depletion region of the upper pn junction. These electrons are then accelerated across the reverse-

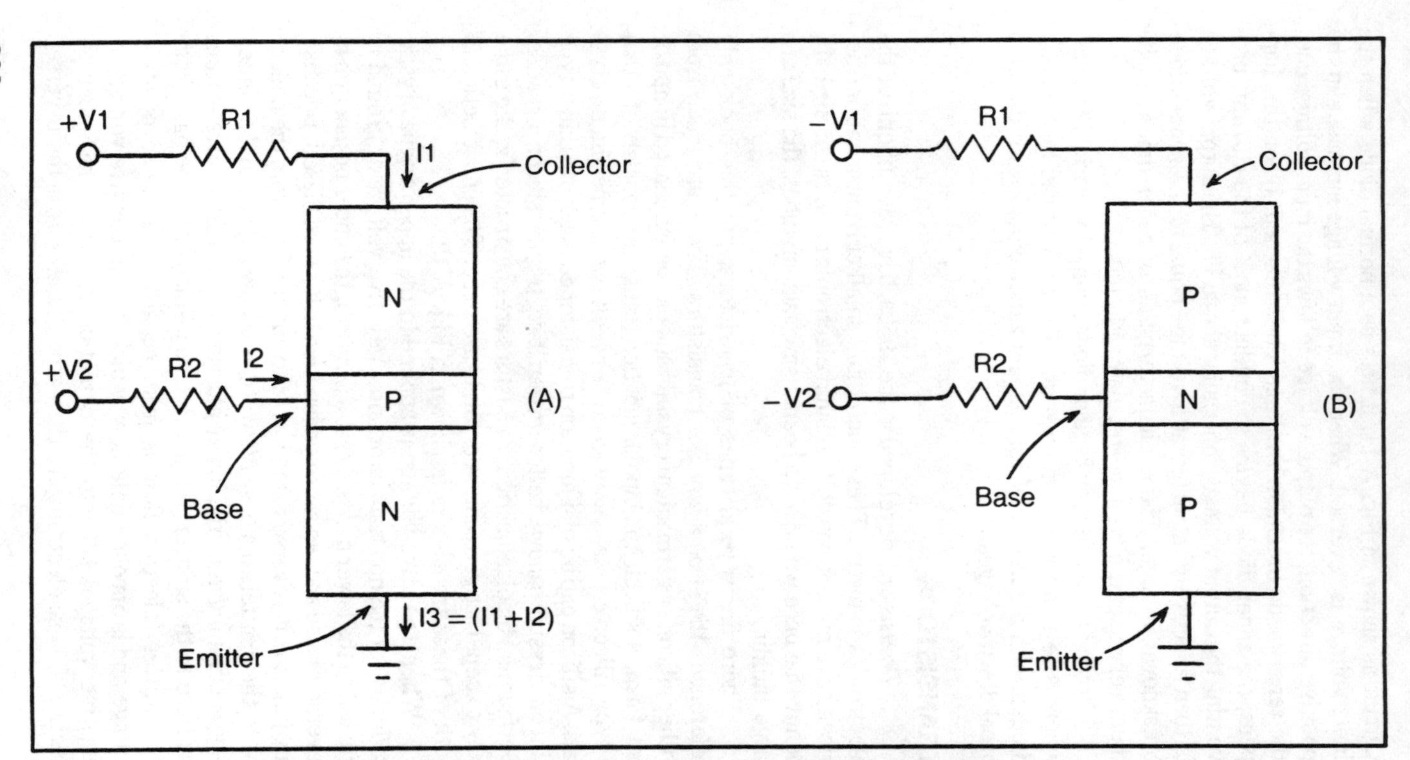

Fig. A-12. Construction of the (A) npn bipolar junction transistor and the (B) pnp bipolar junction transistor.

biased junction by the large positive voltage applied to the collector and constitute the collector current. The magnitude of this collector current is proportional to the base current. The pnp transistor operates in the same way, except that the voltages and currents are reversed.

Figure A-13 shows the symbols for both the npn and the pnp bipolar transistors. The arrow between the base and emitter shows the direction of the hole current. Engineers traditionally use this hole flow when referring to electrical current, while technicians normally use electron flow. It is immaterial which is used because the mathematics is the same. In this book current is assumed to be hole flow.

Figure A-13 also shows mathematically the current gain characteristics of the npn and pnp bipolar transistors. In this figure h_{fe} represents the current gain of the transistor and I1, I2, and I3 represent the collector, base, and emitter currents, respectively. The collector current I1 is equal to h_{fe} multiplied by the base current I2. The emitter current is the sum of the base current and the collector current, as shown in Eq. A-13.

$$\begin{aligned} I3 &= I2 + I1 \\ &= I2 + h_{fe} I2 \\ &= I2(1 + h_{fe}) \end{aligned} \qquad \textbf{Eq. A-13}$$

If $h_{fe} = 70$ amps and $I2 = 0.1$ amps, then $I1 = 7$ amps, and $I3 = 7.1$ amps. Transistors typically have current gains in the range of 25 to 150.

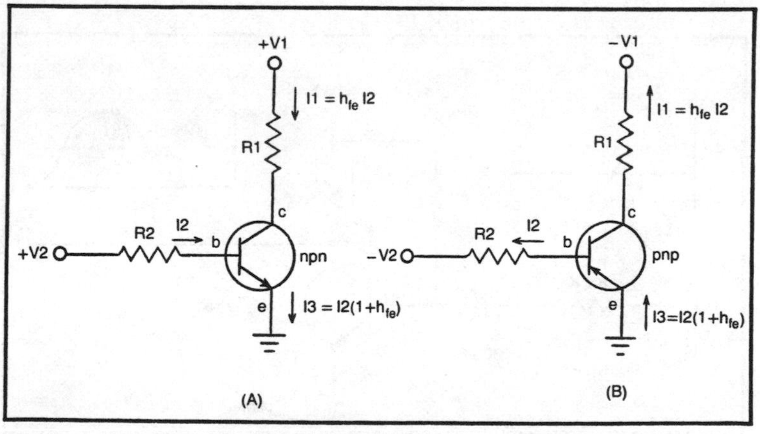

Fig. A-13. Circuit symbols and current flow for the (A) npn bipolar junction transistor and the (B) pnp bipolar junction transistor.

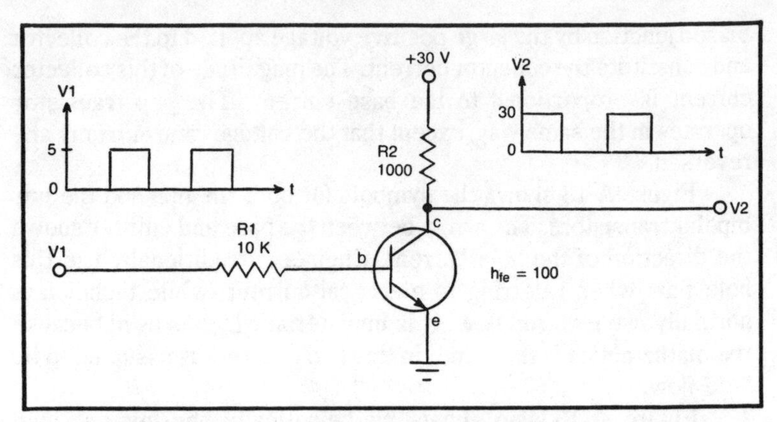

Fig. A-14. An inverter stage made from an npn transistor.

Two typical applications for transistors are inverters and amplifiers. An inverter circuit is shown in Fig. A-14, and a single-stage amplifier is shown in Fig. A-15. The inverter uses the transistor in the switching mode (saturated or off), while the amplifier in Fig. A-15 uses the transistor in the linear mode (always in active region). The switching mode is much more efficient because less power is used. When the transistor is off, no power is dissipated in the transistor. When the transistor is saturated, it conducts as much current as possible through resistor R2 in Fig. A-14, but the voltage drop across the transistor is very low. Power dissipation in the transistor is equal to the voltage across the transistor multiplied by the current through the transistor; therefore, power dissipation is low due to the low voltage drop.

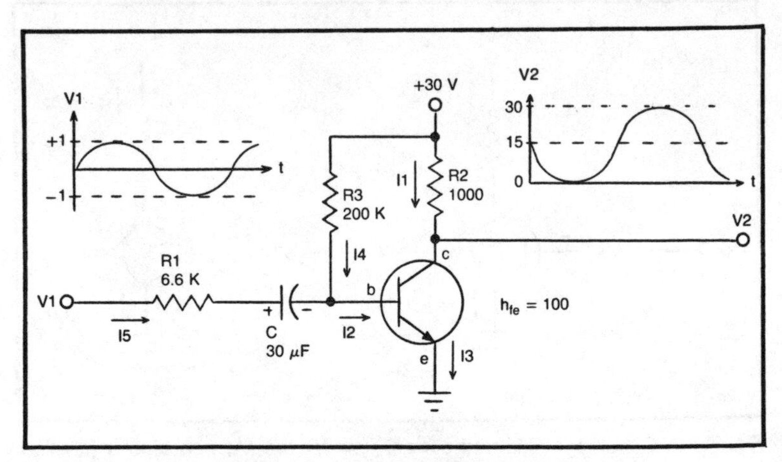

Fig. A-15. A linear amplifier made from an npn transistor.

The inverter in Fig. A-14 changes the 5-volt square wave to a 30-volt square wave that is 180 degrees out of phase. When the input voltage is 5 volts, 0.0005 amps of current is forced through the base to the emitter of the transistor. Since the current gain of the transistor is 100, the transistor would like to have a collector current of 0.05 amps. However, from Ohm's Law the collector current cannot be higher than 0.03 amps (30 volts/1000 ohms), due to the collector voltage and resistor R2. The transistor then, will draw only 0.03 amps, and most of the 30 volts will be dropped across the 1000-ohm resistor R2. The output voltage will be approximately 0 volts.

When the input voltage is zero, the base current is zero, and the transistor does not conduct current. In this case the resistance of the transistor between the collector and the emitter is nearly infinite, and the output voltage is maximum or 30 volts.

These inverter circuits are widely used in digital electronics for changing zero voltage levels to 5 volts and 5-volt levels to 0 volts. For digital electronic applications both the base voltage and the collector voltage are 5 volts.

The linear amplifier circuit in Fig. A-15 can be used to accurately reproduce an input signal with an output signal that is amplified and shifted in phase by 180 degrees. If no phase shift is desired, two of these circuits can be placed in series, and the total gain will be the product of the two gains.

To keep this amplifier operating in the linear mode, a bias resistor (R3) is used to provide a constant current (I4) of 0.15 milliamps into the base of the transistor. If there is no input current to the base other than this, the collector current will be 15 milliamps and the output voltage with be 15 volts. This voltage is the center operating voltage for the 0-to-30-volt output signal.

As shown in Fig. A-15, the input voltage is a sine wave with a peak amplitude of 1 volt. The equation for this input voltage, V1, is the following:

$$V1 = \sin Wt \qquad \textbf{Eq. A-14}$$

where W is the radian frequency of V1 and t represents time. The input current (I5) to the base due to V1 is given by

$$\begin{aligned} I5 &= \frac{V1}{R1} \qquad \textbf{Eq. A-15} \\ &= \frac{\sin Wt}{6600} \end{aligned}$$

The total base current (I2) due to both the biasing resistor R3 and the input voltage is

$$I2 = I4 + I5$$
$$= 0.00015 + \frac{\sin Wt}{6600} \qquad \textbf{Eq. A-16}$$

Remember that capacitors will pass ac signals but not dc signals. The 30-μF capacitor in Fig. A-15 is used to pass the input ac signal and to block the dc bias current from flowing backwards toward the input voltage. The resistor/capacitor combination in the base circuit has a cutoff frequency at which it will attenuate any lower-frequency input signal. This radian frequency is equal to 1/RC or 5 radians per second. Therefore, the input frequency, W, must be greater than 5.

The collector current, I1, is equal to the base current, I2, multiplied by the current gain of the transistor. This is shown by Eq. A-16.

$$I1 = h_{fe}\, I2 \qquad \textbf{Eq. A-16}$$

The output voltage, V2, is equal to

$$V2 = 30 - R2(I1) \qquad \textbf{Eq. A-17}$$

Combining Eqs. A-15, A-16, and A-17 we see that the output voltage is

$$V2 = 30 - 1000h_{fe}\,(0.000015 + \frac{\sin Wt}{6600})$$

or

$$V2 = 15(1 - \sin Wt)$$

or

$$V2 = 15 \sin(Wt + 180°) \qquad \textbf{Eq. A-18}$$

This signal is similar to the input voltage but is amplified by 15 and shifted in phase by 180 degrees.

The gain of a transistor stage can be increased significantly by using two transistors in a Darlington configuration, as shown in Fig. A-16. The gain of this transistor stage can be as high as 20,000. The only disadvantage is that now the base-to-emitter junction voltage drop is 1.2 volts, instead of 0.6 volts, for silicon transistors and 0.4 volts, instead of 0.2 volts, for germanium transistors. This dual transistor configuration can be purchased as a single package if desired.

Besides the bipolar junction transistors, there also exist *field-effect transistors* (FET). These transistors have a very high input resistance and do not draw any appreciable input current. The output current is a function of input voltage instead of input current. This book does not discuss the use of FETs because they are slightly more difficult to work with, and they are usually lacking in current capacity. The simplicity and diversity of the bipolar junction transistor makes it the obvious choice for applications referenced in this book.

OPERATIONAL AMPLIFIERS

The *integrated circuit operational amplifier* is another component that has revolutionized the electronics industry, but not to the extent that transistors have. These single-chip integrated circuits can be used for a large variety of applications, including amplifiers, summers, comparators, buffers, integrators, differentiators, and filters. Because most of these circuits are used in analog computers, the development of the operational amplifier has definitely been a significant technological breakthrough in this area.

The circuit symbol for the operational amplifier (op-amp) is

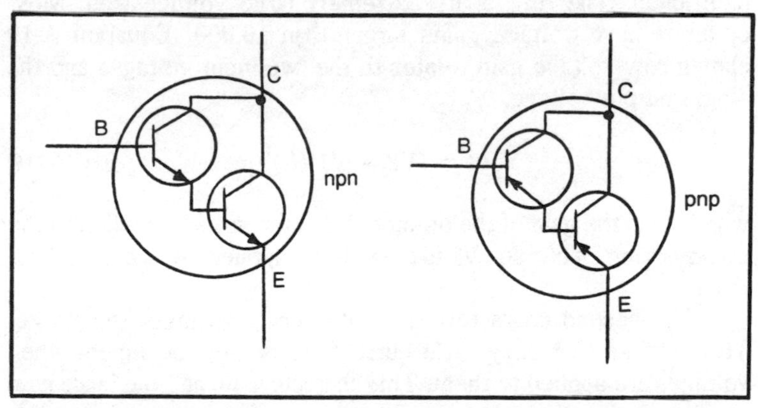

Fig. A-16. Transistors placed in a Darlington configuration for higher current gain.

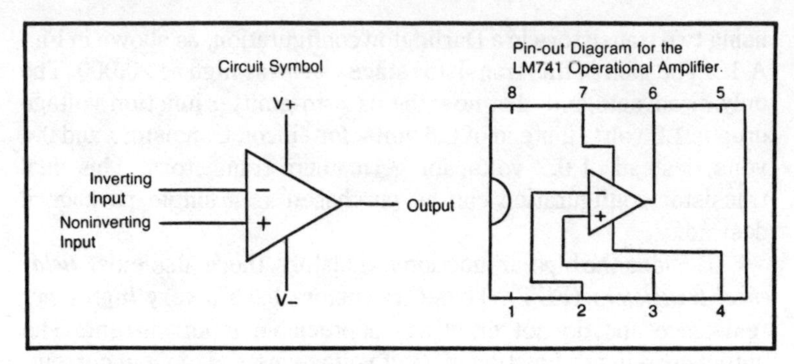

Fig. A-17. Operational amplifier circuit symbol and pinout diagram for an LM741 operational amplifier.

shown in Fig. A-17. This symbol shows two inputs, two power connections, and one output. One of the inputs is *noninverting*, which means that a voltage placed on this input will be amplified with no sign change. The other is an *inverting* input; a voltage placed on this input will be amplified with a sign change. The op-amp also requires two input voltages: one positive and one negative; however, when the op-amp is used as a comparator, the negative power input can be connected to ground. The circuit symbol shown in Fig. A-17 does not show any internal connections, but it actually contains several transistors connected in a differential amplifier configuration. Figure A-17 also shows a pinout diagram of the most popular op-amp integrated circuit, the LM741.

The operational amplifier has three distinguishing characteristics that enable it to be used in all the applications previously mentioned. The first is its extremely large voltage gain. Most op-amps have voltage gains larger than 10,000. Equation A-19 shows how voltage gain relates to the two input voltages and the single output voltage.

$$Vout = (V2 - V1)(G)Vin \qquad \textbf{Eq. A-19}$$

where G is the gain of the op-amp, V2 is the voltage applied to the noninverting input, and V1 is the voltage applied to the inverting input.

The second characteristic is its very high input resistance. This implies that very little current flows into the inputs when voltages are applied to them. This characteristic and the large gain characteristic are used to simplify the equations relating output to input in the circuits discussed in the following sections.

The third characteristic of the op amp is its low output resistance. This implies that, compared to the input current, the output can supply unlimited current to a load connected to the output. Actually this current is limited to about 0.03 amps. This characteristic makes the op-amp ideal for use as a buffer, which will also be discussed later.

INVERTING AMPLIFIERS

The most widely used application of operational amplifiers is the inverting amplifier. The circuit diagram for this amplifier is shown in Fig. A-18. From the op-amp characteristics previously mentioned, assumptions can be made that will greatly simplify the equation relating output voltage (Vout) to input voltage (Vin) for this circuit.

The first step in the process of finding the voltage gain Vout/Vin is sum currents at the inverting input terminal. The input current (Iin) can be found from Ohm's Law:

$$Iin = \frac{Vin - V1}{R1} \qquad \textbf{Eq. A-20}$$

where V1 is the voltage at the inverting input, and Vin – V1 is the voltage across resistor R1. The current Ifb through the feedback resistor R2 is found in a similar manner.

$$Ifb = \frac{V1 - Vout}{R2} \qquad \textbf{Eq. A-21}$$

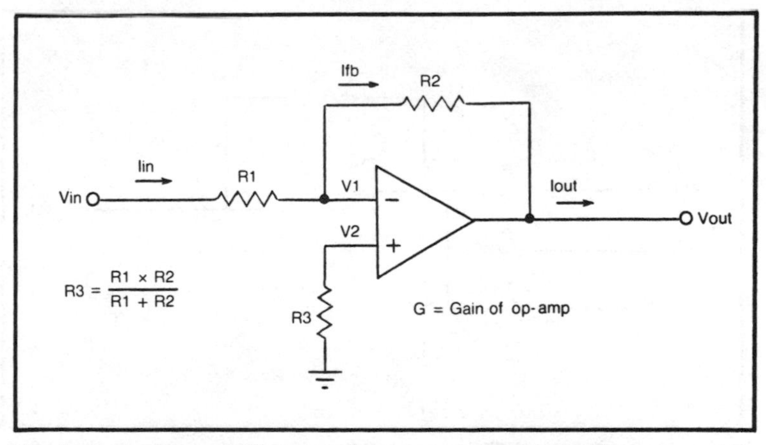

Fig. A-18. Inverting amplifier made from an operational amplifier.

Because the input current to the inverting input can be assumed to be zero, Iin must equal Ifb. Setting Eq. A-20 equal to Eq. A-21 gives

$$\frac{Vin - V1}{R1} = \frac{V1 - Vout}{R2} \qquad \textbf{Eq. A-22}$$

Because the noninverting input is tied to ground through resistor R3, the input voltage V2 is equal to zero. From Eq. A-19, Vout is now equal to $-V1 \times G$, where G is the gain of the op-amp. This equation can be rearranged to give

$$V1 = -Vout/G \qquad \textbf{Eq. A-23}$$

The value for V1 in Eq. A-23 can now be substituted into Eq. A-22 to get, after simplification,

$$Vout\ (R1 + R2/G + R1/G) = -VinR2 \qquad \textbf{Eq. A-24}$$

If G is assumed to be very large, then Eq. A-24 can be simplified to

$$VoutR1 = -VinR2 \qquad \textbf{Eq. A-25}$$

or

$$Vout/Vin = -R2/R1 \qquad \textbf{Eq. A-26}$$

This equation states that the voltage gain of the circuit in Fig. A-18 is the negative of the ratio of the feedback resistor R2 over the input

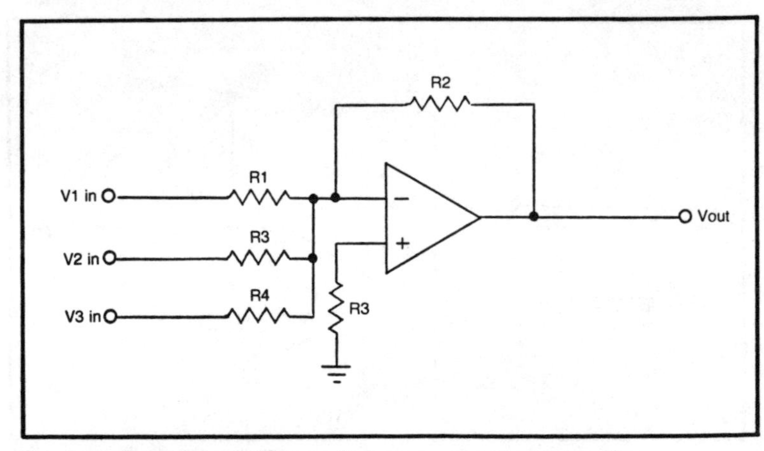

Fig. A-19. Summing amplifier made from an operational amplifier.

resistor R1. This is, then, an inverting amplifier with a voltage gain that can be changed by changing the input or feedback resistor.

Summers

A *summer* is a device used to give an output voltage proportional to the sum of all the input voltages. Figure A-19 shows the circuit diagram of a summer made from an op-amp. This summer is basically an inverting amplifier with more than one input. The output voltage is the sum of output voltages due to each input, as shown in Eq. A-27.

$$\text{Vout} = \text{V1in}\,(-\text{R2/R1}) + \text{V2in}\,(-\text{R2/R3}) + \text{V3in}\,(-\text{R2/R4}) \qquad \textbf{Eq. A-27}$$

If all the input resistors are made to be equal, such as R1=R3=R4, then

$$\text{Vout} = -(\text{R2/R1})(\text{V1in} + \text{V2in} + \text{V3in})$$

If R2 is now made to be equal to R1, then

$$\text{Vout} = -(\text{V1in} + \text{V2in} + \text{V3in}) \qquad \textbf{Eq. A-28}$$

Equation A-28 shows that if all the resistors in Fig. A-19 are equal, the output voltage is the negative of the sum of all the input voltages. These summers are used extensively in analog computers.

Comparators

Operational amplifiers can also be used to compare two input voltage levels and produce an output voltage dependent on the higher input voltage. This type of circuit, called a *comparator,* can be built from an op-amp with no additional components, as shown in Fig. A-20. Figure A-20 also shows the relationship of the two input voltages to the output voltage of the op-amp when used as an comparator.

In Fig. A-20 the input voltage V1 is held at a constant 5 volts, and input voltage V2 is increased from 0 to 10 volts. When V2 is less than V1, the output voltage, Vout, is equal to V−, which is the voltage of the negative input power. When V2 is greater than V1, the output voltage is equal to V+, which is the voltage of the positive input power. The switchover point is at V2 = 5 volts.

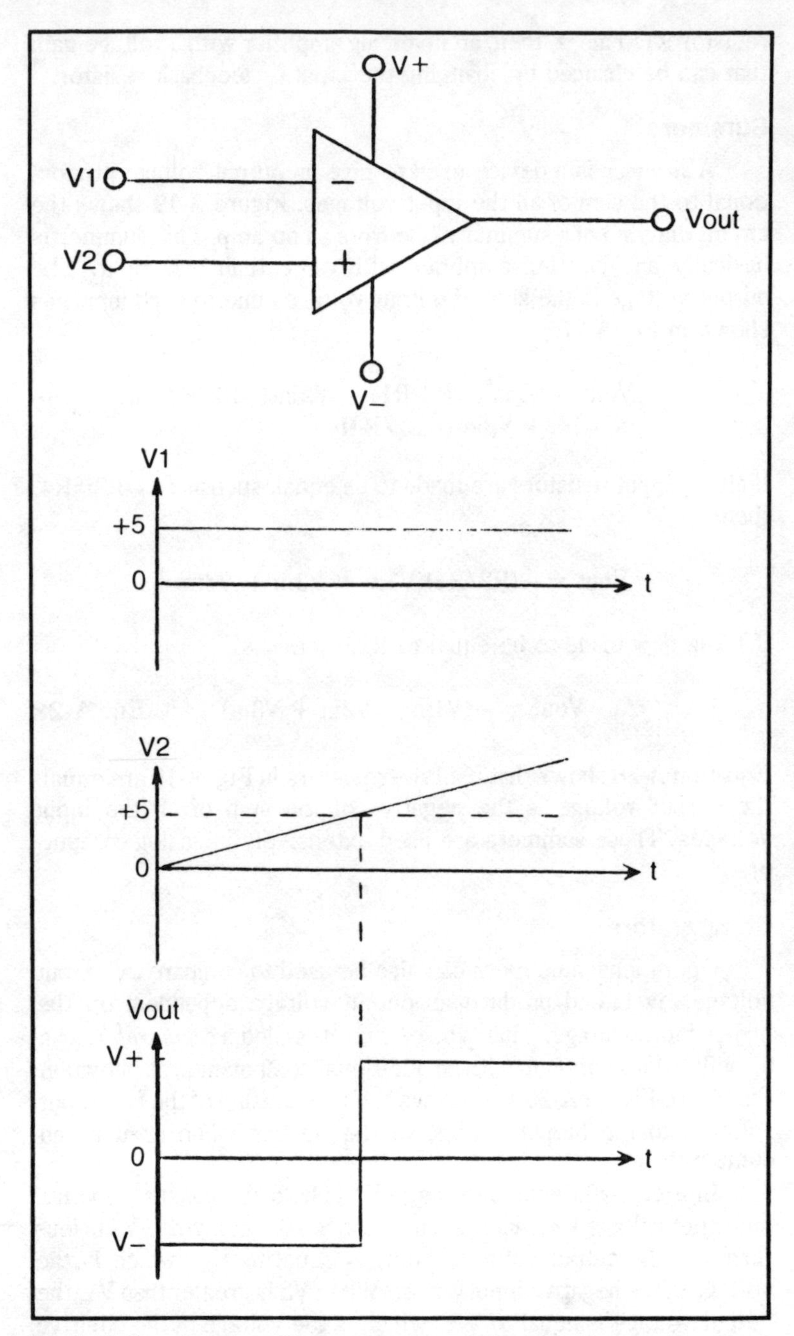

Fig. A-20. Operational amplifier used as a comparator. Also shown are graphs of input and output voltages showing their relationships.

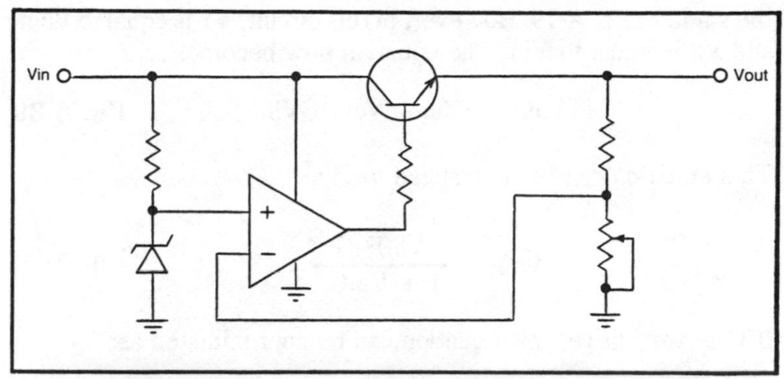

Fig. A-21. Basic voltage-regulator circuit using an operational amplifier.

Comparator circuits are extremely useful in voltage regulator circuits. If used in this type circuit the constant reference voltage can be produced by a zener diode. This reference voltage is then compared to a voltage that is proportional to the voltage to be regulated. The output of the op-amp is then used to regulate this voltage by turning a transistor on or off. Figure A-21 shows a simple voltage-regulator circuit that uses this technique.

Buffers

A *buffer* is a circuit that boosts the current-supplying capacity of a signal. In other words, it takes a low-current signal and changes it to a high-current signal without changing the voltage level.

It was stated earlier that the output of the op-amp had low resistance and high current capacity (relatively). It was also stated that the input of the op-amp had high resistance and would draw only a small amount of current. If the operational amplifier is used as shown in Fig. A-22, it makes a very good buffer with low input current and high output current if the output voltage is equal to the input voltage.

The equation for the output voltage of the buffer in Fig. A-22 is

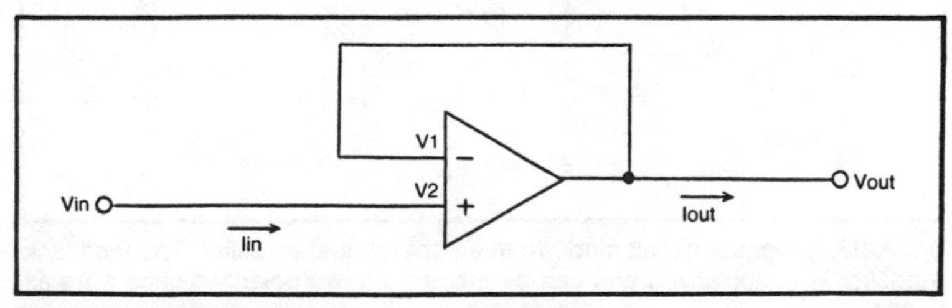

Fig. A-22. Operational amplifier used as a buffer.

the same as Eq. A-19. However, in this circuit, V1 is equal to Vout, and V2 is equal to Vin. The equation now becomes

$$\text{Vout} = (\text{Vin} - \text{Vout})\text{GVin} \qquad \textbf{Eq. A-29}$$

This equation can be rearranged to give

$$\text{Vout} = \frac{(\text{Vin})^2\ \text{G}}{1 + \text{VinG}} \qquad \textbf{Eq. A-30}$$

If G is very large, this equation can be approximated as

$$\text{Vout} = \frac{(\text{Vin})^2\ \text{G}}{\text{VinG}}$$

or

$$\text{Vout} = \text{Vin} \qquad \textbf{Eq. A-31}$$

This analysis shows that the output voltage of the buffer in Fig. A-22 is indeed the same as the input voltage; therefore, it fits the definition of a buffer.

Integrators

Integrators are circuits that mathematically integrate an input

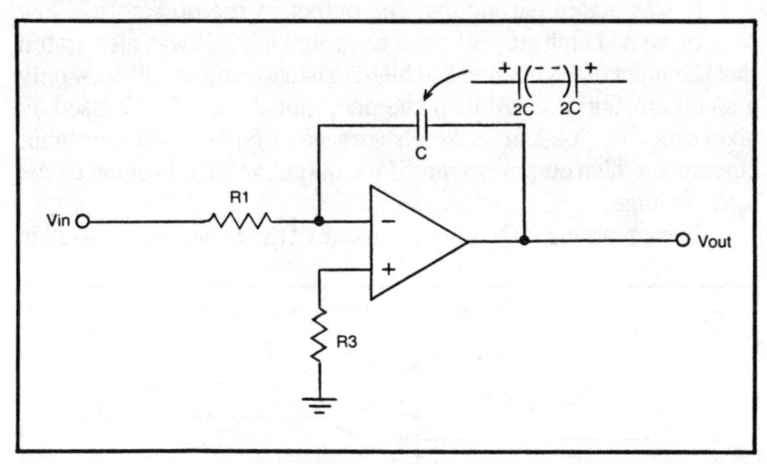

Fig. A-23. Integrator circuit made from an operational amplifier. The feedback capacitor is nonpolarized and can be made from two polarized capacitors as shown.

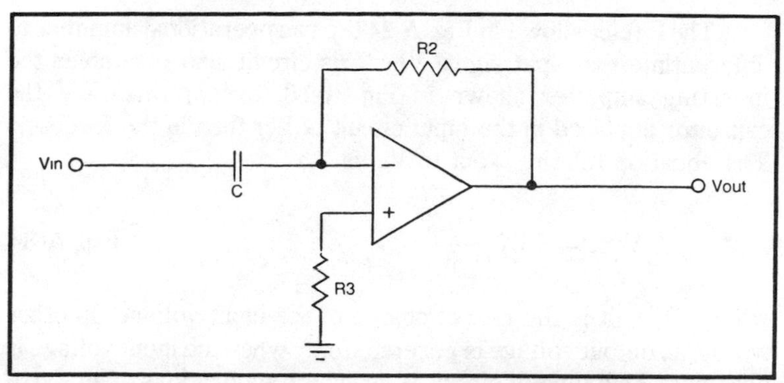

Fig. A-24. Differentiator circuit made from an operational amplifier.

voltage and produce the result on the output. A complete definition of integration here would be useless unless the reader is familiar with calculus. Because this book is written on a technician level and technicians typically do not take calculus, a complete discussion of integration will not be given, but the equation relating output voltage to input voltage for the integrator in Fig. A-23 will be.

The integrator in Fig. A-23 is similar to the inverting amplifier shown in Fig. A-18, except that a feedback capacitor is used instead of a feedback resistor. The equation relating output voltage to input voltage for this circuit is

$$Vout = \frac{-1}{CR1} \int Vin\, dt. \qquad \textbf{Eq. A-32}$$

If the input voltage is constant, this equation simplifies to

$$Vout = (-t/CR1)Vin \qquad \textbf{Eq. A-33}$$

where t is time in seconds from the time the input voltage is applied.

These integrator circuits are especially useful in servo control circuits, such as solar trackers. If one is used between the error detector and the servo motor, the steady-state alignment error can be reduced to zero.

DIFFERENTIATORS

Differentiators are circuits that mathematically differentiate an input voltage and produce the result on the output. Like integration, differentiation is a subject learned in calculus and will not be discussed in depth here.

The circuit shown in Fig. A-24 uses an operational amplifier to differentiate the input signal Vin. This circuit also resembles the inverting amplifier shown in Fig. A-18, except this time the capacitor is placed in the input circuit rather than in the feedback. The equation relating Vout to Vin is

$$\text{Vout} = \text{CR2}\,\frac{\text{dVin}}{\text{dt}} \qquad \textbf{Eq. A-34}$$

where dVin/dt is the rate of change of the input voltage. In other words, an output voltage is generated only when the input voltage is changing. This type of circuit is useful for applications where sudden changes in voltage need to be detected.

FILTERS

Filters are circuits that pass certain frequencies while blocking others. The four types of filters are low-pass, high-pass, bandpass and band-reject.

The low-pass filter will pass only low frequencies. It can be made with an operational amplifier as shown in Fig. A-25. At low frequencies the gain of the circuit is –R2/R1, but once the input frequency reaches 1/CR2 radians per second the gain begins to decrease until it reaches zero at infinite frequency. The plot of –Vout/Vin versus frequency is also shown in Fig. A-25.

The high-pass filter passes only high frequencies. It is shown in Fig. A-26 along with its plot of gain versus frequency. At zero frequency (dc) the gain is zero, but at infinite frequency the gain is

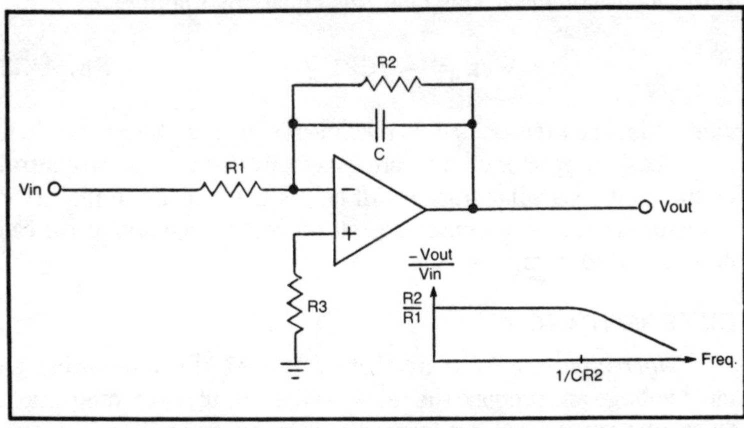

Fig. A-25. Low-pass filter made from an operational amplifer. Also shown is a graph of gain versus frequency.

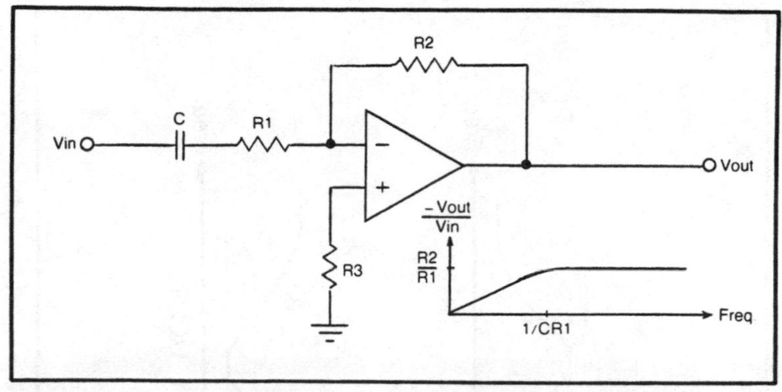

Fig. A-26. High-pass filter made from an operational amplifier. Also shown is a graph of gain versus frequency.

−R2/R1. At a frequency of 1/RC1 the gain has reached 70.7 percent of its final gain of −R2/R1. For both low-pass and high-pass filters, the cutoff frequencies (1/CR2 and 1/CR1) can be set by choosing values for C. The low-frequency gain for the low-pass filter and the high-frequency gain for the high-pass filter can be set by choosing values for R1 and R2.

Filters can also be made that will pass or reject bands of frequencies. Figure A-27 shows an operational amplifier circuit that will do either, depending on what values are chosen for C1, C2, R1, and R2. If the product C1 × R1 is chosen smaller than the product C2 × R2, the filter will be a bandpass filter that passes frequencies between 1/C2R2 and 1/C1R1 radians per second. If C1 × R1 is chosen larger than C2 × R2, the filter will be a band-reject filter that rejects frequencies between 1/C1R1 and 1/C2R2 radians per second. The gain versus frequency plots for the bandpass and band-reject filters are also shown in Fig. A-27.

The gains at the low and high ends of the frequency scale for both filters are equal to −C1/C2. The gain at the logarithmic frequency midpoint of the bandpass or band-reject region is dependent on the distance between the two cutoff frequencies (1/C2R2 and 1/C1R1). This gain is represented by k and is found from Eq. A-35.

$$k = \frac{-R2}{R1}\left[\frac{1+\sqrt{\frac{C1R1}{C2R2}}}{1+\sqrt{\frac{C2R2}{C1R1}}}\right] \qquad \textbf{Eq. A-35}$$

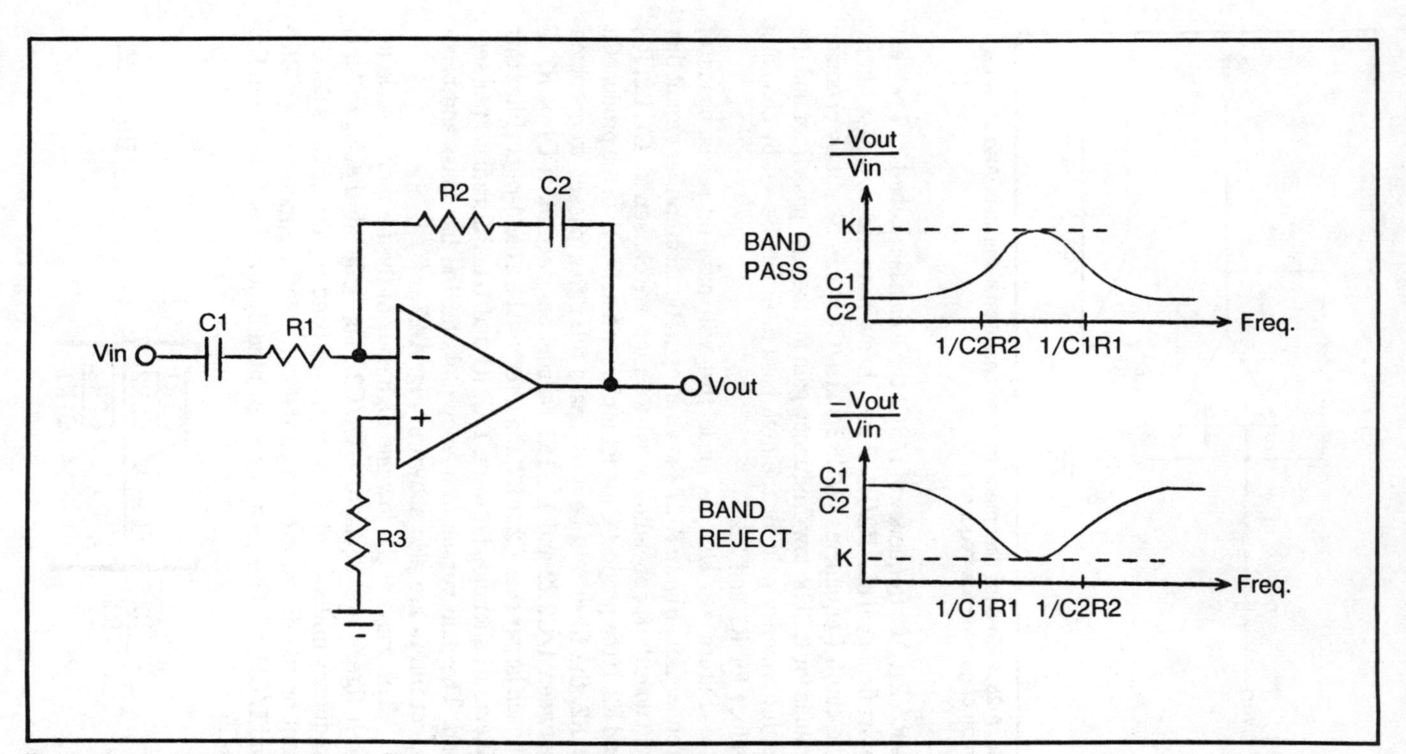

Fig. A-27. A bandpass or band-reject filter made from an operational amplifier.

You can use this equation to calculate R1 and R2 for the desired value of k once C1/C2 has been chosen.

Because of the diversity of operational amplifier applications, it is obvious why they have become so popular. The knowledge of their operation can be a powerful tool for any engineer or technician.

Appendix B
8080A Instruction Set

Octal Code	Mnemonic	Description
000	NOP	No operation.
001 (B2) (B3)	LXI B	Load the following two 8-bit data bytes (B2, B3) immediately into register pair B and C.
002	STAX B	Store contents of register A into memory location addressed by registers B and C.
003	INX B	Increment the contents of register pair B and C by 1.
004	INR B	Increment contents of register B by 1.
005	DCR B	Decrement contents of register B by 1.
006 (B2)	MVI B	Move the following 8-bit byte (B2) immediately into register B.
007	RLC	Rotate contents of register A left by 1 bit.
010	—	This instruction does not exist.

Octal Code	Mnemonic	Description
011	DAD B	Add contents of register pair B, C to register pair H, L and store result in register pair H, L.
012	LDAX B	Load contents of memory addressed by register pair B, C into register A.
013	DCX B	Decrement contents of register pair B and C by 1.
014	INC C	Increment contents of register C by 1.
015	DCR C	Decrement contents of register C by 1.
016 (B2)	MVI C	Move the following 8-bit data byte (B2) immediately into register C.
017	RRC	Rotate contents of register A right by 1 bit.
020	—	This instruction does not exist.
021 (B2) (B3)	LXI D	Load the following two 8-bit data bytes (B2, B3) immediately into register pair D and E.
022	STAX D	Store contents of register A into memory location addressed by register pair D and E.
023	INX D	Increment the contents of register pair D and E by 1.
024	INR D	Increment the contents of register D by 1.
025	DCR D	Decrement contents of register D by 1.
026 (B2)	MVI D	Move the following 8-bit data byte (B2) immediately into register D.
027	RAL	Rotate the contents of register A left through carry by 1 bit.

Octal Code	Mnemonic	Description
030	—	This instruction does not exist.
031	DAD D	Add the contents of register pair D, E to register pair H, L and store the result in register pair H, L.
032	LDAX D	Load contents of memory addressed by register pair D, E into register A.
033	DCX D	Decrement the contents of register pair D and E by 1.
034	INR E	Increment the contents of register E by 1.
035	DCR E	Decrement the contents of register E by 1.
036 (B2)	MVI E	Move the following 8-bit data byte (B2) immediately into register E.
037	RAR	Rotate contents of register A right through carry by 1 bit.
040	—	This instruction does not exist.
041 (B2) (B3)	LXI H	Load the following two 8-bit data bytes (B2, B3) immediately into register pair H, L.
042 (B2) (B3)	SHLD	Store contents of register pair H, L into memory locations addressed by B2 and B3. Contents of L go into memory addressed by (B3)(B2). Contents of H go into memory addressed by (B3)(B2)+1.
043	INX H	Increment the contents of register pair H, L by 1.
044	INR H	Increment the contents of register H by 1.
045	DCR H	Decrement the contents of register H by 1.

Octal Code	Mnemonic	Description
046 (B2)	MVI H	Move the following 8-bit data byte (B2) immediately into register H.
047	DAA	Change the result from the addition of two BCD 4-bit words back into BCD format. This is all done in register A.
050	—	This instruction does not exist.
051	DAD H	Add the contents of register pair H, L to contents of register pair H, L and store result in register pair H, L.
052 (B2) (B3)	LHLD	Load register pair H, L with contents of memory addressed by (B3)(B2) and (B3)(B2) + 1. Contents of (B3)(B2) go into L and contents of (B3)(B2) + 1 go into H.
053	DCX H	Decrement the contents of register pair H, L by 1.
054	INR L	Increment the contents of register L by 1.
055	DCR L	Decrement the contents of register L by 1.
056 (B2)	MVI L	Move the following 8-bit data byte (B2) immediately into register L.
057	CMA	Complement the contents of register A (for example, change 1's to 0's and 0's to 1's).
060	—	This instruction does not exist.
061 (B2) (B3)	LXI SP	Load the following two 8-bit data bytes (B2, B3) immediately into the stack pointer.
062 (B2) (B3)	STA	Store the contents of register A into the memory location addressed by (B3)(B2).

Octal Code	Mnemonic	Description
063	INX SP	Increment the stack pointer register by 1.
064	INR M	Increment the contents of memory location M by 1. The address for M is located in register pair H, L.
065	DCR M	Decrement the contents of memory location M by 1. The address for M is located in register pair H, L.
066 (B2)	MVI M	Move the following 8-bit data byte (B2) immediately into memory location M addressed by register pair H, L.
067	STC	Set the carry flip-flop to logic 1.
070	—	This instruction does not exist.
071	DAD SP	Add the contents of the stack pointer register to register pair H,L and store result in register pair H, L.
072 (B2) (B3)	LDA	Load the content of memory location M addressed by (B3)(B2) directly into register A.
073	DCX SP	Decrement contents of the stack pointer register by 1.
074	INR A	Increment the contents of register A by 1.
075	DCR A	Decrement the contents of register A by 1.
076 (B2)	MVI A	Move the following 8-bit data byte (B2) immediately into register A.
077	CMC	Complement the carry flip-flop.
100	MOV B,B	Move the contents of register B to register B.

Octal Code	Mnemonic	Description
101	MOV B,C	Move the contents of register C to register B.
102	MOV B,D	Move the contents of register D to register B.
103	MOV B,E	Move the contents of register E to register B.
104	MOV B,H	Move the contents of register H to register B.
105	MOV B,L	Move the contents of register L to register B.
106	MOV B,M	Move the contents of memory location M addressed by register pair H, L to register B.
107	MOV B,A	Move the contents of register A to register B.
110	MOV C,B	Move the contents of register B to register C.
111	MOV C,C	Move the contents of register C to register C.
112	MOV C,D	Move the contents of register D to register C.
113	MOV C,E	Move the contents of register E to register C.
114	MOV C,H	Move the contents of register H to register C.
115	MOV C,L	Move the contents of register L to register C.

Octal Code	Mnemonic	Description
116	MOV C,M	Move the contents of memory location M, addressed by registers H and L, to register C.
117	MOV C,A	Move the contents of register A to register C.
120	MOV D,B	Move the contents of register B to register D.
121	MOV D,C	Move the contents of register C to register D.
122	MOV D,D	Move the contents of register D to register D.
123	MOV D,E	Move the contents of register E to register D.
124	MOV D,H	Move the contents of register H to register D.
125	MOV D,L	Move the contents of register L to register D.
126	MOV D,M	Move the contents of memory location M, addressed by register pair H, L, to register D.
127	MOV D,A	Move the contents of register A to register D.
130	MOV E,B	Move the contents of register B to register E.
131	MOV E,C	Move the contents of register C to register E.
132	MOV E,D	Move the contents of register D to register E.

Octal Code	Mnemonic	Description
133	MOV E,E	Move the contents of register E to register E.
134	MOV E,H	Move the contents of register H to register E.
135	MOV E,L	Move the contents of register L to register E.
136	MOV E,M	Move the contents of memory location M, addressed by register pair H, L, to register E.
137	MOV E,A	Move the contents of register A to register E.
140	MOV H,B	Move the contents of register B to register H.
141	MOV H,C	Move the contents of register C to register H.
142	MOV H,D	Move the contents of register D to register H.
143	MOV H,E	Move the contents of register E to register H.
144	MOV H,H	Move the contents of register H to register H.
145	MOV H,L	Move the contents of register L to register H.
146	MOV H,M	Move the contents of memory location M, addressed by register pair H, L, to register H.
147	MOV H,A	Move the contents of register A to register H.

Octal Code	Mnemonic	Description
150	MOV L,B	Move the contents of register B to register L.
151	MOV L,C	Move the contents of register C to register L.
152	MOV L,D	Move the contents of register D to register L.
153	MOV L,E	Move the contents of register E to register L.
154	MOV L,H	Move the contents of register H to register L.
155	MOV L,L	Move the contents of register L to register L.
156	MOV L,M	Move the contents of memory location M, addressed by register pair H, L, to register L.
157	MOV L,A	Move the contents of register A to register L.
160	MOV M,B	Move the contents of register B to memory location M, addressed by register pair H,L.
161	MOV M,C	Move the contents of register C to memory location M, addressed by register pair H,L.
162	MOV M,D	Move the contents of register D to memory location M, addressed by register pair H,L.
163	MOV M,E	Move the contents of register E to memory location M, addressed by register pair H,L.
164	MOV M,H	Move the contents of register H to memory location M, addressed by register pair H,L.

Octal Code	Mnemonic	Description
165	MOV M,L	Move the contents of register L to memory location M, addressed by register pair H,L.
166	HLT	Halt
167	MOV M,A	Move the contents of register A to memory location M, addressed by register pair H,L.
170	MOV A,B	Move the contents of register B to register A.
171	MOV A,C	Move the contents of register C to register A.
172	MOV A,D	Move the contents of register D to register A.
173	MOV A,E	Move the contents of register E to register A.
174	MOV A,H	Move the contents of register H to register A.
175	MOV A,L	Move the contents of register L to register A.
176	MOV A,M	Move the contents of memory location M, addressed by register pair H,L, to register A.
177	MOV A,A	Move the contents of register A to register A.
200	ADD B	Add the contents of register B to register A.
201	ADD C	Add the contents of register C to register A.
202	ADD D	Add the contents of register D to register A.
203	ADD E	Add the contents of register E to register A.

Octal Code	Mnemonic	Description
204	ADD H	Add the contents of register H to register A.
205	ADD L	Add the contents of register L to register A.
206	ADD M	Add the contents of memory location M, addressed by register pair H,L, to register A.
207	ADD A	Add the contents of register A to register A.
210	ADC B	Add the carry bit and contents of register B to register A.
211	ADC C	Add the carry bit and contents of register C to register A.
212	ADC D	Add the carry bit and contents of register D to register A.
213	ADC E	Add the carry bit and contents of register E to register A.
214	ADC H	Add the carry bit and contents of register H to register A.
215	ADC L	Add the carry bit and contents of register L to register A.
216	ADC M	Add the carry bit and contents of memory location M, addressed by register pair H and L, to register A.
217	ADC A	Add the carry bit and contents of register A to register A.
220	SUB B	Subtract the contents of register B from register A.
221	SUB C	Subtract the contents of register C from register A.

Octal Code	Mnemonic	Description
222	SUB D	Subtract the contents of register D from register A.
223	SUB E	Subtract the contents of register E from register A.
224	SUB H	Subtract the contents of register H from register A.
225	SUB L	Subtract the contents of register L from register A.
226	SUB M	Subtract the contents of memory location M, addressed by register pair H and L, from register A.
227	SUB A	Clear contents of register A.
230	SBB B	Subtract the carry bit and contents of register B from register A.
231	SBB C	Subtract the carry bit and contents of register C from register A.
232	SBB D	Subtract the carry bit and contents of register D from register A.
233	SBB E	Subtract the carry bit and contents of register E from register A.
234	SBB H	Subtract the carry bit and contents of register H from register A.
235	SBB L	Subtract the carry bit and contents of register L from register A.
236	SBB M	Subtract the carry bit and contents of memory location M, addressed by registers H and L, from register A.

Octal Code	Mnemonic	Description
237	SBB A	Subtract the carry bit and contents of register A from register A.
240	ANA B	AND the contents of register B with register A.
241	ANA C	AND the contents of register C with register A.
242	ANA D	AND the contents of register D with register A.
243	ANA E	AND the contents of register E with register A.
244	ANA H	AND the contents of register H with register A.
245	ANA L	AND the contents of register L with register A.
246	ANA M	AND the contents of memory location M, addressed by registers H and L, with register A.
247	ANA A	AND the contents of register A with register A.
250	XRA B	Exclusive-OR the contents of register B with register A.
251	XRA C	Exclusive-OR the contents of register C with register A.
252	XRA D	Exclusive-OR the contents of register D with register A.
253	XRA E	Exclusive-OR the contents of register E with register A.

Octal Code	Mnemonic	Description
254	XRA H	Exclusive-OR the contents of register H with register A.
255	XRA L	Exclusive-OR the contents of register L with register A.
256	XRA M	Exclusive-OR the contents of memory location M, addressed by registers H and L, with register A.
257	XRA A	Exclusive-OR the contents of register A with register A.
260	ORA B	OR the contents of register B with the contents of register A.
261	ORA C	OR the contents of register C with the contents of register A.
262	ORA D	OR the contents of register D with the contents of register A.
263	ORA E	OR the contents of register E with the contents of register A.
264	ORA H	OR the contents of register H with the contents of register A.
265	ORA L	OR the contents of register L with the contents of register A.
266	ORA M	OR the contents of memory location M, addressed by registers H and L, with the contents of register A.
267	ORA A	OR the contents of register A with the contents of register A.
270	CMP B	Compare the contents of register B with the contents of register A (condition flags are

Octal Code	Mnemonic	Description
		set as if contents of register B were subtracted from register A).
271	CMP C	Compare the contents of register C with the contents of register A in same way as CMP B.
272	CMP D	Compare the contents of register D with the contents of register A in the same way as CMP B.
273	CMP E	Compare the contents of register E with the contents of register A in same way as CMP B.
274	CMP H	Compare the contents of register H with the contents of register A in same way as CMP B.
275	CMP L	Compare the contents of register L with the contents of register A in same way as CMP B.
276	CMP M	Compare the contents of memory location M, addressed by register H and L, with the contents of register A.
277	CMP A	Compare the contents of register A with the contents of register A.
300	RNZ	Return from the subroutine if zero flip-flop is set at logic 0.
301	POP B	Pop data off the stack and store it in register pair B and C.
302 (B2) (B3)	JNZ	Jump to address (B3)(B2) if the zero flip-flop is set to logic 0.

Octal Code	Mnemonic	Description
303 (B2) (B3)	JMP	Jump unconditionally to the memory location addressed by (B3)(B2).
304 (B2) (B3)	CNZ	Call the subroutine located at address (B3)(B2) if the zero flip-flop is set at logic 0.
305	PUSH B	Push the contents of register pair B and C onto the stack.
306 (B2)	ADI	Add the following 8-bit data byte (B2) immediately to register A.
307	RST O	Call the subroutine located at address 000.
310	RZ	Return from the subroutine if the zero flip-flop is set to logic 1.
311	RET	Return from the subroutine.
312 (B2) (B3)	JZ	Jump to memory address (B3)(B2) if the zero flip-flop is set to logic 1.
313	—	This instruction does not exist.
314 (B2) (B3)	CZ	Call the subroutine located at address (B3)(B2) if the zero flip-flop is set to logic 1.
315 (B2) (B3)	CALL	Call the subroutine located at address (B3)(B2).
316 (B2)	ACI	Add the following 8-bit data byte (B2) and the carry bit to register A.
317	RST 1	Call the subroutine located at memory address 010.

Octal Code	Mnemonic	Description
320	RNC	Return from the subroutine if the carry flip-flop is set to logic 0.
321	POP D	Pop data off the stack and store it in register pair D and E.
322 (B2) (B3)	JNC	Jump to memory address (B3)(B2) if the carry flip-flop is set to logic zero.
323 (B2)	OUT	Output data in register A to output device addressed by (B2).
324 (B2) (B3)	CNC	Call the subroutine addressed by (B3)(B2) if the carry flip-flop is set to logic 0.
325	PUSH D	Push the contents of register pair D, E onto the stack.
326 (B2)	SUI	Subtract the following 8-bit data byte (B2) immediately from register A.
327	RST 2	Call the subroutine located at memory address 010.
330	RC	Return from the subroutine if the carry flip-flop is set to logic 0.
331	—	This instruction does not exist.
332 (B2) (B3)	JC	Jump to memory address (B3)(B2) if the carry flip-flop is set to logic 1.
333 (B2)	IN	Input data to register A from the input addressed by (B2).
334	CC	Call the subroutine addressed by (B3)(B2) if

Octal Code	Mnemonic	Description
(B2) (B3)		the carry flip-flop is set to logic 1.
335	—	This instruction does not exist.
336 (B2)	SBI	Subtract the following 8-bit data byte (B2) and carry bit immediately from register A.
337	RST 3	Call the subroutine located at address 030.
340	RP0	Return from the subroutine if the parity flip-flop is set to logic 0.
341	POP H	Pop data from the stack and store it in register pair H, L.
342 (B2) (B3)	JPO	Jump to the memory location addressed by (B3)(B2) if the parity flip-flop is set to logic 0.
343	XTHL	Exchange data on top of stack with contents of registers H and L.
344 (B2) (B3)	CP0	Call the subroutine located at address (B3)(B2) if the parity flip-flop is set to logic 0.
345	PUSH H	Push the contents of register pair H and L onto the stack.
346 (B2)	ANI	AND the following 8-bit byte (B2) immediately with the contents of register A.
347	RST 4	Call the subroutine located at address 040.
350	RPE	Return from the subroutine if the parity flip-flop is set to logic 1.
351	PCHL	Jump to the memory location addressed by register pair H and L.

Octal Code	Mnemonic	Description
352 (B2) (B3)	JPE	Jump to the memory location addressed by (B3)(B2) if the parity flip-flop is set to logic 1.
353	XCHG	Exchange the contents of registers H and L with the contents of registers D and E.
354 (B2) (B3)	CPE	Call the subroutine located at address (B3)(B2) if the parity flip-flop is set to logic 1.
355	—	This instruction does not exist.
356 (B2)	XRI	Exclusive-OR the following 8-bit data byte (B2) immediately with the contents of register A.
357	RST 5	Call the subroutine located at address 050.
360	RP	Return from the subroutine if the sign flip-flop is set to logic 0 (positive sign).
361	POP PSW	Pop data from the stack and store it in register A and flag flip-flop.
362 (B2) (B3)	JP	Jump to the memory location addressed by (B3)(B2) if the sign flip-flop is set to logic 0 (positive sign).
363	DI	Disable the interrupt input.
364 (B2) (B3)	CP	Call the subroutine located at address (B3)(B2) if the sign flip-flop is set to logic 0 (positive sign).
365	PUSH PSW	Push the contents of register A and flags onto the stack.
366 (B2)	ORI	OR the following 8-bit data byte (B2) immediately with the contents of register A.

Octal Code	Mnemonic	Description
367	RST 6	Call the subroutine located at address 060.
370	RM	Return from the subroutine if the sign flip-flop is set to logic 1 (minus sign).
371	SPHL	Transfer the contents of register H and L to the stack pointer.
372 (B2) (B3)	JM	Jump to the memory location addressed by (B3)(B2) if sign flip-flop is set to logic 1 (minus sign).
373	EI	Enable the interrupt input.
374 (B2) (B3	CM	Call the subroutine located at address (B3)(B2) if the sign flip-flop is set to logic 1 (minus sign).
375	—	This instruction does not exist.
376 (B2)	CPI	Compare the following 8-bit data byte (B2) immediately with the contents of register A.
377	RST 7	Call the subroutine located at address 070.

Appendix C
Keyboard and Bootstrap Machine Language Program

Address (octal)	Memory Content	Command	Comments
000 000 000 001 000 002	061 000 010	LXI SP	Set stack pointer to highest memory address in RAM plus 1.
000 003 000 004 000 005	041 000 002	LXI H	Put address of first RAM memory location into register pair H,L.
000 006 (point A)	116	MOV CM	Move contents of memory location M, addressed by registers H and L, to register C.
000 007	174	MOV AH	Move contents of register H to register A.
000 010 000 011	323 001	OUT	,Output contents of register A to output port number 1 (high address now displayed).
000 012	175	MOV AL	Move contents of register L to register A.

Address (octal)	Memory Content	Command	Comments
000 013	323	OUT	Output contents of register A to output port number 0 (low address now displayed).
000 014	000		
000 015 (point B)	171	MOV AC	Move contents of register C to register A (put memory data back into register A).
000 016	323	OUT	Output contents of register A to output port number 2 (memory content now displayed).
000 017	002		
000 020 (point C)	315	CALL	Call keyboard routine located at 000 315 (wait and input next key closure to register A).
000 021	315		
000 022	000		
000 023	376	CPI	Compare octal code 010 with contents of register A.
000 024	010		
000 025	322	JNC	Jump to 000 044 (point D) if key octal code was greater than or equal to 010 (jump if alphabetical key pushed).
000 026	044		
000 027	000		
000 030	107	MOV BA	Move key code in A to register B.
000 031	171	MOV AC	Move old memory content in C to register A.
000 032	027	RAL	Rotate left through carry by 1 bit.
000 033	027	RAL	Rotate left through carry by 1 bit again.
000 034	027	RAL	Rotate left through carry by 1 bit again.

Address (octal)	Memory Content	Command	Comments
000 035	346	ANI	AND 370 with contents of register A (mask out least significant octal digit).
000 036	370		
000 037	260	ORA B	OR contents of register B with register A (put key code in as the least significant octal digit).
000 040	117	MOV CA	Move new data into register C.
000 041	303	JMP	Jump back to point B.
000 042	015		
000 043	000		
000 044 (point D)	376	CPI	Compare 011 with contents of register A.
000 045	011		
000 046	302	JNZ	Jump to point E if key closure not an L (octal code 011).
000 047	055		
000 050	000		
000 051	151	MOV LC	Move contents of buffer register C into register L.
000 052	303	JMP	Jump to point A.
000 053	006		
000 054	000		
000 055 (point E)	376	CPI	Compare 010 with contents of register A.
000 056	010		
000 057	302	JNZ	Jump to point F if key closure not an H (010).
000 060	066		
000 061	000		

Address (octal)	Memory Content	Command	Comments
000 062	141	MOV HC	Put contents of register C into register H.
000 063	303	JMP	Jump to point A.
000 064	006		
000 065	000		
000 066 (point F)	376	CPI	Compare 013 with contents of register A.
000 067	013		
000 070	302	JNZ	Jump to point G if key closure not an S (013).
000 071	100		
000 072	000		
000 073	161	MOV MC	Move data from register C to memory whose address is stored in H and L.
000 074	043	INH X	Increment address in H and L registers.
000 075	303	JMP	Jump to point A.
000 076	006		
000 077	000		
000 100 (point G)	376	CPI	Compare 012 with contents of register A.
000 101	012		
000 102	302	JNZ	Jump to point C if key closure is not a G.
000 103	020		
000 104	000		
000 105	351	PCHL	Execute program starting at address pointed to by H and L.

Address (octal)	Memory Content	Command	Comments
		10-Millisecond Delay Subroutine (TIMEOUT)	
000 277	365	PUSH PSW	Push contents of register A and flags on stack.
000 300	325	PUSH D	Push contents of register pair D and E on stack.
000 301	021	LXI D	Load 001 046 into D and E.
000 302	046		
000 303	001		
000 304 (point H)	033	DCX D	Decrement contents of register pair D and E by 1.
000 305	172	MOV AD	Move contents of register D to register A.
000 306	263	ORA E	OR contents of register E with register A.
000 307	302	JNZ	Jump to point H if contents of E and D are not 0.
000 310	304		
000 311	000		
000 312	321	POP D	Pop stack and store in register pair D and E.
000 313	361	POP PSW	Pop stack and store in register A and flag flip-flops.
000 314	311	RET	Return from subroutine.
		Keyboard Input Subroutine	
000 315 (point I)	333	IN	Input data from port 0 into A register.
000 316	000		

Address (octal)	Memory Content	Command	Comments
000 317	267	ORA A	OR contents of register A with register A.
000 320	372	JM	Jump to point I if sign flip-flop = logic 1 (if last key not released).
000 321	315		
000 322	000		
000 323	315	CALL	Call TIMEOUT subroutine.
000 324	277		
000 325	000		
000 326 (point J)	333	IN	Input data from input port 0.
000 327	000		
000 330	267	ORA A	OR register A with register A.
000 331	362	JP	Jump to point J if sign flip-flop = logic 0 (if no key depressed).
000 332	326		
000 333	000		
000 334	315	CALL	Call TIMEOUT subroutine.
000 335	277		
000 336	000		
000 337	333	IN	Input data from input port 0.
000 340	000		
000 341	267	ORA A	OR register A with register A.
000 342	362	JP	Jump to point J if key not still depressed.
000 343	326		
000 344	000		
000 345	346	ANI	AND 017 with register A (mask out the key depressed flag).
000 346	017		

Address (octal)	Memory Content	Command	Comments
000 347	345	PUSH H	Push contents of register pair H and L onto stack.
000 350	046	MVI H	Move 000 into register H.
000 351	000		
000 352	306	ADI	Add 360 to register A (where 360 is address of beginning of table).
000 353	360		
000 354	157	MOV LA	Move contents of register A to register L.
000 355	176	MOV AM	Move contents of memory location H, L to register A (key code from table).
000 356	341	POP H	Pop H and L off stack pointer.
000 357	311	RET	Return from subroutine.

Bibliography

Anderson, Edward E. *Fundamentals of Solar Energy Conversion*. Reading, Mass.: Addison-Wesley Publishing Company, 1983.

Cheremisinoff, Paul N., and Thomas C. Regino. *Principles and Applications of Solar Energy*. Ann Arbor, Mich.: Ann Arbor Science Publishers, Inc., 1978.

Crouch, Stanley R., Christie G. Enke, and Howard V. Malmstadt. *Electronics and Instrumentation for Scientists*. Reading, Mass.: Benjamin/Cummings Publishing Company, 1981.

Daugherty, Robert L., and Joseph B. Franzini. *Fluid Mechanics with Engineering Applications*. New York: McGraw-Hill Book Company, 1977.

Engineering Staff of Intel, Inc. *MCS-80 User's Manual (with Introduction to MCS-85)*. Santa Clara, Calif.: Intel Corporation, 1977.

Engineering Staff of Texas Instruments, Inc. *The TTL Data Book for Design Engineers*. Dallas: Texas Instruments, Inc., 1976.

Fitzgerald, A.E., Charles Kinsley, and Alexander Kusko. *Electric Machinery*. New York: McGraw-Hill Book Company, 1971.

Fox, T.R. "Measure the Wind," *Elementary Electronics* 16 (January/February 1976): 46-89.

Halkias, Christos C., and Jacob Millman. *Integrated Electronics: Analog and Digital Circuits and Systems.* New York: McGraw-Hill Book Company, 1972.

Kuecken, John A. *How to Make Home Electricity from Wind, Water and Sunshine.* Blue Ridge Summit, Pa. TAB Books, 1979.

Lindsley, E.F. "New Inverter Gives Wind Power Without Batteries," *Popular Science* 207 (October 1975): 50-52.

Mims, Forrest M. "For Sale: Free Energy from the Sun," *Popular Electronics* 18 (November 1980): 96-100.

Moran, Edward. "Wind-Powered Shop," *Popular Science* 209 (July 1976): 93.

Pierson, Richard E. *Technician's and Experimenter's Guide to Using Sun, Wind, and Water Power*. New York: Parker Publishing Company, 1978.

Ray, Edward D. "Build a Sun Tracker and a Solar Engine with Solar Muscle," *Popular Science* 216 (February 1980): 126.

Rony, Peter R. *The 8080A Bugbook: Microcomputer Interfacing and Programing.* Indianapolis: Howard W. Sams and Co., Inc., 1981.

Smith, Rob. "Designing an Electronic Field Control for a Wind Generator." In *Energy Book 2: More Natural Sources and Backyard Applications*, 36-38. Philadelphia: Running Press, 1977.

Titus, John. "Build This 'Dyna-Micro' and 8080 Microcomputer," *Radio Electronics* 47 (May 1976): 33.

Index

OTHER POPULAR TAB BOOKS OF INTERES

Transducer Fundamentals, with Projects (No. 1693—$14.50 paper; $19.95 hard)

Second Book of Easy-to-Build Electronic Projects (No. 1679—$13.50 paper; $17.95 hard)

Practical Microwave Oven Repair (No. 1667—$13.50 paper; $19.95 hard)

CMOS/TTL—A User's Guide with Projects (No. 1650—$13.50 paper; $19.95 hard)

Satellite Communications (No. 1632—$11.50 paper; $16.95 hard)

Build Your Own Laser, Phaser, Ion Ray Gun and Other Working Space-Age Projects (No. 1604—$15.50 paper; $24.95 hard)

Principles and Practice of Digital ICs and LEDs (No. 1577—$13.50 paper; $19.95 hard)

Understanding Electronics—2nd Edition (No. 1553—$9.95 paper; $15.95 hard)

Electronic Databook—3rd Edition (No. 1538—$17.50 paper; $24.95 hard)

Beginner's Guide to Reading Schematics (No. 1536—$9.25 paper; $14.95 hard)

Concepts of Digital Electronics (No. 1531—$11.50 paper; $17.95 hard)

Beginner's Guide to Electricity and Electrical Phenomena (No. 1507—$10.25 paper; $15.95 hard)

750 Practical Electronic Circuits (No. 1499—$14.95 paper; $21.95 hard)

Exploring Electricity and Electronics with Projects (No. 1497—$9.95 paper; $15.95 hard)

Video Electronics Technology (No. 1474—$11.50 paper; $15.95 hard)

Towers' International Transistor Selector—3rd Edition (No. 1416—$19.95 vinyl)

The Illustrated Dictionary of Electronics—2nd Edition (No. 1366—$18.95 paper; $26.95 hard)

49 Easy-To-Build Electronic Projects (No. 1337—$6.25 paper; $10.95 hard)

The Master Handbook of Telephones (No. 1316—$12.50 paper; $16.95 hard)

Giant Handbook of 222 Weekend Electronics Projects (No. 1265—$14.95 paper)

103 Projects for Electronics Experimenters (No. 1249—$11.50 paper)

The Complete Handbook of Videocassette Recorders—2nd Edition (No. 1211—$9.95 paper; $16.95 hard)

Introducing Cellular Communica... Mobile Telephone System (No... paper; $14.95 hard)

The Fiberoptics and Laser Handboo... $15.50 paper; $21.95 hard)

Power Supplies, Switching Regulato... and Converters (No. 1665—$1... $21.95 hard)

Using Integrated Circuit Logic Dev... 1645—$15.50 paper; $21.95 hard)

Basic Transistor Course—2nd Edit... 1605—$13.50 paper; $19.95 hard)

The GIANT Book of Easy-to-Build Electron... jects (No. 1599—$13.50 paper; $21.95 ...

Music Synthesizers: A Manual of Design an... struction (No. 1565—$12.50 paper; $... hard)

How to Design Circuits Using Semiconductors ... 1543—$11.50 paper; $17.95 hard)

All About Telephones—2nd Edition (No. 1537—$11.50 paper; $16.95 hard)

The Complete Book of Oscilloscopes (No. 1532—$11.50 paper; $17.95 hard)

All About Home Satellite Television (No. 1519—$13.50 paper; $19.95 hard)

Maintaining and Repairing Videocassette Recorders (No. 1503—$15.50 paper; $21.95 hard)

The Build-It Book of Electronic Projects (No. 1498—$10.25 paper; $18.95 hard)

Video Cassette Recorders: Buying, Using and Maintaining (No. 1490—$8.25 paper; $14.95 hard)

The Beginner's Book of Electronic Music (No. 1438—$12.95 paper; $18.95 hard)

Build a Personal Earth Station for Worldwide Satellite TV Reception (No. 1409—$10.25 paper; $15.95 hard)

Basic Electronics Theory—with projects and experiments (No. 1338—$15.50 paper; $19.95 hard)

Electric Motor Test & Repair—3rd Edition (No. 1321—$7.25 paper; $13.95 hard)

The GIANT Handbook of Electronic Circuits (No. 1300—$19.95 paper)

Digital Electronics Troubleshooting (No. 1250—$12.50 paper)

TAB

TAB BOOKS Inc.

Blue Ridge Summit, Pa. 17214

Send for FREE TAB Catalog describing over 750 current titles in print.

Index

R

S

T

V

W

Z